Home
Workplace

A HANDBOOK FOR EMPLOYEES
AND MANAGERS

by Brendan B. Read

CMP**Books**

San Francisco

Published by CMP Books
An imprint of CMP Media LLC
Main office: CMP Books, 600 Harrison St., San Francisco, CA 94107 USA
Phone: 415-947-6615; Fax: 415-947-6015
www.cmpbooks.com
Email: books@cmp.com

ISBN: 1-57820-310-4

For individual orders, and for information on special discounts for quantity orders,
please contact:
CMP Books Distribution Center, 6600 Silacci Way, Gilroy, CA 95020
Tel: 1-800-500-6875 or 408-848-3854; Fax: 408-848-5784
Email: bookorders@cmp.com; Web: www.cmpbooks.com

Distributed to the book trade in the U.S. by:
Publishers Group West, 1700 Fourth Street, Berkeley, California 94710

Distributed in Canada by:
Jaguar Book Group, 100 Armstrong Avenue, Georgetown, Ontario M6K 3E7 Canada

Cover design and watercolor by Brad Greene
Text design by Robbie Alterio
Text composition by Greene Design

Transferred to Digital Printing 2010

DEDICATION

This book is dedicated to CMP's IT Home Support team and IT staff, in particular Jeff, Illysa, Marjorie, and Matthew, without whom this book, and all my other work for CMP, would literally have not been possible.

✪

Table of Contents

Foreword

In *Home Workplace: A Handbook for Employees and Managers* Brendan Read addresses the transition of work from the "Industrial Age" to the "Information Age." While each of us may claim to understand some of the benefits that modern technology provides in our daily work life, few among us can truly bring together a complete picture of the many facets and the symbiotic relationships necessary for the success of the Information Age's "New Ways of Working"—a phase coined by British Prime Minister Tony Blair when he encouraged and supported the work of the 5th European Assembly on Telework, held in Lisbon, Portugal.

During the past few years, we have witnessed numerous detonations and loss of life events in and around commercial / government buildings: A nerve gas attack on a major capital city transit system, rail line explosions in multiple countries, random sniper attacks, and the murderous attacks against the innocent on September 11, 2001. These events provide even more reasons to use technology and geography as ways to protect and defend ourselves.

No longer are we shackled to centralized working locations. With the advent of terrorist attacks around the globe, working from home has become all the more important. In my own family's experience, my wife's stepsister, Becky Bristol, became the first civilian ever awarded the United States' Purple Heart Medal (posthumously) by President Ronald Reagan, when she was killed by a terrorist car bomb while working at children's school on a U.S. Air Force base in Germany.

Whether in the call center industry or the general business world, somewhere between two-thirds and three-quarters of all economic activity is information based and therefore electronically transportable. This economic fact represents an enormous opportunity to reinvent not only what we do, but also how we do it. Our Information Age is the era where we make the most of both new technologies and expanding broadband telecommunications. New technologies are merging with networks. Going forward, "Intelligent Broadband Networks" will be available to all organizations and workers, regardless of geography.

At the same time, the wonders of modern technology and telecommunications are upsetting or causing manifold changes in labor agreements, jurisdictional taxation, and revenue streams. Management perceptions and stylistic trends need to follow the lead of technology and telecommunication innovations. Cities, states, and provinces, desperate for tax revenues, do not want to relinquish traditional income and commuter tax revenues from those no longer required to work within their jurisdictions. Much as the law often lags about fifteen years behind reality, so too does our effective adoption of the virtual work place, home working, and taxation policies.

A review of the many benefits of home working: less congestion and pollution; fewer sick days taken and transportation accidents / injuries; lower risk management and facility costs; improved employee recruiting and retention; increased productivity and job satisfaction; greater rural economic development opportunities; and enabling workers to better balance their work and family responsibilities—makes one wonder why every organization has not adopted an employer "fully sponsored" working from home *modus operandi*? It all has to do with change management and helping decision makers understand the new work models and options available to them.

This book systematically addresses the reasons why employers and employees should enthusiastically bring home working into their business's mainstream. A WIN-WIN for all interested parties, if they are willing to understand, embrace, and manage the changes— changes that are inevitable. Equally, the book provides business case studies, citations, economic models, anecdotal stories, and visions on how to incorporate home working into our day-to-day work experience, individually and collectively.

Brendan has done a superb job of challenging the workplace's status quo, by stimulating thinking and perhaps a bit of revolution in the workplace, and for making us question traditional working relationships, while at the same time offering solutions. He is offering suggestions for each of us to re-examine our roles with vendors, consultants, and analysts. Are we truly doing the best we can for our organizations and ourselves? Are we seizing the opportunities available for innovation in our lives and in our businesses?

It is the responsibility of all of those who have managerial, fiduciary, and profit and loss responsibilities to understand and embrace the dynamics of *Home Workplace* and to *Make It So!* in their organizations. Likewise, every individual with work and family responsibilities has a duty to reduce risks, and reduce costs, while fulfilling their everyday obligations.

I commend to each of you the ideas, methodologies, and insights provided in *Home Workplace*. I have experienced first hand both the old and the new ways of working and foresee only greater opportunities and benefits for increasing numbers of employees to work remotely. Intelligent broadband networks, coupled with "The New Ways of Working," are the future of safe, productive, high-performance work environments for the Information Age.

Jack Heacock
Virtual Office, Organizational Development, and Customer Service Consultant
Call Center Magazine 'Pioneer Award Winner'
Parker, Colorado

Preface

As a newspaper reporter, I've covered fires, but never fought a fire. I've watched doctors save lives, but never saved one myself. As a business editor/writer I've written about electrical installations, steelmaking, transit systems, and call centers, but I've never wired a light fixture, worked in a steel mill, driven a bus or train, or managed a call center. (Though I once worked as a telemarketer—for the King County Democrats in Seattle—in between newspaper jobs.)

Home working is different. It is something I've done for most of my career, beginning in 1975, when I was 17. That was when I began clacking out high school news for the (Peterborough, NH) *Ledger* on my mother's portable manual Remington typewriter a short walk from a gorgeous view of Mount Monadnock. I've come full circle, writing this book on an IBM ThinkPad laptop from my home office on Vancouver Island, Canada, again with a view of the mountains (and a glacier) outside of my house.

I'm passionate about home working. It works for more people than you may think. If more people worked from home we could have cleaner air, a better quality of life and lower taxes.

That's why I've written this book, to provide information to help you decide whether to deploy this workplace method. If home working will work for you I wanted to provide advice on how to implement it.

✪ THE HOME WORKPLACE REVOLUTION

You may not know it but you're in the middle of a revolution.

The signs of it are not protests in the streets. Or picket lines inches away from employer-contracted strikebreaker thugs.

Instead the signals are the vehicles parked all day long in the carport or on the street; the men and women with the cellphones and ear buds in the local office supply store; and the neighbors in sweats with the cordless phone dangling out of their front pockets while rolling out their garbage can and waving a cheery "hi." You half-heartedly wave back, twisting the arm of your itchy suit and gunning the air-killing engine of your freshly-dented tank-like SUV as your radio warns of 45-minute backups on the I-whatever.

Then there are the e-mails from colleagues saying "I'm working from home today" and the employees who show up now and then with their laptops and cellphones and camp out in a vacant cube.

You might think these people are a little odd. Why aren't they in the office? How can they work like that? Yet you get the same e-mails. You hear from your counterparts that these people are great workers: always available, there on time, willing to work early or late, rarely do they call in sick, and that they're usually cheerful and rarely fatigued.

Then you look at the employees who show up daily in their cubicles: too often bedraggled, stressed-out and worn out—and the workday's just begun—mumbling something about traffic or late trains. Then you get the requests to leave early to pick up a daughter at the soccer field or go to a doctor's appointment: before the traffic.

Could those people in the sweats and cordless phones, who send the "working from home today" e-mails, and their employers, be onto something?

Let's admit it. Haven't you ever taken your laptop home, booted up your desktop, and donned your jogging suit instead of your monkey suit? Turning over your coffee mug instead of your vehicle's engine? Kissing your loved ones good morning instead of greeting them with a grunt of exhaustion sometime late in the night after a long day?

You are witnessing a *home workplace revolution*. Where people are or want to carry out their employment tasks from where they live in order to avoid long, stressful commutes that cost them $4,000 to $5,000 a year in work clothes, car expenses and carfare. Where employers avoid providing expensive offices: at up to $35,000 per employee/workstation including the real estate, furniture, cabling, heating/air conditioning, power, share of amenities and parking.

Where both employees and employers save that priceless commodity—*time*—which is wasted in commuting. Where employers and employees avoid getting infected and paying the price for illnesses and other disasters that afflict workplaces.

And when disaster strikes, wouldn't you and your employees prefer to be at home? I know all too well *that* feeling: I was one of many who evacuated Manhattan after the terrorist attacks on September 11, 2001, along with my wife and sister-in-law.

You may know *home working* through other titles: telecommuting and teleworking. But *only* home working defines the location: employees' domiciles. The other terms often include working from satellite offices and from the road.

The Revolution Grows (and Pays Off)

Home working is growing. The International Telework Association and Council (ITAC) reported that the number of employee teleworkers nearly doubled: to about 23.5 million in 2003 from 14.4 million in 1999. That is on top of the millions of self-employed teleworkers (which this book doesn't cover), whose numbers increased at a slower rate: from 19 million in 1999 to about 23.4 million in 2003.

The 2003 American Interactive Consumer Survey, conducted by the Dieringer Research Group revealed that the number of employed Americans who telework at least one day a week has increased by nearly 40% since 2001; 22% of those work from home daily or nearly daily. The META group reports that the number of full-time or exclusive home workers has doubled since 2000.

Yet this is a small number—about 20 million or 15% of the total workforce including the self-employed—who work at home at least once a week in 2001, reports the US

Bureau of Labor Statistics. But the percentage is larger in the information or knowledge workforce: people that create, enter, assemble, file, and communicate *knowledge*; i.e., numbers, words, data, ideas, and messages.

These occupations include accountants, administrators, bookkeepers, civil/public servants, consultants, call center customer service reps, support and sales agents, data entry clerks, managers, professors, journalists, nurses, programmers, teachers, and trainers. Because knowledge is intangible it can be communicated, from anywhere to anywhere, using high-speed voice/data links. You don't have to be in a formal premises office to be a knowledge worker.

AT&T: At the Revolution's Forefront

Communications giant AT&T is in the vanguard of this revolution. It develops and markets many of the products and services that enable this change, beginning with the invention of the telegraph in 1844 (the first "dot" in dot.com) and accelerated with the invention of the telephone in 1876.

AT&T and its scientists developed what became today's landline and cellphone networks. They supply the high-speed DSL connectivity that home workers increasingly rely on. That, in turn, has fostered home working growth by enabling employees to receive and transmit data practically at the same speeds as if they were working in employer-supplied or *premises* offices.

Not only that, AT&T walks the walk on home working. It is reaping the benefits internally as well as externally. It supplies the voice and data services for home workers, e.g., virtual private networks. And in 2002/2003 17% of AT&T managers report that they worked from home full time—up from just 9% in 2001. The average number of days these managers worked from home grew to 9 per month in 2003 from 6.7 per month in 2002.

Home working, plus satellite working (working from nearby satellite offices), has paid off for AT&T, a company that is in a highly competitive marketplace and needs all the savings and productivity gains it can muster.

The telco generated over $150 million in home work benefits in 2003. These gains come from lower real estate and other overhead costs, higher productivity and improved employee retention and recruitment, which lower staffing costs. How many other revolutionary new practices can show such bottom line results?

Out of sight out of mind? Not according to AT&T. Home-based managers worked *7.6 hours* per workday compared with *6.8 hours* for all managers. Home workers gained about *1 hour* in extra productive time. The time saved from commuting isn't spent sprawled in from the TV. Out of 80 minutes each day home working managers avoided commuting, *60 minutes* are redirected to work.

These employees are happier and more loyal. Over 63% of home/teleworking managers report increased job and career satisfaction; 47% who had received competing job offers factored in working from home into their decision to stay within the company.

Has there been any other way that is more effective at getting more out of your employees, voluntarily and willingly? Could home working be the key to attracting and retaining employees without busting budgets on big pay/benefits packages, cozy offices, and shiny gyms?

Home working also hasn't hurt promotability. Managers in these "virtual offices" are more likely to be promotable than managers in traditional premises offices. Nearly one in three managers reports that teleworking (including home working) has had a positive effect on their career.

AT&T says that all these benefits, however monetized, are small when compared to the benefits that *all of us* receive from home working. These include saving millions of gallons of gasoline and avoiding thousands of tons of pollutants from being spewed into the air. Not to mention a better work/family/life balance.

"Rhetorically we ask what other management practice can generate such a success rate?"—AT&T

✪ HOME WORKING DRIVES VIRTUAL WORKING

The home workplace revolution is the largest gear in a bigger movement, the *virtual workplace revolution*—working virtually, anywhere, anytime, anywhere, freed of being face-to-face with colleagues, clients, customers and supervisors.

Virtual working is the carrying out of tasks from home; from the road, i.e. *mobile working;* and even from employer-supplied or *premises* location through calls and chats and through audio, data, Web and video conferences, collectively known as *conferencing,* without face-to-face interaction with others. Virtual working eliminates time, money, and performance lost in commuting and business traveling and in not working when traveling.

In the Industrial Revolution, steam power, canals, and railways broke the chains of inefficient home-based production and enabled low-cost mass production of goods and services conducted in centralized urban premises. In the Automobile Revolution, paved roads, personal vehicles and motorized trucks made living and working independent of rail lines and urban centers possible: people could live and employers could locate workplaces anywhere.

In the "Virtual Working Revolution," employees and employers are freed of workplaces altogether. The revolution has been enabled by technology that has automated production and made communication and information processing cheap, reliable, and feasible.

The two most widespread means of virtual working, the phone and e-mail, have come at the direct and indirect expense of retail, door-to-door, and field sales. Call centers handle calls and text and Web contacts that would have been done in person. Self-service—via the Web and interactive voice response (IVR)—has been made more user-friendly with speech recognition and thus is taking customer service and sales to the next level by eliminating the need for people, in any location.

There are more tasks and jobs that were once handled face-to-face that are now handled virtually. And as technology improves, costs drop and people become used to these different methods of communications.

For example, health maintenance organizations (HMOs) now offer "dial-a-nurse" services to provide information that otherwise would have forced an in-person visit to obtain. One such company, IntelliCare (based in Portland, ME), also has many of its workers take contacts at home.

Education and training, which was once thought to be as personal as you can get, is going virtual: Web conferencing and online, interactive lessons and simulations, which can be supplemented with CDs and videos. With prepackaged lessons, students and employees can learn when and where they want.

Virtual education is popular in colleges and universities. Colleges like the University of Phoenix offer degree programs at hundreds of satellite campuses through a blend of online and in-person classes. Today's students are extremely computer-savvy and most modern information is available online, so why make students and their professors go to lecture halls and old-fashioned libraries?

Home working and the home workplace revolution snaps the bonds to the old-fashioned office by providing an alternative fixed location for non-traveling employees. Home is a static locale—workers are already there. It is a place from where they can conference and base their mobile working.

○ WHY THIS BOOK

This book is a guide for managers and executives who are curious about home working, want to understand which functions this new phenomenon is best suited, want to find out what is entailed in implementing home working, why they should consider doing it, and need advice on how to do it. That includes making the case to supervisors and superiors alike.

This book is also for employees who are currently working from home, or wish to do so in the future. It offers them excellent advice on voice, data, equipment, locations, facilities, and furniture. It contains tips on how to cope with distractions from various sources (and species). It also raises legal issues that they should be aware of and plan for. Because this book is written from the executive and manager perspective, it allows the employees to understand management's thinking, which will help employees to comply better with a company's home working program, making it more of a success.

Because every organization is unique, I recommend those readers who want to pursue home working consider (especially for large-scale projects) contacting professionals who are experienced at putting together and assisting clients in implementing home working programs. I list key consultants and organizations that I've worked with (when writing an article, at trade shows, and authoring this book) in the Resources Guide, which can be found at the back of this book.

One of those consultants is Jack Heacock, who has kindly assisted me with this book. He is a decorated Vietnam veteran: a former US Army signalman who knows all too well that having your people distributed can improve your chances of survival.

That same kind of thinking was what led the US Department of Defense to create the Internet: by distributing computers on a network it would reduce the risk that a single attack could wipe out vital information stores.

The DOD's premise worked on September 11, 2001. When the World Trade Center and a good chunk of the Pentagon buildings collapsed the Internet didn't. People, like me, who evacuated from the scene and had laptops could hook them up and stay in touch.

This book focuses primarily on home working by employees, chiefly in the information/knowledge industries. It does not look specifically at types of home-based businesses, except self-employed contractors, covered in Chapter 11. Chapter 1 touches on the differences and similarities between home working and home-based businesses.

However, home-based business entrepreneurs still will find much advice on voice, data, facilities, disaster protection and ergonomics, all of which is applicable to any type of home office. For instance, Chapter 10 has two sections—face-to-face at home and employees working from other employees' homes—that the entrepreneur will find helpful, e.g. they should pay especially close attention to issues like parking and zoning.

I took the employee tack because I've found that many employers are understandably skeptical about why and how they consider having employees work from home. It is a revolutionary way of working that requires a greater degree of trust and reliance on performance assessment than traditional "line of sight" and "hands on" management.

This book is intended to take away much of the mystery of home working by laying out what it is, what is entailed in deploying it, the issues that arise and how to cope with them, and options for home working. Only after the reader understands what is involved do I then discuss the advantages and downsides of home working. For those that decide to implement a home working program I then discuss how to do that, including cost-benefit analyses, and an example of a home working policy. I've also provided a home working checklist.

From my experience as a home worker, a premises worker, and a mobile worker there is little for employers to fear. As shown by the AT&T example, most studies show that home workers are more productive and loyal than premises workers.

I know that I've been happier and healthier since I've begun working from home. I live where I want to while obtaining higher productivity for my employer and a better quality of life for my family.

Home working is also better for society. The fewer the people who have to commute the less pollution and oxygen-giving greenspace that is chewed up for roads and parking lots, and the lower the needless injuries and deaths from motor vehicle accidents. Less money is wasted on subsidizing roads and transit, including the unaccounted-for paving over of productive, property-tax-paying land for highways, for bus or rail transit, and on emergency services from accidents—lowering our tax burden. That means more money in our pockets.

Equally, if not more importantly, by eliminating the stress-inducing money-draining drudgery of commuting, home working enables you to spend more time with your family, friends, and loved ones, and in your community. To enjoy and take part in life instead of just existing.

Dorothy, from *The Wizard of Oz*, is right. There is no place like home.

✪ WHAT THIS BOOK COVERS

This book is intended to outline the details of home working so you can consider and analyze the business case of having some, if not all, of your employees work from home, some or all of the time, and how to implement and manage a home working program.

The book looks at what is home working, the types of home working, and its components, i.e. voice and data technology, equipment, home office locations, facilities, environment, and furniture. It looks at what employers need to have in place to enable home working including network access, temporary or hot-desks for when home workers come into the office, the means of communicating and staying in touch with home workers, and IT support needs.

There are unique administrative and management matters with home workers, including how to measure home working program performance. The book examines how to select, train, and manage home workers including assessing home offices, monitoring, sign in/sign out, and career development. There also are legal issues to be addressed, like complying with data privacy laws, workplace regulations, liability insurance, and taxation.

There is a chapter on special circumstances. They include international, small community and face-to-face (with customers and clients) home working that have unique issues, requirements, responses, and solutions.

Once you know what is involved with home working, you are provided with the necessary information to help you decide whether to instigate a home working program, and how to implement it. The book explores the advantages and challenges of home working, its alternatives such as adjuncts like satellite offices, home worker contracting, mobile working, and conferencing. There are suggestions on how to plan a home working program and how to market it with colleagues in other departments—especially Facilities, HR and IT—and to senior management. There is advice on rolling out, and if need be, terminating the program.

At the back of the book are: a Resources Guide of associations, consultants and suppliers; a checklist prepared by Jack Heacock, who is one of North America's leading virtual working experts and who wrote the Foreword; and a telework (home working) agreement prepared by John Paddock, attorney with Denver, CO law firm, Hale Hackstaff Friesen. There is also a guide on Getting A Life, i.e. how to overcome the isolation from working at home and how to make the most of this excellent opportunity to enjoy the additional time that working at home provides.

Note: *The names, offerings, locations and contact information of suppliers including consultants and organizations may change over the lifespan of this book. For example, during the final editing Siebel acquired Ineto and Citrix acquired ExpertCity. Best bets: use the old names for Web searches. Also stay in touch with new products, services, technologies, applications and case studies by visiting CMP's websites, reading its print and online publications, and attending its conferences and trade shows.*

You will find that there is repetition in parts of the book—it is deliberate—to minimize the annoying flipping pages back and forth. Where needed, I've put in references to other chapters.

To follow up on topics such as voice/data I recommend: *The Telecom Handbook*, by Jane Laino, Newton's *Telecom Dictionary* by Harry Newton, *A Practical Guide to DSL* and

Going WiFi by this book's editor, Janice Reynolds. For mobile working: _Going Mobile_ by Keri Hayes and Susan Kuchinskas. And for conferencing: _Videoconferencing: the whole picture_ by James Wilcox. For more help to incorporate home working in disaster planning read _Disaster Survival Guide_ by Richard 'Zippy' Grigonis. All of these books are published by CMP Books.

The material for this book comes from many articles, interviews, stories and anecdotes, cited where and when available. Much of the information also comes from my own experience as a home, premises, and mobile worker.

I wrote this book on my own Dell OptiPlex desktop, supplemented by a CMP-supplied IBM ThinkPad. Both machines are linked to my very reliable cable broadband provider, Shaw Cable through a Linksys [Cisco] hub. The CMP machine uses a Nortel Contivity VPN to access our network. Telus is the local and residential long distance phone carrier where I live. My telephone is a very reliable, if sometimes annoying (to my wife), Panasonic Caller ID-fitted cordless speakerphone.

I also received superb assistance from CMP's IT Home Support department. For their patient and always-there professional attitude, I dedicate this book to that group.

I hope you find the information and advice helpful. If you have any comments, please e-mail me at **bbread@shaw.ca**. I'll pick up your message—from home.

Brendan B. Read

Acknowledgments

Journalists, like myself, are often little more than "glorified voyeurs." We hear, see, and report on what others are doing. But we've rarely done these acts ourselves.

My experience and my own writing even on home working, only goes so far. To complete the picture, I sought outside expertise. I especially want to thank home working consultant, evangelist, and inspiration, Jack Heacock who reviewed and made very helpful suggestions for the copy, and who prepared the checklist found at the back of the this book.

Also assisting me with the research by patiently answering my questions has been John Edwards, TELEWORKanalytics; Todd Tanner, The Tanner Group; David Smedley, Wave, Inc.; John Paddock, Hale Hackstaff & Friesen; Chuck Wilsker, the Telework Coalition; Bob Fortier of the Canadian Telework Association and InnoVisions Canada; Jeff Furst, FurstPerson; the International Telework Association and Council (ITAC); Joanne Pratt; Eddie Caine; Gil Gordon; Michael Amigoni, ARO; Basil Bennett, Willow CSN; Tim Houlne, Working Solutions; Reg Foster, Alpine Access; Dave Bjork, IntelliCare; Carla Meine, O'Currance Teleservices; King White, Trammell Crow; John Boyd, The Boyd Company; John Vivadelli, Agilquest; Robert Camastro, Virtual-Agent Services; Rick Frye, Gryphon Networks; and Richard 'Zippy' Grigonis.

I have also sourced direct and background information from several leading home working employers and/or suppliers. These include AT&T, Aspect, Citrix, Convergys, Envision Telephony, GemaTech, MCK, Microcell (Fido), Nortel, Procter and Gamble, Siebel, Siemens, Spectrum, Sprint, Teltone, Telus, UCN, Verizon, West and Westjet.

You can find most of these consultants, firms and organizations in the Resources Guide. If you are interested in setting up or enhancing your own home working program please contact them.

Making the book happen has been the assistance and support from CMP's IT Home Support department, in our Manhasset and San Francisco offices. They are there when CMP home workers need them. They were there for me when I worked there, from home. I also want to thank the staff of *Call Center Magazine*, including group vice president Chris Keating, for allowing and supporting me to work from home in New York and later in Canada when I was employed as the magazine's Services Editor.

I owe the editing of this book to my fellow home worker and book editor Janice Reynolds. She has contributed her knowledge on DSL, VPNs, and Wi-Fi (she is author of *A Practical Guide to DSL,* and *Going Wi-Fi,* both published by CMP Books). She worked with me very patiently on this tome, on *Designing the Best Call Center for Your Business* and *The Complete Guide to Customer Support,* the last book written with *Call Center Magazine* Chief Technical Editor Joseph Fleischer; both books have sections on home (tele)working.

I also owe the putting together of the book, again very patiently, to Robbie Alterio who also designs *Call Center Magazine* and put together my other two books, and to Matt Kelsey who have been very patient with this book. And I owe unlimited time and love to my wife, Christine, whose common sense advice has kept the book (and me) on track.

Brendan B. Read
bbread@shaw.ca
June, 2004

Introduction: The Impending Working Crisis

There is an impending multifaceted crisis facing organizations that could affect their (and your) ability to function and deliver products and services effectively. It is a crisis that will impact the quality of lives of each of us. Unfortunately, the situation, represented by data and reports, will very likely get worse over the lifespan of this book.

Here are the facets:

Worsening (and costly) commutes

Commuting times are getting longer and delays worse. The non-profit group, The Road Information Program (TRIP), reports that commute times increased nationwide by 14% from 1990 to 2000—from 22.4 minutes to 25.5 minutes. Cities with big commutes include Atlanta, GA (31.2 minutes), Miami, FL (28.9 minutes), Orlando (27 minutes), and Jacksonville (26.6 minutes).

The average urban driver now spends 62 additional hours annually—the equivalent of 1.5 working weeks stuck in traffic. That's up from 44 hours in 1990, reports the Texas Transportation Institute *2002 Urban Mobility Study*.

The same study, cited in a published 2003 report for the American Public Transportation Association (APTA) entitled, "Critical Relief for Traffic Congestion," estimated the total cost of congestion in terms of lost hours and wasted fuel was *$68 billion* in 2000. Nationwide, the total annual cost may now approach *$100 billion*.

Commuting costs eat into paychecks

The APTA report also found that each peak-period road user lost $1,160 in wasted fuel and time in 2000, including time shared with family and friends.

The report, citing an article by Philip Reed entitled, "Your Car's Total Cost of Ownership" (published by Edmunds.com in April 2002), says the cost of owning and operating a vehicle can run as high as $6,000 or more a year.

The percentage of household income eaten by commuting range from 15.3% in New York, where there is easily accessible mass transit to 23% in Houston, TX, where mass transit is less available; the APTA reported cited an article by Anthony Downs "Can Transit Tame Sprawl?," that appeared in the January 2002 *Governing Magazine*.

Lives lost, injuries incurred during "drive time"

With automobile-based commuting there are accidents that incur costs as well as pain and suffering. The National Highway Traffic Safety Administration reported that *42,815* people, motorists and passengers, plus pedestrians cyclists and other non-motorists lost their lives in traffic accidents in 2002, or over *117 per day*.

To put that in perspective, 50% more Americans die on the roads each month: over *4,500* compared with the approximately 3,000 who lost their lives in the terrorist attacks on 9/11/01. Perhaps we should end terrorism on the roads?

On top of that, over *2.9 million* Americans are injured in traffic accidents, or over *8,000* a day or *240,000* per month. The APTA report, citing a study by the National Traffic Safety Council, entitled "Injury Facts," which found that in 2000, $71.5 billion was lost in wages and productivity due to motor vehicle injuries.

Natalie Litwin is a board member of Transport 2000 Ontario: part of Transport 2000, Canada's leading transportation alternatives advocacy organization. Writing in *The London* (Ontario, Canada) *Free Press* March 20, 2004, she points to some other disturbing statistics:

* The Ontario Medical Association estimated that in the year 2000, pollution costs to the health-care system and losses to employers and employees in that province were more than $1 billion.
* An Ontario Ministry of Transportation 1994 study estimated the social costs of highway crashes: health care, property damage, policing, insurance and lost earnings at $9 billion in one year.
* According to Smartrisk, a non-profit injury-prevention organization, vehicle-related injuries in Ontario cost $567.1 million in 1996.
* Car and motorcycle crashes caused 54.7% of Canadian spinal-cord injuries in Canada, according to the Canadian Paraplegic Association.
* Motor vehicle crashes caused approximately 50% of all degrees of traumatic brain injuries in the US.

High (and hidden) costs to "remedy" commuting

To "remedy" congestion costs money. The American Association of State and Transportation Officials estimates that the US needs to spend $92 billion—in tax dollars—just to maintain current congestion levels.

The costs to construct a highway is $15 million and up per lane-mile or $30 million for a lane in each direction, based on a recent suburban highway widening project in Victoria, British Columbia, Canada. But each lane can carry only 1,500 cars or 1,680 commuters; there is an average of 1.12 passengers per car in each direction. Bus or rail rapid transit systems can cost about $20 million to over $100 million a mile (the higher end is for elevated lines or subways in large, costly metropolitan areas).

Those "remedies," especially highways, eat land. To transport 15,000 people per hour in private cars on an expressway requires 7 lanes per direction or 167 feet in width; the APTA study cites a paper entitled "Transportation for Livable Cities," by Vukan Vuchic, published by Center for Urban Policy Research, Rutgers University, New Brunswick, NJ,

1999. If those commuters took buses the land and lane-use shrinks dramatically to 36 feet or 1 lane in each direction; if they rode trains that figure drops to 24-26 feet or 1 track in each direction.

Traffic engineering isn't the answer

Don't count on engineering to solve the congestion mess. Ironically engineering roads—to accommodate more traffic—has made congestion worse. The faster the traffic, the easier it is to drive, but often too the more injuries and deaths. That leads to more delays; it also diverts police, fire, and paramedics from less preventable tragedies.

The *New York Press* published, on March 8, 2004, a revealing article on pedestrian tragedies and traffic congestion. It pointed to a pair of pedestrian fatalities: Juan Estrada and Victor Flores, two schoolchildren killed by a gravel truck that could have been prevented by a simple traffic-calming device called a "leading pedestrian interval," or LPI.

An LPI, it says "lights up the pedestrian signal about three seconds before vehicular traffic gets the green. This gives pedestrians a head start into the intersection and forces turning vehicles to be less aggressive as they drive through the crosswalk." LPIs might have prevented the type of "right turn conflict: that killed Juan and Victor. The downside of an LPI is that a few less vehicles may be able to move through the intersection at each cycle of the light.

The story pointed to success with experimental LPIs, however the New York City Department of Transportation is apparently resisting, says the paper. A former DOT director of planning told the publication, under condition of anonymity, how it works. There are no formal "warrants" or requirements when engineers look at intersections, "they are primarily looking to see that an LPI won't degrade vehicular 'level of service.' DOT's attitude is, 'We will do pedestrian safety, but only when it doesn't come at the expense of the flow of traffic.'

"To the traffic engineers, 'it's all about big maps and traffic counts. Guys in Lexuses stuck in traffic jams are simply more important than Mexicans crossing the street.'

"The real reason there is no LPI at the intersection where Juan and Victor died is because the traffic engineers who control and run New York City's Dept. of Transportation fundamentally disagree with the entire concept of traffic calming. In the world of the traffic engineers, taking away five seconds of green time from trucks heading west to the Battery Tunnel is a serious risk. It's a major sacrifice.

"The way that DOT operates is scandalous. But the scandal is not so much about incompetence or corruption. The DOT is controlled and run by an insular and widely discredited group of professionals called "traffic engineers." The more effectively the traffic engineers do what they perceive to be their job, the more choked and immobilized New York City's streets become."

The newspaper also interviewed former DOT Traffic Commissioner and Chief Engineer Sam Schwartz.

"'Traffic engineers have failed,'" Schwartz says. "'If you compare the accomplishments of our profession [in this country] over the last 50 years to the medical profession, our performance is equivalent to millions of people still dying of polio, influenza and other minor bacterial diseases that have been cured.'"

In contrast to traffic engineers' and highway lobbyists' claims reducing traffic supply actually helps traffic congestion. John Kaehny, executive director of Transportation Alternatives, told the *New York Press* about a British study that examined cases where roads were taken out service and found that "a significant amount of vehicular traffic simply 'disappeared.'"

When it wasn't convenient to drive anymore, commuters took a different mode of transit, traveled at a different time of day or made fewer, more efficient trips. They concluded that if you tighten roadway capacity and make a city less accessible to the automobile—particularly a city that offers good transit options–there will be less traffic congestion, higher quality of life and significant economic benefits."

Property taxpayers subsidize cars, buses, trucks, highways

Land consumption, especially for highways, has a nasty hidden price not covered in fuel taxes, tolls, or fares—the recurring lost economic activity and property taxes when land is buried by pavement and tracks. Taxpayers, especially property taxpayers take the hit.

Here's one example from the City of Burlington, Ontario, Canada, a suburban city near Toronto. A typical 4 or 6 lane highway in their area needs about 150 metres (500 feet) of right of way; for 2 kilometres in length (about 1.2 miles) and some extra land for an interchange the total area would be about 38 hectares. If the same area were developed for modest single-family homes, we could achieve about 16 units per hectare. Each household would pay about $3,500 per year in taxes.

Here's the conversion for Americans: 38 hectares for 2 km of 4 to 6 lane highway with interchange converts to 93 acres for 1.243 miles (hectares to acres: 2.4710, kilometers to miles .6214). 2.375 houses per hectare, 5.869 houses per acre.

16 modest single family housing units per hectare amounts to 608 houses on 93 acres. Each house pays $3,500 CDN or $2,625 US annually in taxes or a total $2,128,000 CDN at an exchange rate of $1CDN = $0.75 US, $1,596,000 US in ANNUAL foregone property taxes.

150 meters width of straight stretch equals 492.1 feet (meters to feet, 3.2808) 1 mile of straight stretch equals 25,983,936 square feet or 59.65 acres. That works out to be 350 homes. That amounts to *$1,225,000* CDN or *$918,750* US in ANNUAL foregone taxes.

Mass transit not always viable

As the above studies show, mass transit consumes less land: 24 feet compared with 167 feet.

Mass transit also is far less polluting than automobile traffic, although buses do get snarled in the same traffic as private vehicles. But mass transit can't compete with private cars on most commutes, i.e. suburb-to-suburb, since offices have moved out or located to car-served and badly congested "edge cities."

The APTA study says only 49 percent of Americans live within one quarter mile of a transit stop. The US Census Bureau reports that only 4.6% of Americans take mass transit, compared with 75.7% who drive alone.

Mass transit requires operating subsidies from 30% to 75% or more of costs. Mass transit operators can't recover all their costs from their fareboxes because they are competing with subsidized private vehicles.

Environmental damage

Commuting in North America means an individual using their private vehicle for most work trips, but there is a steep price for such convenience.

The Surface Transportation Policy Project (STPP) reports that motor vehicles are the largest source of urban air pollution, generating more than two-thirds of the carbon monoxide in the atmosphere, a third of the nitrogen oxides (which react to form smog), and a quarter of the hydrocarbons (which also form smog).

Some pollutants emitted by cars and trucks are known to (or are likely to) cause cancer, including toxic substances such as soot (fine particulates), benzene, arsenic compounds, formaldehyde, and lead. In the 1996 National Toxics Inventory, the Environmental Protection Agency estimates that mobile sources such as cars, trucks, and buses release about 3 billion pounds of cancer-causing, hazardous air pollutants each year.

The STPP reports that according to the Intergovernmental Panel on Climate Change (IPCC), the 1990s was the hottest decade of the 20th century. The IPCC further predicts that the earth's average temperature will increase by as much as 10° F during the 21st Century, leading to record heat waves, droughts, an increase in frequency of severe storms, rising sea levels, and the migration of insect-borne tropical diseases like malaria.

Carbon dioxide (CO_2) is the largest contributor to climate change and the transportation sector is one of the largest sources of CO_2. Cars and light trucks emit 20% of the nation's CO_2 pollution. Each gallon of gas consumed pumps 28 lbs of CO_2 into the atmosphere—19 lbs from the tailpipe and nine pounds from upstream refining, transporting and refueling. The US transportation sector, as a whole, is responsible for about 32% of American CO_2 emissions, and almost nine percent of the world's total CO_2 emissions.

Commuting consumes oxygen-producing greenspace that no one has set a price tag on; and few are willing to pay for its use. Yet, unless we quit treating the environment as a free lunch, we will pay for it with our lives.

Oil/gas shortages

Chicken Little wasn't wrong, only a little early. Respected *Vancouver* (BC, Canada) *Sun* journalist, Barbara Yaffe, cited several studies from well-respected organizations in her Feb.7, 2004 column. They include:

* The Oil Depletion Analysis Centre (London, UK) issued a press release in Feb. 2004 "Oil Supply Shortages Likely after 2007." The full report was carried in the Jan.2004 *Petroleum Review*, published by the Energy Institute (London, UK).
* Association for the Study of Peak Oil and Gas predicted in 2003 world oil production would peak around 2010 or slightly earlier.

* The US Geological Survey reported also in 2003 that oil and gas production would peak between 2011 and 2015.

Such predictions are difficult, said Yaffe, because of uncertainties surrounding politics, exploration results, conservation efforts and alternative energy discoveries.

"While a flood of new production is set to hit the market over the next three years, the volumes expected from anticipated projects thereafter are likely to fall well below requirements," says the report.

"'The rate of major new oilfield discoveries has fallen dramatically in recent years,' Yaffe cites author Chris Skrebowski. Mega-field discoveries numbered 13 in 2000, six in 2001, two in 2002 and none in 2003.

"'From 2007, the volumes of new production will likely fall short of the combined need to replace lost capacity from depleting older fields and satisfy continued growth in world demand.'

"Conclusion: 'The world may be entering an era of permanently declining oil supplies in the coming decade.'"

Concludes Yaffe: "It's difficult for most of us to stand back and contemplate the big picture as we go about day-to-day living. But someone had better start acting on this news, because it won't be long before the associated environmental consequences start intruding on the daily lives of every last one of us."

Here's something else to think about: much of the existing and future oil comes from parts of the world not exactly known for their political stability or from being free of armed conflict: Saudi Arabia, Kuwait, UAE, Iraq, Iran, Libya, Russia, the former Soviet republics, Indonesia, Nigeria, Venezuela, Brazil, and Mexico. Their existing and wannabee political leaders know that it doesn't take too many twists of the shutoff valves to bring back the blocks-long gas pump lineups, no matter how many wells are drilled into the Arctic National Wildlife Refuge and wildlife sacrificed in the process.

Many have advocated the use of clean-burning hydrogen in vehicles. But that is some ways off. There are no affordable, viable hydrogen engines on the market; nor there is an efficient distribution and storage system in place as there is for oil and gas.

And, barring a miraculous turn to renewable energy sources like solar and wind power, to produce and distribute hydrogen requires, ironically enough, fossil fuels and metals such as palladium and platinum, as revealed in an article published January 6, 2004 in *The Village Voice* (New York, NY).

Labor shortages/rising costs

With the baby boomers retiring there may well be a labor shortage that will cost employers higher wages while employees will have to pay higher taxes to cover the Social Security shortfalls. Those "boomers" are living longer, which means more government expenditures on Social Security and Medicare.

* A study published by the Employment Policy Foundation (EPF) (Washington, DC) in August 2003 predicts demand for labor will outstrip supply by 22 percent in the US over the next 30 years with most of the unfilled jobs likely to be in highly paid

managerial and professional occupations. Every job unfilled will cost the economy $100,000 per year in lost output and ultimately $3.5 trillion in annual output in current equivalent GDP.

* The National Association of Manufacturers forecasts a skilled worker gap that will grow to 5.3 million workers by 2010 and 14 million in 2020.

* A McKinsey study entitled, "The War For Talent," predicts that the demand for talented employees will rise by 33% over the next 15 years. It also predicts a 15% drop in supply.

Employers are now analyzing what is going to happen with their workforces. For example, *Business 2.0* reported in September 2003 that a quarter of Cigna's IT workers will pass 55 in the next 10 years. More than a quarter of software maker SAS will retire by 2010.

Employees want more time for a life. They want to spend more time with their families, participating in outside activities, and partaking in career development opportunities.

More employees don't want to commute. A July 2002 report by the Information Technology Association of America on its "Anytime, Anyplace, Anywhere" survey found that 54% of American voters felt home working would improve their quality of lives. Moreover 36% would choose home working over a pay raise, 43% felt they would be a better spouse or parent if they were able to home work, and 46% think that the quality of work would improve if they were able to work at home.

Diminishing labor quality

There are growing reports that American labor quality is declining, requiring employers to look farther afield for workers and/or spend more on remedial training. The alternative is moving their businesses offshore to lower-cost developing countries where labor is, sadly, often of a higher quality.

A study released Feb.9, 2004 by the American Diploma Project (ADP), and reported widely in media outlets like *The Washington Times,* said that more than 60% of employers rate high school graduates' skills in grammar, spelling, writing, and basic math as only "fair" to "poor."

The report also reveals 53% of college students take at least one remedial English or math class.

Employers often have had to pick up the slack. One study estimated remedial training cost one state's employers nearly $40 million a year.

In a Feb. 11, 2004 *New York Times* op-ed, Nicholas Kristoff called for more math and sciences education to keep American jobs onshore. He cited the Trends in International Mathematics and Science Study that placed the US at 17th, just ahead of Latvia, in a recent study of eighth-graders.

Jeff Furst, president of staffing firm FurstPerson, reports that in many communities, call center applicants are less skilled than in the past. Reading, writing, and comprehension abilities are diminishing and so is the work ethic.

In a typical community, only 30% to 45% of candidates tested will meet the abilities and behaviors to do the work, "which really impacts the ability to hire employees that meet the job requirements," he says.

Out of 100 interested job candidates about 10 will ultimately be hired. This takes into account pre-screening, selection testing, and background checks. On the other hand those new hires will stay longer and perform better.

Many applicants do not know how to multi-task or to problem-solve. Those are key requirements in today's multichannel multipurpose call centers.

"While there are many job candidates in most markets, it is becoming more challenging to find qualified candidates," Furst points out. "For many call centers, that means increasing the recruitment budget and changing their hiring strategies."

There is a bad joke from northern England—where the Industrial Revolution began—that illustrates this dilemma.

A man walks by the local textile mill and sees a sign on the sooty red bricks posted "Handyman wanted. Apply within." So he tucks his thumbs into his suspenders steps into the office, and asks about the job. The receptionist rings up the managing director, who then asks the applicant to be sent in. The director then asks the man to sit down.

"I see you've come for a handyman's job," says the managing director through his wire-rimmed spectacle. "Are you good at carpentry, banging in a few nails, keeping the frames and woodwork fixed?"

"Sorry guv, never touched a hammer in me life. Me dad smashed his thumb once with one. Lost his hand from the gangrene."

"Hmm, I see. Are you good with plumbing? Fixing the steam pipes and the indoor lavatories?"

"Steam pipes, indoor lavatories? You got that there? Fancy place you got here, guv. Sorry."

The managing director frowned. "What about electricity, wiring, fixing the dynamo and the lights?"

"Electricity? Electrocution like they do down in Florida? All them electric and magnetic fields causing leukemia?!"

The managing director had had enough. He stood up and glowered. "You've come in for a handyman's job. What's so bleeding handy about you?!"

"Me?" smiled the applicant. "I live just around the corner!"

Employment risks

Most workplaces are unsafe. People working in close proximity spread illness. Premises offices are poorly ventilated, with minimal air changes, keeping airborne diseases and toxins inside. This helps to spread the new, dangerous, contagious, drug-resistant disease strains that are increasingly showing up.

For instance, the Seattle *Post-Intelligencer* reported in March 2003 that up to 40% of the strains of Streptococcus pneumoniae, which also causes meningitis, sinusitis, and ear inflections, could be resistant to penicillin and erythromycin in the near future. The Associated Press, in a story posted on the (Toronto, Ontario, Canada) *Globe and Mail's* website Sunday, February 29, 2004, states that British deaths from an increasingly drug-resistant superbug, methicillin-resistant Staphylococcus aureus, are 15 times higher than they were a decade ago. The article went on to say that some strains of staph have also acquired resistance to vancomycin, a drug that is considered by medical professionals as the "last line of defense" when all other antibiotics have failed.

"Although new antibiotics are constantly being developed, some experts fear it is only a matter of time until virtually every drug is useless," warns the article.

The cover story from the June 5, 2000 issue of *BusinessWeek* titled, "Is Your Office Killing You?," reports that some buildings draw in only 5 cubic feet of fresh air per person per minute. "That is almost enough to keep people alive," the article quotes New York architect Robert F. Fox Jr. The American Society of Heating, Refrigeration and Air Conditioning Engineers, recommend 20 cubic feet, below which sick building syndrome increases.

Then there are the fears of deadly respiratory disease pandemics, spread by human-to-human contact. The SARS outbreak in 2003 practically paralyzed Toronto, Ontario, Canada. In response, many employers had their employees work from home and used conferencing techniques instead of commuting or traveling. There are unnerving concerns in some medical quarters that the bird flu virus, which ravaged Asian poultry in the winter of 2003/04, could mutate to enable human-to-human transmission. Researchers had found scary similarities between bird flu and the Spanish flu virus that wiped out over 20 million people worldwide in 1918-1919.

Disaster risks

There is still the danger of terrorism. Workplaces are vulnerable, especially high-profile office buildings, both directly from bombings and indirectly from mass evacuations.

But there also are the more common disasters, e.g. power blackouts. The August 2003 blackout that hit the northeast US and Canada trapped people in elevators and trains and paralyzed entire cities. The same may well happen again.

The location of offices and facilities also has made matters worse. Those with offices near embassies and consulates face demonstrations, or worse. Firms that locate their offices near chemical plants and freight tracks with grade crossings increase the risk that their employees will face fires, gas leaks, or be trapped in an accident.

Cost squeeze in the workplace

At the same time, organizations are looking at ways to cut costs, even as the economy improves. Corporations want profits while government departments and other non-profits face budget reductions or want to do more. Key among the costs: healthcare and absenteeism.

* A 2003 survey of CEOs by PriceWaterhouseCoopers reveals that healthcare benefits costs companies nearly $5,000 per full-time employee.
* A February 2003 report by Mercer Human Resource Consulting and Marsh reported that employees took an average of 12 unscheduled days off annually. Assuming $40,000 in wages/benefits per employee working 5 days a week they are nominally

available, 240 days a year 50 weeks, 10 legal holidays each employee/day costs approximately a $167/day, or about $2,000 a year.

Significantly more people are taking time off for personal reasons. A study by CCH published in 2002 and reported by the *Associated Press* showed that "personal" as a cause for absences jumped to 24% from 20% in 2000 and due to stress to 12% from 5%.

As a result organizations have been deferring raises, cutting back or eliminating benefits, consolidating departments, laying off staff, and automating functions. Yet these moves may backfire as labor shortages begin to bite.

In an increasing number of cases, organizations are moving their work outside of the US to low-cost countries in Asia, Africa, Central/South America, Caribbean and Eastern Europe. Forrester Research estimates that at least 3.3 million white-collar jobs and $136 billion in wages will shift offshore by 2015. But there are consequences of offshoring.

There are reports of customer service, customer retention, and internal help desk and communications problems caused by poor cultural affinity between offshore employees and Americans. There are also security risks from having American data handled in less-safe developing countries, and a talent drain that could harm American products, services, and research and development. It is not clear what productive jobs—if any—will replace those moved offshore.

Cost squeeze at home

When and where there are rising real estate prices and taxes people often must buy or rent homes farther away from traditional workplaces. People also are sometimes forced to find lower-priced houses and apartments that are further out if one of the family members loses their job, or takes a less-paying one.

Either way increasing the distance between home and "work" drives up commuting times and costs, hikes stress and risks, causing a loss in productivity from tardiness, absenteeism, and accidents. The growing demand means more public money, i.e. more taxes, must be spent on improving and maintaining roads and transit systems. At the same time, more tax revenues will be lost through the consumption of additional land to accommodate the commuting populace's transportation needs. More transportation demand also leads to more pollution, illness, and economic waste that threaten to culminate in a lower quality of life for all.

♻ THE WAY OUT OF THE CRISIS

There is a way out of this crisis: home working. Home working minimizes commuting needs, including the need to extract more taxes from individuals and organizations to pay for road and transit infrastructure and operations. It also lowers environmental damage and costs, and in doing so helps us to keep our health. The Telework Coalition cites AT&T research that shows their home workers cut driving by 100 million miles a year, eliminating 50 million tons of CO_2.

Home working ironically enables mobility, by removing commuters from the traffic stream. That leaves roadspace—for those who need to "be there" in person, and for trucks and emergency vehicles.

The results are far quicker to realize with home working than with major transportation investments. Compare: 6 months to a year for home working decision to implementation compared to 6 to 12 years for new roads or transit systems.

By lowering commuting demand home working keeps more of the oil and gas in the ground —petrochemicals are still the most efficient and safest internal combustion fuel sources (they are not making dinosaurs like they used to anymore).

Home working has proven to attract and keep workers. It is attractive to retirees, enabling them to stay productive and to keep contributing to the economy, forestalling any Social Security "crunch."

Home working minimizes health risks; people working at home can't spread diseases as easily. It also mitigates disasters. Being trapped at home is safer than being trapped in a building. Dispersing your workforce through home working means some employees are still able to work when disaster strikes.

Home working is also the last cost-saving frontier. Real estate expenses are 5%-10% of operating costs. At the same time home working has proven to lower staffing and turnover costs and boost output.

With home working your employees can live for less in outlying areas, enabling them to save and spend more, and enjoy a higher quality of life—just as long as those areas have the communications networks and transportation access needed for their jobs. House prices, and taxes in such locales can be 25% to 75% less than in neighborhoods lying in so-called "commuting distance" of traditional workplaces.

Because many of these smaller communities have few other employment opportunities, offering home working pays off for you by keeping them loyal to you. You get more out of your investment in your employees and you lower your turnover costs.

With home working you can afford to maintain and grow American (and Canadian) jobs in North America. You are not limited to the labor pool that exists "in commuting distance". Depending on the job, the functions, and the communications requirements your employees can work from anywhere. That keeps your customers, employees, elected officials and the public satisfied. Without a thriving work force who will buy or create demand for your products or services?

✪ BUT WE NEED A REVOLUTION

To enable home working requires a revolution in how we look at and manage employees. There is now such a revolution underway. And that's what this book is about...

Chapter 1: What Are Home Workplaces?

Home working is as the name says, working from a home, i.e. one's or one's employees' permanent residence. The home workplace is an alternative or adjunct to the employer-supplied "traditional" workplace—traditional only since the mid-19th Century—i.e. the office, which I call, *premises offices* or *premises workplaces*. I put quotes around the word *traditional* because prior to the mid-19th century most people worked from home.

Home working refers to employees carrying out their job tasks someplace at their homes, full-time, *exclusively,* or *occasionally* (more later on those distinctions). The employees have wired or wireless phones: they also have computers, fax machines, and printers—depending on the need and frequency of home working—which they or their employer supply and maintain.

Home workers can be your employees, but they also can be self-employed contractors. In addition, they can be employees of outsourcers or service bureaus that you hire to handle functions such as customer service, sales and support, internal help desk, accounting/bookkeeping, employment screening, programming, proofreading, transcription and translation. These options are examined in detail in Chapter 11.

Home working is a method of *virtual working*. Virtual working is the carrying out of employment-related tasks that are not totally dependent on employees being at premises, or being face to face with others. Mobile working, and audio-, data-, Web- and video-conferencing, which I label *conferencing*, are examples of virtual working. So is data collaboration, through the use of groupware. *Virtual workplaces* are the space, stationary or mobile, where the work is being carried out.

Exclusive home workers can be mobile workers. They can, and often do use conferencing. I am an exclusive home worker, although I occasionally mobile work, and I often conference. Many of my colleagues who occasionally home work also mobile work and conference.

I explore mobile working and conferencing in Chapter 11. For greater detail on these topics read *Going Mobile* by Keri Hayes and Susan Kuchinskas and *Videoconferencing: The Whole Picture* by James Wilcox. Both books are published by CMP Books.

But home working is usually but not always virtual. The work sometimes requires home office employees to meet with customers and clients at their homes. Chapter 10 explores this special circumstance in depth.

○ HOME WORKING CHARACTERISTICS

Home working has the following key characteristics:

Employees handle some or most of their job tasks at home

Obvious and simple, but this can be challenging to implement. That is what this book is about.

Employees have adapted their homes to become workplaces; the newest condos and houses now are designed for "live/work"—space is set aside for work purposes. In other abodes, the spare desk or the spare bedroom becomes the office; Junior does the home-work in his bedroom and the mother-in-law sets up her fortress on the futon. When Junior moves out (and learns the hard way from neighbors that keeping the sound down may not be a bad idea after all), or when the mother-in-law moves to "a warmer climate" (i.e. Florida or someplace *real* "down south" that is unlikely to freeze over) you may have the space for a home office.

The tasks do not require other individuals to complete in-person

Working from home, practically by definition, is working alone, almost always without another person in the same work area. As the explained later on in detail in this chapter, there are many tasks that can be handled without face-to-face interaction with colleagues or with clients and customers.

Notice the terms: in-person and face-to-face. Home working can entail employees directly interacting with colleagues, clients and customers by communications i.e. wire-line, wireless, telegraph, Telex, fax, 'snail mail,' e-mail, instant messaging (IM), and/or conferencing.

Employees are self-starting, self-disciplined

Successful home working employees are highly responsible and mature, show up and leave when they are supposed to, are good at following organization policies and per-form their tasks well and on time. They do not need someone staring over their shoul-ders or checking up on them every few minutes.

Employers manage by performance

Successful home working arrangements have employers—from supervisors on up—who manage by performance, which is what employers are paying for, rather than warming seats and taking up real estate. Managing by performance entails examining how well employees meet goals that line up with organization objectives.

These managers trust their employees to do what they are supposed to do. They check up on home workers—by calling, e-mailing, instant messaging, or in specialized cases such as call centers, monitoring calls and e-mails—the same as they do in prem-ises offices.

○ WHAT HOME WORKING ENTAILS

Home working involves your employees setting up and working from offices in their

HOME WORKING
NOT TELEWORKING/TELECOMMUTING

I have avoided where possible the terms "Teleworking" and "Telecommuting" for a number of reasons:

1. Teleworking often includes mobile working as well as home working. But mobile working is quite different from home working—in physical requirements, functions, investments, legal issues, and benefits.

   To accommodate home working there needs to be a fixed home office premises with furniture, equipment, and connections. Mobile working has no such requirements. Mobile workers can literally be baseless or have a home office, a premises office, and/or satellite office bases. Home workers can be connected via wireline or wireless; mobile workers are line-free. While there are no legal restrictions on working while stationary there are increasingly strict laws regulating work while traveling, e.g. non-handsfree cellphone usage is illegal in many locales.

   Also home workers are not pumping emissions into the air; the same is not true for most mobile workers—they also take up costly road space, adding to traffic congestion and costly delays.

2. Telecommuting is not technically accurate. The only "commute" is connecting to the employers' servers, which is the same as if the home worker worked in a premises office; the servers are often not in the same building. Home workers typically get their calls, faxes, and e-mail directed to their home offices.

3. The terms "Teleworking" and "Telecommuting" raise red flags with many senior managers. They connote slacking off, playing "Terminate the boss" on their laptops—as if that doesn't already occur behind the boss's back in premises offices.

But home working doesn't have such negative baggage. Who can argue with "home work"? You do your work at home and do it well just as you were expected to do in school.

homes—in accordance with requirements that you, as the employer, set out—enabled by investments that both you and your employees make. Home working is, therefore, a *partnership* between employees and employers because as employers you, at minimum, are taking up some amount of space in your employees' homes.

Employees must make changes in their personal lives and environments to accommodate home working. Like arranging for childcare during the day, telling dearest daughter when she comes home from school not to crank up the "blessed" stereo, and keeping Poopsie and Hairball from doing their "thing," any of which can interrupt your work.

Employers also need to adapt to home workers. Those adaptations range from faxing rather than hand-delivering papers, conference calls rather than in-person meetings, online log-ins rather than punching a time clock, and by managing and making review and promotion decisions on employees' performance rather than what is seen by shoulder-surfing.

Home working specifications can range from simply requiring the employees to inform you and your colleagues when they are going to work from home, to formal login procedures. The investments can range from setting up web e-mail accounts so home workers can access e-mail via the Internet, to sophisticated off-premises extensions, to phone switches that transfer calls and screen-pop customer files.

The home office can be a basement, bedroom, back closet, or a desk at the back wall. As an employer, and as a condition of permitting employees to work at home, you can also specify that the office be in a separate room, that the computer screens not be seen readily from the street or by neighbors, that the computers be lockable, and that the office have lockable filing cabinets to protect data.

You stay connected, via a wired or wireless connection, with your home employees for work, communications and performance assurance. That includes monitoring their calls, e-mails, instant messages, Web activity, and work tasks. These employees can also deliver the work phone-only; and/or they can fax, Telex ™, or semaphore the material. Additionally, they can courier, snail mail or dog-cart their work to you on "tangible media," i.e. floppies (remember them?), CDs, MemorySticks ™, or old fashioned paper.

To supervise these employees you pick up the phone or send them an e-mail or instant message as well as grade and remark on their output. You can require them to come to your premises, say once a week or once a month. You can also visit them at their home. Chapter 9 examines communicating with and managing home workers.

Home workers can hot-desk when they use temporary space at your premises offices. That space can be a spare computer, dedicated computers and cubes or desks in shared spaces for home or mobile workers, or cubes and desks for their laptops. Chapter 7 looks at hot desks.

Home workers can also satellite work from mini-or satellite offices near their homes and on the road, rails, seas, in the skies and in hotel rooms, cafes and in waiting areas as mobile workers. They can audio/data/video/Web-conference from their homes (and from satellite offices) instead of making business trips, which saves you money and prevents productivity losses caused by travel: not even seamless wireless coverage on planes can make up for time wasted going places and in message tag. Chapter 11 looks at these alternatives and adjuncts.

Because home working is a partnership between employees and employers there must be policies and agreements on who pays for what in home offices: voice/data, equipment and sometimes even furniture. Chapters 2-6 cover the issue for voice, data, and facilities, respectively, including power backup.

✪ TYPES OF HOME WORKING

Home working comes in several different strands with their own benefits and costs. You can mix them with mobile working and premises (including satellite) working to get the work environment that best meet your needs and budget.

You will need to review the components: phones, computers, connections, and furniture that all strands use; who pays for what; and specific issues that arise from such work environments. For example, who gets to work at home, how to communicate with staff, how to implement data protection and how to devise a home working policy that covers everything.

Exclusive Home Working

Exclusive home working refers to those employees who are based at home and handle the work for their employer primarily from their home office. Exclusive home workers can work either full-time or part-time, but the vast majority of their work is performed from their home office. These employees do not need to be on *your* premises, consuming *your* organization-subsidized space, furniture, washrooms, parking spaces and yes, coffee machines. Exclusive home workers therefore can *boost* your profits and *free up resources* enabling you to better meet your goals.

By the same subway token, exclusive home workers do not regularly work on premises and are typically not assigned desks and equipment there. However, employers often supply exclusive home workers with equipment (phones and computers) and pay for their communications costs—but not always.

Exclusive home working employees can be asked to work at premises office occasionally, i.e. once or twice a week, once a month, or as needed. They are accommodated at premises by assigned desks or by hot-desking. But they have their principal base of operations at home; rarely will (or should) employers supply them with two sets of equipment.

Exclusive home working employees have the same way to communicate with customers and clients as exclusive premises workers. Employers may choose to directly forward inbound calls to employees, such as through "extension" switches, give out home workers' phone numbers, or have employees pick up voice mail from mailboxes (for more discussion see Chapter 2). Whatever method chosen, employers do not have to pay for idle phones and cabling.

Faxes and "3-D" or "snail mail" and packages may be sent to the premises and forwarded or sent directly to the home-working employees' homes. In some companies, however, the home workers pick up their faxes and mail when they come into premises for regular meetings or part-time work.

These employees receive most, if not all, instructions and supervision and stay in touch with colleagues and other departments by communication methods such as phone, faxes, e-mails, instant messaging, Telexes, and carrier pigeon rather than in-person. Employers also have the option of seeing their home workers directly at their employees' home offices—with the written permission of the employees.

Occasional Home Working

Occasional home working employees are those employees who are based at premises offices, but who work occasionally from their home. Allowing employees to work from home on occasion is arguably the best first step towards permitting them to work from home exclusively. Having employees work from home occasionally gives an organization vital experience and information in developing an exclusive home working program.

Examples of occasional home working are where a premises-based employee will ask a supervisor if they can work from home on a specific day: they could be waiting for the plumber or their cousin Godot to arrive. Or they could say they are feeling a little sick.

Sometimes employees will occasionally work from home if there is a storm that blocks roads or knocks out the employers' premises. Or in the case of the 9/11/01 terrorist attacks they will work from home if they cannot get to the premises... or if there are no premises left.

Occasional home-working employees may use their own personal computers, but sometimes companies supply them a computing device (e.g. a laptop) that can be used at-home and at-premises. If they use their personal phones, employers may accept phone expenses.

Such employees typically have their calls transferred from their at-premises machines to the landline phones at their homes or to their cellphones. They usually pick up their snail mail at-premises, but if urgent, they can have faxes, packages, and mail sent to their homes.

Typically, occasional home working employees don't have secure access, especially if they use their personal computers. Employers avoid loading corporate applications and software on personal machines that they have no control over because such access could pump into a reeking stew of hidden computer viruses onto the network, and onto everyone else's machines.

Instead, employers supply occasional at-home workers and others with e-mail or Web mail accounts so their home working employees can access their e-mail over the Internet. These programs have limited utility; they often don't support calendar or database functions, and they often time out after 30 minutes with little or no warning. Still these accounts allow occasional home workers to stay in touch.

Note: *First, the tools you select for your home workers must enable them to deliver equivalent if not superior performance than at a premises office. Second, whether exclusive or occasional, you should own the home workers' equipment. Only by owning the tools can you control what is on them, which in turn, allows you to protect your network, data, and customers from computer viruses and data theft. You also better protect your outfit from legal action resulting from alleged violations of privacy and computer protection laws.*

✪ HOME WORKABLE TASKS

There are many tasks that can be home worked exclusively or occasionally. Examples include:

Editing and writing
Graphics design
Proofreading
Transcription
Translation (spoken and written)
Legal
Credit/background checks
Recruiting/interviewing/testing
Research (chances are that most material produced since the mid/late 1990s has been archived electronically)

Business/program analysis and planning
Computer programming
Data and report entry
Database management
Website hosting
Customer service, reservations, support and sales
Internal help desk
Accounting and bookkeeping
Programming
Engineering and design
Sound recording
Teaching (including lecturing, and correcting students' work)
Tutoring
Training
Management
The list goes on...

The common characteristics of most of these tasks are that they require *information processing or knowledge work*. Information processing is the receiving, handling, manipulating and communication of non-tangible objects: data. That data can be product orders, service requests, orders to employees, questions from employees to other employers or supervisors, instructions (which is what computer programming ultimately involves—instructing machines, i.e. computers) complaints, advice, articles, music, photos, graphics, and yes this book.

Information processing is also known as knowledge work because information is knowledge in some form or another. Knowledge is communicated through the same means as information: spoken, written, or visually through graphics and images.

Face-to-face interaction with customers, clients, and/or colleagues is usually not essential to complete such tasks. An accountant, attorney, bookkeeper, secretary, or transcriptionist could be working at-home and receive visits from superiors, colleagues, and customers to go over work details. But it is an option.

To expound on the point made in explaining the characteristics of home working, working at home *does not mean* working in isolation. Communication: from messengers and dispatches to instant messaging have enabled home workers to stay in contact with colleagues, clients, and customers.

Yes, you can make and manipulate tangible objects at home. That arguably predates written information processing: you don't need to write to learn how or to bake bread, make implements, or sew garments. Where I live, in central Vancouver Island in British Columbia, Canada there is a very active home-based business association representing nearly every manner of business.

But there are many more information processing applications that can be home worked at the same, if not greater efficiency levels, than at premises offices. The customer who dials a toll-free number can't tell if the agent is in a call center located in Indiana, India, or in their home in Indiana or India.

My corporate laptop and my personal desktop machines are of the same power and capability as those of my premises-working colleagues. But I can't easily have a printing press in my home and if I did, it wouldn't possess the same capability as the one that printed this book.

✪ NON-HOME WORKABLE TASKS

There are many tasks that can't be readily worked at home. Examples include:

* Where manipulation of material, equipment and machinery is involved. For example, assembling items, repairing by hand and with tools, driving.
* Where you must physically interact with people. For example, medical, retail sales and service.
* Seeing clients and customers at their premises, at yours, or at events such as trade shows.
* Training when direct touching and operation of equipment and the handling of tangible products is involved, including training others face-to-face.
* Teaching, where teachers need to read their students' in-person behavior.
* Negotiations and dealmaking, where it is vital to read others' full body language.

You will need to look at the mix of home workable and non-home workable tasks and how much of the employees' workday or workweek is devoted to each. You may be surprised to find out just how many of those tasks are home workable and how much time your employees spend on those tasks.

That, in turn, may open some opportunities for deploying a home working program. Consultant Jack Heacock uses the example of dockworkers that need to handle paperwork, i.e. information processing at the end of their shifts—which could be after midnight and on weekends. Instead of having to keep offices open at such odd hours, why not have the employees file the information electronically from their homes? The longshoremen would certainly appreciate it, especially after a long day on the cranes, trucks, loaders, and docks—sometimes in lousy weather.

Yet when looking at what jobs can and cannot be handled at-home, keep in mind that in a natural or man-made disaster executives and managers suddenly discover that employees can and are doing many of those "can't-be-done-at-home" functions either at-home or while mobile. Life, and work, goes on. People, like their employers, find ways to cope.

✪ THE HOME WORKABLE TREND

Keep in mind that more and more tasks are now home workable thanks to rapid advances and dropping prices in computers and communications. The growing web of broadband services: cable, DSL, wireless LAN a.k.a. Wi-Fi, are enabling data flow to homes at speeds that were possible only a few years ago exclusively at premises offices.

Today, architects, customer service reps, engineers, financial analysts, managers, nurses, and sales reps can pull up and work on massive volumes of data from the convenience of home. Technicians located anywhere can now diagnose and fix many computer problems. In my trade our blue lines, or page proofs, which were once printed out in blue and checked and corrected by hand, are now in portable document files (PDF) that can be opened, read, and corrected from anywhere.

Any task that is done principally by operating a computer can be done at home. If all the information inputs to the individual operators are data, and all that they do is operate a computer why have them on the scene? For example, sound recording can be done at home. The Associated Press ran a story July 28, 2003 on technology that enables musicians to lay down tracks from home computers instead of having to haul their cool cat tails into recording studios.

Computers are continuing to displace humans at a rapid rate. The automation trend is unrelenting for the simple fact that in developed and developing nations capital is cheaper than labor.

The people-mover systems in many airports and cities are completely automated, as are many industrial processes. Driving trains and working in mills have gone from blue-collar to white collar, from denim to Dockers ™. Even in the mid-1990s, when I was visiting steel mills as part of my job reporting on the metal industry, I was astounded just how much of the people-work took place far off the shop floor in glassed-in, sound-proofed, and wired-up pulpits.

The automation trend is swallowing home workable jobs like those in call centers. Organizations have deployed interactive voice response (IVR) systems, often with speech recognition that enables people to talk to machines. Many also employ Web self-service and ATMs. All are faster, cheaper and, given the sad state of American education, often better quality than having the same service and sales handled by live people.

As more people get broadband, the easier, more convenient, and less expensive it is to get news online (rather than to buy the paper version). *The New York Times* is delivered to me online via e-mail, not on my doorstep. It is worth noting that automated system monitoring, maintenance and upgrades, like for websites, more often than not, are performed by home workers.

Teaching, once thought as the ultimate in-person task, is home workable too. Computers are quickly becoming a key teaching tool in classrooms and colleges. With the growing mass of information being deposited in databases, which are easily linked together, why should students be required to commute at taxpayers' expense (including the roads), to taxpayer-owned buildings on untaxed land? My alma mater, the University of Victoria, plans to accommodate enrollment increases partly through a home working program.

Admittedly, you can't adequately teach chemistry unless students learn how to experiment in-person. But even they can do their research and read results online.

The generations that are in school now and which will be entering school are communication and computer-savvy. They will find even more ways to use technology to learn, without brushing up against the crumbling ivy walls.

Think about it. The savings offered by home working, including that of security in today's sadly violence-prone public schools, may well be more than enough to pay to equip every student with a workable desktop or laptop computer. The cost reductions, such as by eliminating school bus runs and school buses, and cutting back and selling/leasing unwanted buildings for tax-paying development could help lower tuition costs to give more students a chance at a college education.

Don't worry about socialization. Children and young students will socialize before and after-hours as do adults, but their hangouts will most likely be closer to home. Eliminating travel to and from a school reduces injury and death from motor vehicle accidents and from assaults and fights; it also could lower the odds of abductions and sexual assaults. It definitely would lower pollution and congestion caused by parents dropping off and picking up their children, and from teenagers rushing to and from school.

Virtual schools also limit the need for your employees and you to trudge in late and rush out early—on your time—to transport kids to and fro. The end result is higher productivity, for them and for you.

✪ COMPARING HOME AND PREMISES WORKING

At premises employees must travel from their homes to facilities selected, set up, maintained, and secured at *your* expense to carry out their tasks alongside other employees. Premises workers utilize employer-supplied, managed, and maintained equipment and communications.

Premises office employees have to dress according to an employers' dress codes, which at this writing have swung back to being more formal (ties and hosiery) from the ultra-casual t-shirts and jeans of the late 1990s. Even in traditionally casual offices, employees dress up to meet with clients, customers, and senior management.

Employers, with employees who work at premises offices, decide upon amenities (sometimes with employee "consultation") such as what brand or type of coffee and coffee maker goes in the break room The employers also often subsidize those amenities. The fact that it may taste like week-old dishwater is irrelevant especially if the nearest over-priced trademark-paranoid espresso bar is 15 or 20 minutes away.

With at-premises offices supervisors, senior managers, colleagues, and clients have the option of communicating in-person with the employees. But they can also use e-mail, instant messaging, fax, phone, Telex, telegraph, semaphore, and Pony Express.

The irony of premises offices is that most internal communication and external communication with clients and customers are not done in-person. The reason: it takes time away from your job and your co-workers, and when you do speak to them in-person it can derail their train of thought and it can take a while to re-rail those cars after you've spoken to them.

This change was borne out in a 2002 study of Canadian executives by International Communications Research that said 94% of managers often send e-mail rather than meet one-to-one; 67% said very often.

The Calgary Herald, which carried the story in February 2003, quoted Bob Schultz, professor of strategic management at the University of Calgary's Haskayne School of Business saying that e-mails "give managers the ability to respond to more people than before. Managers don't have enough time to do face-to-face meetings with everyone."

When I worked at-premises for *Call Center Magazine* I could go for days without seeing my supervisor or colleagues. Instead I would get e-mail or phone calls even though my cube was next to theirs. When I had to speak to them face-to-face I got polite: "what do you want but I'm busy" looks; the same looks I would give to them when they came by my cubicle.

Home working employees do not have to travel to another location to carry out their tasks. Instead they work at home, in an environment that they selected, set up, maintain, and keep secure—at their expense. Employers may or may not have had a say in the home office set up or paid for the equipment and services.

Home working employees can dress as they see fit—yes they even can "work naked" if there aren't too many prying eyes and if their workstation's upholstery is soft and sanitary enough. I wouldn't advise it, though. You never know who might walk in...and yes I wear sweats unless I have to go out. I live in a small casual West Coast city at the edge of the woods where wearing a tie is equivalent to donning a top hat, tails and white gloves.

Home employees decide and pay for their choice of caffeinated or decaffeinated mud. They buy it, they drink it, and if they hate how it's made and want to string up the person who bought the beans they can blame the doofus in the mirror.

Communication with home working employees is usually electronic (jargon-addicted high-tech marketers notwithstanding, telephones—wired and wireless—are "electronic" too) and courier. But most employers have the right to visit their employees' home offices, and will do so if the at-home offices are reasonably close to the at-premises offices

Home workers must take a greater responsibility than premises workers to ensure they have the tools and technologies they need to do their jobs: like broadband connections and phones. Home workers also must arrange for a functional work environment, free of noise and other distractions, and have methods and procedures to secure computer data, such as locked doors and filing cabinets, if employers require it.

Home workers must be self-starters and fixers. There's no one there to make sure they show up and finish on time, to look over their shoulders or to run to their cubicles.

✪ HOME WORKPLACE MANAGEMENT

Employers and managers like you are familiar with the two other virtual working components: mobile working and conferencing. But most employees who work on the road and who conference do so from premises office bases. You probably don't think twice about going on the road, or an employee taking to the highway or the runway. You think little, if anything, about employees taking conference calls, instead of making business trips.

Like many companies, you may outsource many of your functions, such as customer service/sales, drafting, programming, tech support, back office processing, and transcription to outside firms, done by people you will never see, hear, and chances are with whom you will never otherwise communicate. Increasingly these operations are being based in low-cost regions and countries like Argentina, Africa, Canada, the Caribbean islands, Central America, Chile, China, Eastern Europe, India, Malaysia, Mexico, Russia, and The Philippines.

You may also have operations in other parts of the country, or in other countries. The colleagues or employees working in those locales most likely will rarely be seen or visited by you.

But you probably think much harder about home working for you or for the employees on your premises. Sounds strange, doesn't it?

The reason is part psychological, part reality. When an employee is based in premises offices—even if they are on the road most of the time—you feel you have more control. Yet the words: "you're doing a great job" or "you're fired" have equal import and consequence whether expressed in person or electronically.

The biggest fear employers appear to have about home working is their perceived lack of control. Still, employers have little control when their employees are on the road, are located in other premises, or if they outsource to vendors located in other premises, including outside of the country.

Home working forces employers to trust their employees more to behave as adults: to come to work on time and leave when they are supposed to. Home working requires employers to evaluate employees more on what they did rather than on how often they warmed the $800 chairs you bought them.

The reality comes in deciding on how to set up and manage employees from home workplaces with the same results as if they were working from premises workplaces. In other words: home workplace management.

Ask yourself, how do you manage your traditional, office-based employees? Is it via management by wondering around or with annual performance evaluations?

Consultant Jack Heacock's Number One Rule for virtual workers (home and mobile): ***Employers must treat, measure and pay home office workers the same as traditional, centralized premises office workers.***

In my case, I am graded and evaluated on how well I do my stories, whether I get them in on time, whether the stories serve my reader base, and on any other work, such as online that I am assigned or agree to take on. Also factored in are how often I come up with new and interesting story ideas, how well I communicate and listen to others, and how I work with my editorial team. Nearly all of these factors are location-independent.

The last is a little different because I don't see my colleagues and they don't see me. But that should not, and in my case does not, make any difference in my performance or my performance rating. If my colleagues place an inquiry via a call, e-mail or instant message, or I do likewise to them, the inquiries have the same impact and meaning as if my colleagues knocked on my old cube wall, or I knocked on theirs. The contact and the information in the contact and interaction is what counts, not the physical presence of the parties.

You can (and should) screen *all* employees for attributes that enable successful home working, e.g. self-discipline, problem-solving abilities, and ability to work independently. You should apply how employees deploy those abilities in subsequent evaluation practices and procedures.

Jack Heacock suggests that when employing home working strategies, home workers become *more* measurable than premises office counterparts. The reason: the vast majority of all efforts (on the part of the home worker) are electronic and thereby quantifiable.

"Just as the ACD (automatic call distributor) measures everything worthwhile that a call center agent accomplishes, much of that technology is now available via 'Intelligent Networks' for the non-call center work at home employee," says Heacock. "An employer can frequently measure a home or mobile worker better than an employee next to the manager who uses out-of-date industrial age measurement models."

Home working is not investment-free. It requires unique "components": voice and data hardware, software, networks, tech support, furniture, procedures to cover employee selection, training and supervision, ability to minimize background noise, and information security to enable home working. Many of these components require buy-in from other departments such as HR and IT.

Employers must set out separate home working policies for employees and supervisors to follow. If employers require exclusive home workers to come on premises, i.e. hot desking, they need to set up policies and facilities.

Yet many of these methods, investments and policies have already been made, or are needed to be made in more a limited scale with home workers. For example, you can (and should) specify the furniture requirements for home workers to prevent carpal tunnel syndrome and other injuries. But your employees may be required to pay for that equipment—which often is considered a bargain when compared with commuting and "work clothing" expenses.

You also can stipulate that your employees have locks on their home office doors and require them to shut and lock the doors when they are inside and after they leave, to prevent prying eyes and hands. Try that with a typical office cubicle! I've visited many call centers with live screens; and when I worked as a security guard in my late teens and 20s, I saw plenty of documents lying around.

For those psychological and real reasons home working is the most revolutionary of the virtual workplace methods. With the exception of complete automation or moving the entire business to a developing country—as many businesses have done—home working, especially coupled with conferencing, promises to deliver the highest costs savings and productivity gains of all the management methods.

✪ HOME WORKING COSTS AND BENEFITS

To join the revolution, however, employers need to examine the benefits and costs of home working and make a business case for it. They must determine what tasks they want to have worked from home, and how often: exclusively or occasionally, and then devise implementation strategies. Those strategies include laying out how to manage people—by their performance—and in winning over other departments and the head office.

Equally, if not more importantly, home working must be shown to work. There are many case studies that you or a home working consultant can look at to see which are the most comparable to yours.

More critical is if you informally home work, i.e. employees already occasionally work at home, those experiences *must be positive*. If there are any problems, the sources need to be found and the issues corrected. Otherwise, why should any executive approve a full-scale home working plan if you can't manage the informal home working that now happens?

In deciding whether to implement a program, planning for such a program, and managing home workers, the most important consideration is ensuring that the processes, tools, and procedures selected and implemented enable your employees working at home to deliver **equivalent if not superior performance than at your premises offices**.

The last part is very important: superiors expect better results for the investment and the cultural change home working entails.

I cannot stress this point enough. Customers, clients, colleagues, and your superiors should not detect any difference in how your home working employees carry out their responsibilities or in the quality of their work. They don't care and it shouldn't matter where the employee is as long as the tasks are accomplished to their satisfaction.

Jack Heacock and Associates put together an excellent model a few years ago for a 100-employee operation that has proved itself highly accurate and predictive of bottom-line results with clients, with less than 5% variance from their live data worldwide.

Keep in mind that many economic development agencies provide tax-free buildings, land, and other subsidies, or sometimes take the risk and guarantee new buildings to attract businesses to their communities. Those savings offset the property portion of premises office costs.

But any organization that is offered such benefits must also calculate how much it costs to outfit the new premises office. Heacock points out that while the total cost is approximately $35,000 to put in a workstation, real estate only accounts for $7,000-$8,000 of that. That leaves $26,000 to $27,000 you have to fork over.

"When you subtract $9,000 that you need in hardware, software, voice, and data regardless of where the agents are you are still spending $17,000 to $19,000 per workstation unnecessarily by having them in a premises office," Heacock states. "That doesn't include premises-only costs like maintenance and moving workstations to form new teams."

Jack Heacock's model has employers paying home workers' furniture and equipment. But you can save more money if you asked your employees to pay these costs themselves.

The flip side is potentially higher integration problems between your software and what the agents have on their desktops, and possible security risks, because you don't know what's on their computers (see Chapters 3 and 7). If you don't own it, you don't control it. Also, if you don't take the proper ergonomic steps, either by buying or requiring employee agents to buy ergonomically-sound furniture, your agents run the risk of carpal tunnel, tendonitis, and other injuries (see Chapter 6).

Not surprisingly, self-employed home agent contracting (Chapter 11) is becoming a popular option. It is the self-employed workers' responsibility to look after their equipment, facilities, and ergonomics. It also avoids some legal issues, such as taxation (Chapter 8).

Start Up Cost per Teleworker [Home worker]

Furniture, Equipment and Telecommunications	$5,900
One time charges, Set-up & Installation	$1,200
Total per Teleworker	$7,100

Ongoing Monthly Costs

Telecommunications	$ 110
Network & home office/teleworker support	$ 110
Total Average Cost per Year	*$2,640*
Year One Average Cost	*$9,740*
Year Two Average Cost	*$2,640*
(Numbers based on average typical costs.)	

Based upon the above assumptions the following results are expected:

	Recurring	*One-Time*
Savings from unneeded office buildout	$1,500,000	
Onetime start-up Costs		(710,000)
Annual Real Estate Cost Avoidance	$500,000	
Productivity ($50K salary x .15 x 100)	$750,000	
Retention ($50K salary x .20 x 100)	$1,000,000	
Ongoing annual costs		(264,000)
Net Annual Savings (1st Year)	$1,276,000	$790,000
Net Annual Savings (2nd - 5th years)	$1,986,000	
Total Five Year Savings	*$10,010,000*	

Allen Bourne, principal, Contact Knowledge, based in Plano, TX points out that it costs $24.85/hour for a self-employed home worker, in compensation, facilities, and equipment costs versus $35.86/hour for a premises employee agent. Jack Heacock says the self-employed home agent option will give you greater flexibility and may cost less than setting up your own home working program.

The downsides of using the contractor solution are less control and loyalty, especially if the self-employed workers are handling several different clients' programs. Lower employee loyalty results in higher turnover and lost productivity costs.

Employees also are generally more loyal than self-employeds, especially those who are self-employed only because they are out of work and/or who need employer-supplied medical benefits. In-house employees, on the whole, are more loyal than outsourced employees.

"However you do your home working, you're better off, in most cases, than setting up hugely expensive premises based offices," says Heacock.

Yet even passionate home working advocates such as Jack Heacock agree that you need some premises facilities, for example, a showroom office. If your organization does a lot of training, like a call center, there should be a training area. If you do architecture or engineering, it may be a good idea to have a few employees working on premises, even though much of the work has gone virtual too—with amazing computerized 3-D models.

With home workers you can reduce your building/facilities footprint by approximately 50% to 65%. There are sometimes cross-border taxation feuds between states, counties, and cities that may compel you to have small storefronts in the same states where you have a number of home agents. That may help you avoid authorities double-taxing your home agents: where they live and where your business is located (see Chapter 8).

"With functions like call centers you will always need a place where you have in-person training, especially of those who have never worked in [say] call centers before, and especially for complex programs like customer support and transportation that take months to train agents," says Heacock. "You will also need a place to show clients and senior executives, at least for now, to demonstrate what it is that you do."

❂ HOME-BASED BUSINESSES VERSUS HOME WORKPLACES

While this book is intended for managers and executives, and their home working employees, it is also very useful for home-based businesses. That is because although there are many differences between home working and home-based businesses, there are also similarities.

Differences

The key difference between home-based businesses and home workplaces (as delineated in this book) is that a home-based business is not limited to a home workable function, i.e. primarily knowledge-based functions (described earlier). Basically any economic endeavor, including hands on: architecture, carpentry, contracting, crafts, electrical, fishing, forestry, healthcare, landscaping, manufacturing, mining plumbing, repair, sales, services, tourism, transportation, wellness, and yes writing...you name it...can be home-based, even if only for administrative purposes. For example, Janice Reynolds, who edited this book (and others) runs a home-based business from her home in Pennsylvania. She also writes books and does consulting work.

The other differentiators are corporate and legal requirements. Home-based entrepreneurs don't have to connect into another organization's voice and data networks. They are freed from having to communicate and stay in touch with colleagues, customers, and supervisors through a premises office. They do not follow another's corporate policies and guidelines.

Instead, it is their responsibility to set out how they will do business, the hours they work, and how best to comply with laws and regulations. By definition home-based businesses have to do their own accounting, legal, and marketing. They also are responsible for business taxation. If they have employees, they must look after their paperwork; their home office becomes a workplace in the fullest sense of the word. Such business owners must comply with employment standards laws and regulations including workplace safety and workers' compensation.

Home-based businesses can pick the communication modes, and equipment, that suit their needs. It is up to them to select the space, fit it out, and choose the furniture. No one is going to check on them: except possibly insurance companies and building inspectors if they had substantial renovations and wiring done.

"Home-based businesses may also have zoning issues to contend with, which home workers generally do not. Some communities ban home-based businesses in some areas," reports *The Vancouver* (B.C., Canada) *Sun* in a March 13, 2004 story. Other communities have restrictions, such as the number of employees on site. Local governments are most concerned with building and especially code violations, traffic (from patrons, deliveries and employees) parking, signage, and appearance (such as materials on the front lawn, says the *Sun*). Condominiums, co-ops, and gated developments may have their own restrictions. Home-based businesses (along with home workers) may generate additional garbage, which means both must recycle whenever possible.

But governments are beginning to recognize the benefits of home-based businesses. Vancouver city councillor Peter Ladner pushed for home-based business liberalization in his city, reports the *Sun*. He pointed out the advantages: reducing traffic congestion and pollution (by eliminating commuting) as well as providing jobs and income to those who must stay at home.

Moreover today's home-based business may be tomorrow's big business. The *Sun* reported that Ladner cited a Canadian Federation of Independent Business study that showed 42% of business owners had at one time or another operated their business from their residences.

Similarities

Yet, there are many similarities between home-based businesses and home working. Which is why, if you are a home-based business entrepreneur, you may find this book very helpful.

For instance, successive chapters (Chapters 2-4) cover voice (wireline PSTN and VoIP, wireless and services such as Caller ID), equipment (phones, computers and peripherals), and data (including Wi-Fi). Chapter 5 looks at locations and facilities requirements, including power wiring, electrical codes, and backup. It also discusses how to cope with "occupants" i.e. family, friends, visitors, and pets. Perhaps most importantly, Chapters 5 and 6 examine how to correctly set up home workplace environments and ergonomics: space, lighting, temperature, noise, and furniture to maximize home working comfort and productivity.

Entrepreneurs may find Chapter 11, which discusses self-employed home workers, particularly insightful. That chapter touches on taxation and types of self-employment arrangements. For instance, you may find that it is profitable to enter into a business relationship with clients that see the benefits of home working, and prefer to work with home-based businesses.

The succeeding chapters in this book will help you put together a hopefully successful home working plan. They may assist you in refining your current program. Either way you will have made home working work.

Chapter 2: Home Voice Components

To provide home working, whether exclusively or occasionally, you will need to arrange, specify, or make investments in hardware, software, and networks. Some of these "components" will be at your premises, or on your network, others will be at your employees' home. You also will need to have IT staff and technologies to support them.

This chapter covers arguably the single most universal IT requirement: components to handle voice. The following chapter, Chapter 3, covers equipment: computers, peripherals and backing them up; Chapter 4 examines data, which includes fax and video.

Yes, voice is technically data—audio-supplied data—and voice is increasingly being carried as data, in packets (more about that later). But voice is a different form of data: it must be spoken and heard and that information, in the form of sound must be converted into electrical impulses and converted back to sound at the receiving end.

✪ ISSUES

There are several issues to keep in mind when considering supplying voice to home workers. These are:

Employee/employer needs

You and your employees need to have the right voice tools for the right job. Remember the mantra: enable home workers to deliver *equivalent if not superior performance than at premises offices.*

If your premises office workers, like customer service reps, receive calls direct from outside customers, those home workers should have the tools to do the same at home. If employees only call into the premises for messages, like computer programmers, you don't have to directly route calls to them when they home work.

The only possible exceptions to these rules are in emergencies, for which you need to plan for ahead of time. Having someone there and connected is more important than the details, like noise on the line, and background sounds. People will understand in such circumstances if there are kids screaming, or dogs barking in the background, or if they're put on hold because the other line is ringing, or the call sounds scratchy because the employee is on a cell. Typically, a quick explanation and apology at the beginning of a call is all what is needed.

Existing carrier relationships

Many organizations have preferred vendor relationships with telcos on long distance that offer substantial discounts. But home workers may have long distance carriers that are not the same as their employers' carriers. This is not an issue if employees only make local calls. However, it is an issue that comes when employees make long distance calls.

You can't reasonably expect home workers, especially occasional home workers, to switch carriers to suit your needs on their personal lines. That is an unfair intrusion on their privacy. So you have several choices. They include supplying employees with their own lines (more about that later), issuing calling cards and/or paying employee-submitted long distance bills. If your employees are working out of another country you may not have a choice in long distance carriers unless you go wireless (more later on that too).

Technology changes

Voice technology is changing rapidly, and organizations must keep up with it for all of their workers, regardless of where these employees are located. Remember the mantra...

Up until very recently the only voice choice has been PSTN a.k.a. plain old telephone service (POTS) that comes into homes and premises, typically over copper wires. But quality improvements to voice over Internet protocol (VoIP), which delivers voice over data lines, has made that technology more feasible. Wireless also is quickly becoming an alternative to residential and business lines.

So when looking at the components, and at the vendors and products, make sure you select those that give you the option of adding new features that could cut costs and enable your home working employees to work more efficiently and effectively. But remember—those features must enable you to deliver a similar service quality for the tasks your workers are undertaking.

○ CHOICES

You will need to communicate with your employees, and they will need to communicate with you, your colleagues, and with customers and clients. Chances are that means you will need to arrange and supply some means that your employees can perform such tasks by phone—wireline and/or wireless.

No means of communication has come close to voice, in the form of a phone call in real-time interactive interactions from anyone anywhere to anyone anywhere. Online text communications: telegraph, Telex, fax, e-mail, messaging and chat are slower because nobody can type faster than they can speak. Also you have to wait until you receive the entire message before you respond. But with voice you can jump right in.

Yes, it is possible to home work by means other than phone. Your home workers can receive and transmit assignments, information, and remarks by mail, fax or written phone messages taken at a local business or a friend's place, or by courier. After all, it is only within the past 30 to 40 years that phone service has become near-universal in North America, the past 20 years in Europe.

But those techniques—mail, faxes and left messages—are not real-time, and they can be expensive (I will deal with couriering later), and an imposition on others, which risks curtailment if they're not there or they're annoyed with you. There also are fewer opportunities to use these methods. Declining first-class mail volumes (who writes letters especially business letters anymore?) have forced post offices to close outlets and reduce deliveries to save money. How many businesses perform faxing duties for a fee? Many people have switched to e-mail or have their own fax machines.

Besides who lives in "Mayberry" anymore where they know, let alone trust, their neighbor for such favors like taking phone messages? Where there is a friendly local grocery—which hasn't been replaced by an eat-here-and-get-gas outlet owned by a big oil company and/or by a *****Mart in the next town? Automatic phone switching did away with the operators: I saw an old "cordboard" at a local museum—it was last used in the early 1960s.

There are several choices that you and your home working employees face in deciding how to be connected by voice: in the connections, technology, switching, and home-side hardware. Here is a look at them:

Wireless

We will first look at wireless technology. Wireless is just that, transmitting and receiving calls via radio signals.

For wireless the still-standard technology is cellular. In cellular, wireless transmitters/receivers known as cells interface with the PSTN to handle wireless-to-wireless, wireless-to-wireline, and wireline-to-wireless calls. Cell systems handle many calls on the same frequency. To find out more about wireless I highly advise reading *Going Mobile* by Keri Hayes and Susan Kuchinskas.

The other option is satellite, which is used mainly in remote areas; attempts to successfully launch widespread satellite-based wireless networks literally fell to earth in the late 1990s. High costs and risks (if you think finding locations to erect cell towers is expensive and painful, try watching multimillion-dollar rockets carrying your satellites blow up on launch pads), an inability to carry data, transmission delays, and cellular competition worked together to bring down satellite-based communication for the masses. But that doesn't mean satellite has been grounded forever. Look for someone to come up with a better, cheaper means of providing this communications mode to the general public in the future.

There are technologies that enable integration of wireless voice with premises office phone switches. Richard Grigonis, editor-in-chief of *VON* magazine and author of several CMP books, points to the MBX (Mobile Branch Exchange) software from OnRelay that integrates a PBX (private branch exchange) or other office phone system with employee mobile phones.

The MBX "Office Phone" software client installs on the mobile phone while the server software integrates with the switch, or PBX. MBX then routes all incoming and outgoing mobile calls through the office phone system, making the mobile phone operate as an exact duplicate of the desk phone, delivering PBX dial tone to you and your mobile phone anywhere in the world. This saves up to 40% on mobile bills and gives you access to your PBX's advanced features (call barring, conference calling, class of service, call forwarding, hunt groups), as well as any voice processing systems accessible by the PBX.

MBX works over both GSM and CDMA wireless networks and integrates to Avaya and Nortel phone systems. Grigonis says OnRelay also intends to support systems from Cisco, Siemens, Ericsson, Mitel, Alcatel, and those with CSTA interface.

OnRelay told *VON* that it's a relatively straightforward matter for developers to extend the MBX solution to support mobile data technologies. Those include CDPD, USSD, GPRS and 3G, as well as mobile voice technologies such as CDMA, WLAN, Bluetooth and VoWiFi. The OnRelay architecture keeps its calling intelligence and algorithms at the server and client endpoints, not in the mobile network.

Your employees or you can buy wireless phones separately, and then subscribe a separate service plan, or pay-as-you-go. But typically people buy phones and service plans together because that lowers the hardware costs, although in doing so you are locked into a long-term contract.

Keep an eye over the horizon for a new wireless option: voice over wireless fidelity (VoWi-Fi), where VoIP calls are transmitted over wireless LANS (WLANs) to end users, typically in business "hotspots."

I review WLANs and LANs in the following two chapters: on equipment and data. But to get in-depth information on Wi-Fi I highly recommend reading *Going Wi-Fi* written by Janice Reynolds, who edits all of my books published by CMP.

VoWi-Fi has much less range than cell; 350 feet compared with the range of more than 25 miles that can span a cellphone and the base station with which it is communicating. But you or your employees can plug a headset into a so-equipped laptop, link to a Wi-Fi network (instead of fiddling about with a separate cellphone and cell connection) and begin communicating. Still, VoWi-Fi hasn't replaced cellular—at least not just yet. But as Wi-Fi "hotspots" multiply and overlap, especially in large urban areas, expect to see it grow. Taking and making long distance calls for free anytime, anywhere is a very attractive proposition, unless you are a long-distance carrier.

Wireline versus Wireless

For most employees PSTN wireline is all they need. The voice quality is excellent, it is cheap, ultra-reliable, secure, and easy to install, use, and repair. Wireline gets its power from batteries, which are located in the telcos' central offices (COs). These batteries, which typically have backup generators to keep the batteries charged, enable wireline service to ride out power outages.

Wireless service costs about 10%-20% more than wireline, at this writing. But as competition grows expect those prices to plummet.

The instrument costs more also. You can buy a wireline PSTN phone for $10 ($5 at a flea-market) compared with $100-plus for a wireless. Break or lose a wireless phone and your employees are out big time. Especially if someone else picks it up or "borrows it" when they break into a home or car and then calls their aunt in Antarctica. But wireless handset prices have begun to drop.

Unless your employees keep the cellphone on their desk or strapped to wherever is comfortable, your workers risk playing "where's the QWERTYUIOPing cell?!!" And if your employees don't keep their cellphones charged the devices won't work: no matter how many times they bang them.

Cellular wireless networks still can't support data effectively or as inexpensively as wireline; however, the industry is working feverously to change the current paltry data transfer rates. There are growing networks of 100K-plus cell data networks. You can also do CTI screen pops—transferring customer identification and data from computers with the call—with wireless if you have a separate data line.

Still, wireless does provide benefits. A pure wireless voice set-up means no more juggling of landlines and cellphones. You also don't have to deal with two different sets of numbers and payment plans. Having all voice traffic coming in on one wireless phone means no more missed calls and playing phone tag because your employees are on the wireline while calls comes in on the wireless, and vice-versa.

More people are getting rid of spare wirelines or dropping their wireline service altogether in favor of wireless, especially with wireless number portability that lets you take your number wherever you go. Why? Because many users found that once they were comfortable with their cellular service, they rarely used their wireline. One call reaches them wherever they are. If those ranks include your employees you may not have a choice but to use wireless.

Another possible benefit of wireless is fewer telemarketing calls. It is illegal, in the US at least, for telemarketers to call your wireless because you are charged when you answer. But because you can't tell by the number if it is wireless (*Call Center Magazine* was assigned a 917 exchange, used by cellphones/pagers when it 'switched' PBXes)—the authorities may give marketers a break if they truly accidentally called cellphones.

Note: *The Wall Street Journal's March 17, 2004 issue had an interesting article entitled, "Viruses Lurk as Threat to 'Smart' Cell Phones." To paraphrase that article, the wireless industry is bracing for the first major cell phone virus attacks. These viruses could cripple smart cellular phones (cellular-connected hand-held computers) and even shut down portions of wireless networks. As these smart phones become more powerful, they have become more susceptible to viruses, mainly because their operating systems have become more complicated and more open to developers of third-party applications such as games. Any exchange of data leaves a smart phone potentially vulnerable to viruses. And these devices are exchanging more and more data, as users send text messages and e-mail, browse the Web, and download applications.*

Wireless Recommendations

For home working employees that also heavily mobile work and/or often go to premises offices, and/or need real-time access (e.g. support reps that repair gear on-site, salespeople, senior managers), consider wireless especially if the price is comparable to your PSTN rates. Many such employees and professionals no longer have their own premises office or home office landlines; their calls are routed to their cell numbers directly or the company's phone switches reroute their calls to their cellphones.

Also look at wireless as second or third lines, especially for part-time or occasional at-home working. I sometimes used my wireless for local when I had Telus's Talk North America plan, it saved money over long distance calls.

Wireline

The principal choice, at this point of time, for voice communication is wireline. The most common wireline method is circuit-switched a.k.a. PSTN a.k.a. POTS, "Ma Bell", "telco" and "the !#@$%^&*() expletive-deleted phone company!" The other wireline choice is packetized voice, a.k.a. Voice over Internet Protocol (VoIP), which will be discussed later.

PSTN phones are still everywhere. Whether corded or cordless, they work the same way (cordless phones ultimately relay signals to corded bases)—when you dial the phone, it sends a signal with the number you are trying to reach. When the party answers it completes a circuit, hence circuit-switched.

Calls can reach employees directly via their home number or rerouted off either a long distance, local phone company, or your own phone switches. Alternatively, calls can be routed into voice mail where employees call in to retrieve them from mailboxes.

Sophisticated phone switches enable specific types of calls to reach employees. For example, if your workers are support techs and you have a team working from their homes that are Mac specialists, those switches will route mainly or only Mac-related calls to them. This is known as *skills-based routing*. Those switches also enable transmission of caller data, e.g. their phone numbers and data records to be transmitted with the rerouted calls. That is known as *computer-telephone integration* (CTI).

You or your employees lease a line (business or residential, more about that later), or perhaps you or they buy the equipment from the jack to the ears. All your employees essentially need (besides phone service) are phones to connect into jacks; the phone company takes care of the rest and bills you for it.

VoIP

Wireline isn't just PSTN and POTS; there is now VoIP where voice is treated as data and transmitted over the Internet. VoIP promises to liberate users from long distance and additional line charges; it also promises to provide *real* competition for voice as well as data to the hidebound telcos.

Here's how VoIP works. You "dial" the phone number or plug in the IP address of the party you are trying to reach either by using software on your computer (a.k.a. softphones) or by purchasing and using specially equipped Internet "phones" (IP phones). Or you or your employees can buy "VoIP boxes" (which is about the same size as your Cable or DSL modem) into which you hook your regular wireline phone(s). The box acts as a gateway between your phone and your Internet connection.

No matter the device, when the party answers, the packets take different routes over the Internet from you to the called party. Those routes can be "naked" or encrypted through a virtual private network (VPN).

The analogy is a railroad that sends a switching locomotive to pick up a shipment of 6 cars, breaks them up in a yard and put them on three different trains, leaving at different times, and on two routes. The cars arrive at the yard closest to your customer, are put in the right order, and delivered to them, at the same time.

VoIP usually provides the same types of mandatory and optional services as PSTN. That includes caller ID, call waiting, directory assistance, and enhanced 911.

Wireline VoIP is becoming an alternative to PSTN because it avoids long distance charges: voice packets are local. VoIP also potentially simplifies wiring: if an employee needs voice and data VoIP permits them to be carried on one wire instead of two. That could save $29/month to $39/month on residential, more on business lines. There are carriers such as Primus Telecommunications that provide VoIP-based long distance to/from the US and Canada.

VoIP also provides employees (and you) much needed competition to the local telco on the "last mile" (or kilometer) to their home, saving them (and you) money. In the US, VoIP has enabled "Ma Bell" a.k.a. AT&T to sneak back into the homes. The federal government had evicted "Ma" when it broke up the AT&T local and long distance "family," i.e. monopoly; AT&T provided long distance, in competition with other carriers like MCI and Sprint while the "Baby Bells" took care of the last mile over PSTN.

In Canada, which has greater broadband penetration, but less residential phone competition, than in the US Primus also provides local service. Meanwhile the cable companies are rolling out VoIP to go after the telcos' local and long distance business.

My other beats are outsourcing and site selection. Much of my calling is overseas: to Europe, Asia and occasionally to the rest of the Americas. VoIP would save CMP a bundle on my outbound calls.

Another VoIP plus: It may enable you to survive some disasters—if you have circuit-switched/VoIP interfaces (i.e. gateways) at your switches that are located far from ground zero. When terrorists attacked New York City's World Trade Center several phone company (COs) that switch calls to premises collapsed with the twin towers. But the Internet, with servers distributed nationwide, continued to operate, albeit slowly due to the additional traffic it took on after the disaster.

PSTN versus VoIP

With VoIP you need specialized hardware to convert PSTN calls to IP and back. The standard equipment includes H.323 terminals (common LAN-based end points for voice transmission), gateways that interface between the LAN and the PSTN's circuit-switched network, a gatekeeper that performs admission control functions and other chores, and possibly a multipoint control unit (commonly known as an "MCU") to provide conferences between three or more endpoints.

There's some debate between what's best: H.323 or SIP. One phone switch vendor, Interactive Intelligence believes SIP is superior; it is faster and easier to implement. Others may disagree. Check with your IT/telco staff.

Some organizations may want to or need to add VoIP carriers to their vendor roster. If so, you may have to draft up RFPs to sift applicants out and, once a carrier is selected, put in place methods to continually review their performance.

At this writing, VoIP phones are much more expensive than PSTN phones; VoIP boxes add to the costs (more about this in Chapter 3). But prices are coming down.

Some VoIP carriers such as Vonage offer VoIP boxes in their rate plans. This is akin to wireless carriers offering cut-rate phones. The practice harkens back to the old Bell days where you got a phone hardwired into your PSTN circuit when you ordered phone service.

Also some VoIP services provide poorer quality of service (QoS) than PSTN. That mainly comes from delayed packets, known as latency, which makes people sound like they're talking into a long tube, like the transoceanic cable phone calls my parents made to our family in the UK in the 1960s. A condition known as jitter also can occur, when the packets don't arrive in order—jitter means that the conversation can be garbled with clicks and pops.

There are greater risks of disconnections with VoIP. VoIP is also dependent on the electrical grid so if it goes down, so does your employees' Web-enabled phones unless they have battery-powered uninterruptible power supplies (UPSes) or, in extreme cases, onsite generator backup (more in Chapter 3 about that). Both disconnections and power outages could cause serious problems where maintaining connections is vital for the work, like in call centers, professional services, and telemedicine.

In addition to occasional poor call quality, network contention, dropped packets, and delays can also, in some cases, disrupt signaling, which result in interference with the transmission of touchtones or speech recognition.

Most VoIP switch vendors recommend having a separate wireline—or wireless service—as a backup. Some switch manufacturers offer an option so that their phone systems can automatically failover to PSTN or wireless. But like any backup, it doesn't offer full functionality. You might lose accompanying information such as automatic number identification (ANI) and Caller ID, which have the same rough function: identifying the callers' numbers. But, if your firm has a call center and uses computer-telephone-integration to screen pop data to your agents' screens, PSTN failover should bring that data over.

Some, but not all (at this writing), VoIP boxes have inputs for PSTN lines as well as for Ethernet cables from cable or DSL modems. That permits failover from VoIP to PSTN or restoration from PSTN to VoIP without plugging and unplugging hardware or having multiple phones or boxes. Some IP phones, like Net2Phone's *Max IP10* also can handle PSTN calls for failover.

But the failover process is *not* automatic, says VoIP expert and consultant Todd Tanner, The Tanner Group. You must set it up with your carrier—if your carrier is doing the call routing—or with your telecom department for the home worker to notify the appropriate party that the VoIP has tanked and they need to switch to PSTN.

"There are some switch manufacturers that allow users like home workers password access to the administration, to change the routing" says Tanner. "But I strongly advise against doing that. That opens the door to employee abuse."

There are many people who are willing to tolerate VoIP flaws. Some industry observers say that increased wireless use has conditioned many to accept strange noises and disconnections; the callers or called parties might think the VoIP user is on a subway platform in Boston rather than a home office in their basement.

On the other hand VoIP technology is rapidly improving and becoming more reliable. Many of the QoS problems it faces are being resolved or contained at an acceptable level of performance: approaching that of PSTN.

Because VoIP is so cheap you can expect telcos to offer it as the base service to customers and then charge extra for "dedicated-line" circuit-switched PSTN as a "premium feature." While the telcos eventually will have competition from the cable guys on VoIP, they will still maintain their de facto natural monopoly on wireline PSTN especially for the last-mile to the home.

If your employees receive a high volume of calls that come mainly from outside of your company, such as customer service, sales, IT, and those calls come to your employees via a phone switch, either at your premises or at your carrier's, then VoIP may be worthwhile. Many phone switches now have IP gateways, eliminating added long distance charges to at-home workers, and from at-home workers to anywhere.

PSTN/VoIP Recommendations

Check very carefully the state of VoIP technology when making your decision to use VoIP in lieu of PSTN—CMP's magazines and websites are great information sources. Some organizations with stringent connectivity and quality requirements e.g. call center outsourcer O'Currance, based in Salt Lake City, Utah already feels that VoIP is reliable enough to not need PSTN failover—especially for their home workers. Weigh up the long distance and possible second line avoidance cost savings versus the expense of providing gateways at your end and at your at-home workers'. Do some tests.

Determine which home workers' have voice quality, truly mission-or customer-critical tasks (e.g. emergency services, medical information, customer service, support and sales). VoIP may improve enough over the lifetime of this book to limit PSTN use only as a backup: like dial-up Internet access is now for broadband users. But if VoIP is to be the only phone system you support, consider requiring, and buying for your home workers, small UPS systems.

But make sure you back the critical users with PSTN or wireless. Ask occasional home workers and non-critical users to use their personal PSTN lines and reimburse them for long distance or a cellular plan that includes roaming.

Consider VoIP as an emergency strategy—it can provide an additional voice channel to occasional at-home employees or those who are forced to work from home because of a disaster. VoIP can also free vital circuit-switched and cellular bandwidth for emergency services.

If you go with VoIP you will have to treat it like PSTN and data services. You seek, assess, decide, and monitor VoIP providers. Or you rely on your employees to select carriers. You choose whether to reimburse them or have them write it off as business expenses.

Note: When deciding your voice choices check with the employees to see what is available to their homes. There may be limitations to the number of PSTN circuits. Consequently your employees, or you, may have to pay more for the additional lines. Alternatively you may consider either VoIP for voice if there is DSL or cable (and the cable carrier offers VoIP), or in extreme cases wireless.

Residential or Business Connections?

When supplying voice you will need to consider if the employees require residential or business wireline or wireless services. There are points to consider for both.

Residential versus Business Wireline

If you're going to supply the line, specify residential unless the service is critical. Residential lines cost far less—about 60% of the monthly charge—of business lines. But some employers with home workers report better repair time commitments from their local carrier when the employees are provided with business lines.

Recommendations

Most home workers have other options, however. For example, a personal line extension in their office, a spare cordless phone on the same or another home line, or a cellphone. So for many at home workers, it isn't overly important if a few calls on the out-of-service line are missed. Your employees can always see if the calls can be forwarded or rerouted to the personal line or check for messages, if need be.

Residential versus Business Wireless

Typically most home working employees who use wireless use it to supplement their wireline when they are mobile. They are responsible for their own equipment. But with such arrangements, you, as the employer, may not be getting the best rates.

Recommendations

For those workers who are mobile a lot, or if you decide to go wireless instead of wireline for all home offices, you may want to supply the wireless service. In such situations, you still could make the employees responsible for the phones. Employer wireless phones often include walkie-talkie services that enable instant conferencing between supervisors and colleagues; missed cell calls go into voice mail.

Basic Features

You can have your at-home employees served "straight," wireline or wireless, with no features like voice mail. They can pick up the calls with $10 corded phones, $2 at the "Sally-Ann Boutique." If your employees are on another call, the callers get busy signals. If the employees are busy or otherwise not around, the calls don't get answered.

If your firm or department tolerates busy signals and unanswered phones, having no features is fine. However, many companies do not operate that way. Most employers have at the very least voice mail to handle another call when the line is busy or if the employee is otherwise unavailable.

Remember the mantra: to enable home workers to deliver **equivalent if not superior performance than at premises offices**. So what are the features and hardware available?

As the previous section explained, there are no major differences to home working employees between residential and business line services vis-à-vis what they enjoy at premises offices. The employer, though, pays through the fingertips for business lines to the home.

There is a set of services available that you should consider for all home working employees; employer-supplied or employee-shared. If employee-shared and the employees don't have these services already, have the employees expense them.

Before we discuss this set of services understand:

* Wireless essentially offers the same features as PSTN wireline; you can program your switches to route circuit-switched calls to wireless. With text-enabled wireless phones your employees can message other employees, if you are set up for that and your customers and clients have that capability.

* VoIP voice services have most of the same functionalities as PSTN voice services. However, some of the functionalities are proprietary to the device. Standard items, such as transferring a call, forwarding a call, and conference calling, though are available with most VoIP services. But VoIP phones also have their own unique features, among them are instant messaging, voice/sound file e-mails, and buddy list calling to those PCs or phones in the same network.

Here are the basic features available by wireline (PSTN and usually VoIP) and wireless. They can be supplied by the carrier, manufacturer, or government.

Voice mail

Voice mail can be hardware, i.e. an answering machine or a phone with an answering machine, or software/service hosted by the carrier. It has the same function: to record incoming messages.

I prefer the carrier-hosted voice mail; I can't remember when I last had a traditional answering machine. There are no up-front costs for equipment, no desk or wire clutter, one less gadget to plug in, the capacity is near limitless, there are fewer things that can wrong, and the voice mail can be easily accessed from nearly everywhere.

With carrier-hosted voice mail you can use cheap $10 phones. If the power goes out—a risk with answering machines—you still have a phone and your voice mail service.

The downsides are additional recurring charges, though telcos include them in bundles with other handy services. Also the service goes down now and again when the carrier makes changes to it.

There is an option to voice mail: answering services. They are more expensive than voice mail but they provide a personal touch and can provide key services, like paging with an urgency that voice mails can't. Consider them for highly paid professional employees like managers and top-level support reps.

Caller ID

Caller ID identifies the phone number and the caller who is calling. It is advantageous to know who's calling, so you can decide whether to open your line to the call, send them to "voice mail jail" or "busy" them off. It also gives you time to prepare for that caller.

The US federal government bans telemarketers from blocking their phone numbers. But political, charitable, and research calls are exempt from that provision. Many companies that do not have to comply with the "blocking" law and which dial out through T-1 lines do not show their numbers.

Caller ID comes separately, through devices built into the phones, answering machines, or in skimpy little boxes plugged into your phones. Telcos typically offer Caller ID in a bundle with other services like and call waiting and call forwarding.

Caller ID is not perfect. It does not work when a call is placed by someone using a business phone switch or a predictive dialer (a specialized call-making computer that dials numbers and then connects with the callers with the called parties). It also doesn't identify the source of outbound voice message calls. People also can block caller ID, although it is illegal for outbound telemarketers to do so.

Call Waiting (with or without Caller ID), Call Waiting Alert

Call Waiting, available only in telco bundles, enables your at-home workers to provide your callers nice (and sometimes frustrating) "virtual waiting rooms" where they are "shown to" when they call your premises.

Call Waiting is almost always coupled with voice mail so callers can leave a message after shuffling their feet for 30 seconds or so. Alternatively, many home residential as well as business phones have HOLD buttons.

Call Waiting with Caller ID lets your employees know who is calling and that they're there, in case the call is from someone that your workers need to talk to right away. Some Caller ID systems also can announce the call verbally to you with Call Waiting Alert.

Call Forwarding

Call Forwarding is just that; the ability to forward calls to another number, like the transfer buttons on your premises phones. That way part-time and occasional at-home workers can transfer calls they are expecting to the premises, or to their wireless when mobile working.

Three-Way Calling

Three-Way Calling is a limited form of multi-party conferencing (see Chapter 11 on conferencing) that saves at-home workers the cost and hassle of setting up conference bridges. It also enables easy teamwork amongst three workers.

Do Not Call

Do Not Call (DNC) is a service offered by, at this writing, the US federal and many state governments. People place their home numbers on these registries that telemarketers must scrub their business-to-consumer calling lists against.

The benefits for home workers are fewer interruptions and greater productivity—great reasons for having Caller ID. There are exemptions, e.g. companies that your employees do business with, political parties, and charities, so there will still be unwanted calls.

The DNC laws apply strictly to business-to-consumer calls that reach consumers on residential lines. Richard Frye, vice president-corporate development, Gryphon Networks explains that these rules apply to phone lines that are used primarily for personal purposes; they do not apply for lines primarily used for business—including employment—purposes, even though they may be "residential" lines.

But some individuals have been placing their business numbers on DNC registries. At this writing some states have been mulling allowing small businesses to place their numbers on DNC lists. Telemarketers may resort to calling employees to flog consumer wares: a move that is sure to bring down even tougher legislation against them.

"The problem is that there is no way for the government agencies like the FTC to identify which of the numbers on the DNC are fraudulent: that the numbers are actually business numbers or residential numbers being used for business purposes," explains Frye.

DNC recommendations

Your exclusive home workers should not put their numbers on the DNC lists, regardless of who is paying for them. But if you have occasional home workers and their phone lines are primarily used for personal purposes—then they are on stronger legal ground.

If your home workers, whether exclusive or occasional, are having problems with telemarketers then you should suggest that they complain to HR, and HR should take it up with Legal. Legal should complain to regulators and lawmakers: telemarketers are wasting employees' time and your wages and costing you productivity.

Since I've covered federal, state and international lawmakers and their efforts to curtail telemarketers since the mid-1990s, I am reasonably comfortable in saying that if telemarketers calling home workers becomes a problem that there will be legislation passed very quickly to stop it, no matter the type of line or who pays for the service. Nobody likes being bugged, especially when they're working. The media and political sympathy will be with the person making an honest living in their own house, not the outfit trying to sell them wax sunshades or pay-per-view with free commercials from call centers in some low-cost country like India or Mexico.

In the meantime, you can certainly get your home office numbers unlisted if you are supplying them, or in extreme cases change the numbers and ensure the new ones are unlisted—and insist that your employees not give out those numbers unless they are for work purposes. Your employees can do the same if they supply the lines.

Your employees should be especially careful where they place their home office numbers, like in product or service orders and registration forms. Many firms sell personal information like phone numbers unless you specifically tell them *not* to do so.

Premium Features

For employees who get a lot of phone calls and/or need to be reached by unknown people outside of your firm, such as customers and prospects, there are what I call *premium features*. Premium for the added value they offer and for the added investments in hardware, software and support.

These premium features are hosted on the organization's owned/leased or contracted-for business phone systems.

Business phone systems are complex topics by themselves. To get fully familiar with them I highly recommend reading several top books: *Next Generation Phone Systems*, by David Krupinski, Brian McConnell and Charlie Schick, *Newton's Telecom Dictionary*, by *Call Center Magazine* (and Telecom Books) "flounder" Harry Newton, *The Telecom Handbook*, by Jane Laino, and *Which Phone System Should I Buy*, by Lyle Deixler. All are published by CMP Books.

Premium features work with VoIP and with wireless. If your feature requires data along with voice on the same interaction between callers and employees you will need second separate lines with circuit-switched, you also may need the same type of line as a backup with VoIP and as your employees' only wireline for wireless.

There are also premium features that enable you to monitor at-home workers' calls and "whisper" to your employees when they are on calls.

If you supply multiple voice lines or have one voice line plus VoIP on your data line, employees can transfer those calls—and accompanying data—to other employees. Here's an example: a Level One help desk rep that can't solve a user's problem and refers the call to a Level Two rep.

Most premium features, like routing calls with skills-based routing and CTI, are justifiable for high-volume and/or high-value calls such as call center customer service, support, and sales. However, you may then find it necessary for your home workers to have business rather than residential lines to support the calls.

In brief, premium features are call handling, routing methods, and enabling technologies that provide services over and above those available on residential and straight business lines. The key premium features are:

Advanced routing

Advanced routing is the taking of inbound calls on your organization's phone number(s). It can include skills-based routing (discussed earlier). It typically supports automatic number identification (ANI), the bigger cousin to Caller ID. ANI cannot be blocked, unlike Caller ID. ANI also enables CTI screen pops that carry customer data (like how many times you complained that your Mac was rotten, and why) to that employee.

Advanced routing often, but not always, supports call monitoring. Call monitoring enables supervisors to listen in both sides of conversations on inbound calls.

There are several routing options, including:

Network routing, also known as "intelligent networks"

Network routing refers to calls that are destined for the organization and reaches their carrier's phone switch, which then reroutes the calls to anywhere the organization sees fit: to premises offices, mobile workers, or to home workers. The organization tells the carrier where to route the calls through a network automatic call distributor (ACD). The calls can pick up caller data (CTI) or be routed to employees with specific skillsets (skills-based routing).

Network routing has its advantages and challenges. Carriers own and upgrade the technology: switches plus optional interactive voice response and CTI, which saves up-front costs amounting to hundreds of thousands of dollars in expensive switches that keep working long after the technology has become obsolete.

With network routing there is no disaster-vulnerable phone room. You also avoid the headache of moving and/or disposing of the dustbunny-breeding boxes when your lease is up and/or senior management decides to decamp your operations to Outer Podunkistan where PhDs happily work for $1/hour on 25-hour shifts.

But network routing places you at the mercy of the carrier; customization is limited because carriers have to please other customers and you are on the carrier's timeframe, not yours, for upgrades. You are tied into using that carrier.

You pay the same amount to reach a home worker's phone as you would to a premises office worker's phone. But you incur long distance charges when the home worker lives outside of local calling distance (unless you use VoIP) from the carriers' point of presence.

While in the long run you may pay more for network routing/intelligent networks, compared with buying your own phone switch, network routing/intelligent networks including to the home may be the way to go for many companies. Network routing/intelligent networks supply sophisticated call routing, monitoring, recording, and reporting via the ACD that heretofore has been only affordable and justifiable to call centers. Home working consultant Jack Heacock points out there are some vendors that offer network routing for the same cost as handling "unintelligent" inbound long distance calls.

Some of these networking providers supply a range of carrier connections. UCN, which offers the inContact network routing, resells services from leading national carriers and incorporates their products into its network.

"Making available the special features of an ACD to the non-call center community of teleworkers [home and mobile] enables organizations to better measure, hence manage the home office employees in a non-intrusive manner," says Jack Heacock.

Centrex routing

Centrex routing is a generic term for contracting with your local exchange carrier to handle the inbound call routing, rather than using the long-distance carrier's or switches on your premises. Centrex routing supports ANI, CTI, and skills-based routing.

As with network routing, you avoid the up-front investments and risk of obsolescence from your owned phone switches (not to mention the space to house them or the personnel to operate and maintain them), lugging them around or disposing of them when you move. But as with network routing, you are locked into the telco's technology platform and upgrade timetable.

You also may want or need to invest in *off-premises extensions (OPXes):* separate boxes located on your premises and depending on the technology on your home workers' to reroute Centrex-switched calls to home workers, whether wireline or wireless. If these employees live outside of the local calling areas you may need to pay long distance charges: those inbound calls became outbound calls.

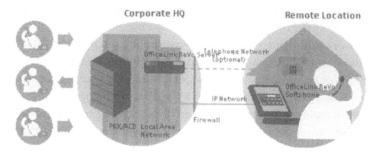

Off-premises extensions (OPXes) link phone switches with home workers and satellite premises offices. They avoid having to invest in new phone systems. This diagram illustrates how Teltone's OfficeLink ReVo OPX makes those connections between corporate headquarters and remote locations like home offices. Credit: Teltone

PSTN-enabled premises routing

PSTN-enabled routing refers to the routing of your inbound calls to your employees on phone switches that you own, modify, upgrade (however you wish), and have located on your premises. With premises routing you can add intelligence like ANI and CTI and skills-based routing.

But with premises routing, as with Centrex routing, you will need additional technologies to reroute calls to home workers. That technology can be OPXes: in the form of special cards built into phone switches or separate aftermarket boxes attached to the switches.

The aftermarket hardware and software, like Teltone's *OfficeLink ReVo*, have the same functionality as PBX/ACDs; they can also support wireless. There are other OPX vendors including GemaTech and MCK.

GemaTech's suite includes, for example, secure voice recording, with technology that makes that feature surprisingly affordable: at approximately $500 per seat. The firm says recording typically costs $1,000 to $1,500 per seat. GemaTech bases its prices on recording all 24 channels of each T1 or 24 concurrent calls at an end-user price of $15,000 per T1, without the need for any end user licenses calculated per seat. Call centers, brokerage firms, and healthcare providers need or, in some cases, may be required to record calls to ensure performance and protect themselves in case of legal action.

The routing technology to home workers also can be software that you develop and program onto the switch, with thin-client applications at the home workers' computers. For example, Alpine Access, a call center outsourcer wrote such software to enable its Avaya and Eon switches to route calls to home workers. Employees log into the switches from their PCs, which they own and the switch activates their lines: no different than in a traditional premises call center. The employees take or make calls with standard phone sets that they own.

VoIP-enabled routing

VoIP-enabled routing for premises and Centrex is a popular newer option that avoids

SWITCH FEATURES FOR HOME-BASED EMPLOYEES

There are features in network and premises-based switches that are exceptionally useful for supporting home workers. Telephony@Work only charges for its multimedia CallCenter@nywhere switch when the call center agents are interacting with customers not by selling per-seat licenses or running the meter when the employees login. When the employees end the interactions the licenses are freed up.

The payment method avoids being charged twice for employees who split their time between home and premises call centers. CallCenter@nywhere is marketed to carriers for network routing and directly to call centers.

"Our method keeps down the cost of supporting home agents, especially those that are part-time," explains ceo Eli Borodow. "Why pay for something you're not using?"

The switch has "being there" supervision by tracking agents' absence from their desks. If the agents do not answer contacts routed to them it automatically logs the agents out and sends reports to the supervisors; it then routes customers to other agents.

CallCenter@nywhere's seamless database switching at the server end avoids training agents to toggle between different systems for different tasks. That also provides greater security—a concern with home-based employees—because only the data relevant to customers is pushed to the agents' desktops.

CallCenter@nywhere has monitoring which removes supervisors' fears that home agents may be rude to customers if they feel no one is listening to them. It also enables supervisors to listen for background noise, which can also be an issue in home offices.

The supervisor can make the agent aware of the problem by whispering to them, and if the problem isn't resolved supervisor can join or take over the call. If the agent is misbehaving, however, the supervisor can lock them out of the communication system altogether.

"The supervisors actually have more control in a virtual call center with home agents than they do in a traditional premises-based call center," says Borodow.

the two cost cripplers: long distance charges and additional voice as well as data lines. For those reasons, many employers with at-home employees are looking at or deploying VoIP.

Most major switchmakers have by now installed IP cards in their phone switchgear, i.e. "IP-PBXes." There also are PC-based phone systems that have been designed from the get-go as IP-enabled. In addition, there are IP-enabled OPXes that run off IP. And there are VoIP-equipped aftermarket OPX hardware and software.

But because VoIP is still not as reliable—yet—as circuit-switched voice, many (and soon all) of these switches will have circuit-switched failover. However, the perceived quality of service of IP telephony is improving rapidly. A growing number of firms, including connections/quality-sensitive call centers, are doing away with the failover. This truth in the technology comes from increased bandwidth and gradually improving Internet, data-switching, and ISP networks. IP phones, PBXs, and gateways are also steadily improving their ability to compensate for latency, jitter, and lost packets, while maintaining good audio quality.

But is it good enough? Check out some VoIP installations and hear for yourself.

Hosted routing

An alternative to network, Centrex, and premise-based routing is hosted routing, where third parties: applications hosting firms, vendors, and outsourcers supply virtual network ACDs for voice over PSTN, and increasingly over VoIP and wireless, and inevitably data communications (e.g. e-mail, chat, and co-browsing). As with network and Centrex routing, you avoid the up-front capital costs and you avoid being tied down to a particular carrier, which protects you if their rates go up and service deteriorates.

For example, Contact OnDemand from CRM software vendor Siebel is a hosted infrastructure that is telephony-independent and supports PSTN, VoIP, and wireless phone connections. The app can be voice only or include data, supporting dial-up through broadband. That gives considerable flexibility in matching call and data handling needs with the VoIP reliability and size of dialup or cable/DSL pipes at the home agents' end.

Contact OnDemand also has been tightly integrated with both the Siebel CRM OnDemand hosted application and its on-premise licensed product. But Contact OnDemand can work with any CRM package.

West, a large global outsourcer has a hosted virtual routing platform, with well-integrated IVR functionality, plus predictive dialing for outbound calls. The technology is used for West's premises call centers, its self-employed home workers (not for outbound by them as of press time), and for clients with their own employees at-premises and at-home.

Prosodie hosts the phone network for WillowCSN, a call center outsourcer staffed with self-employed agents working from home (more about them in Chapter 11). That enables the firm to arrange for and connect home workers in many parts of the country without additional long distance charges.

White Pajama's Hosted VoIP Contact Center offers the same features as its PSTN service. It supports VoIP agents with either a softphone running on the agent's desktop, a VoIP physical phone, or through a VoIP box with an analog phone.

"The Internet as transport means is cheaper than PSTN but it is not as reliable yet and not ready for all applications," explains president and ceo Mansour Salame. "However our customers can switch with a simple mouse click from VoIP to PSTN and back so there is very little risk to them."

But as with network and Centrex routing, you're tied to the host's technology and upgrade program. Check out the latest releases, see if it will work for you and get references.

Transferring

Transferring calls is just that; the calls are transferred when they reach the premises—either off the network, Centrex, or PSTN-enabled switch. The home working employee or the receptionist can punch in the worker's home number and all calls are then rerouted to that number.

The advantage of transferring is direct routing to employees (with only slight time delays) with no additional hardware or software costs. Depending on the equipment, you might not be able to ANI or CTI info—with the right equipment, this should be no problem.

Transferring is an especially handy feature if your home workers are occasional and they switch between locations. They can then transfer the calls when they leave the premises to the number they pick, then knock off the transfer function when they return.

Voice mailboxes

Voice mailboxes on your phone system are premium features that enable callers to leave messages for employees, giving callers indirect access to them. All you have to do is set up voice mailboxes on your switch—premises, Centrex, network—then your employees have to create greetings and check for messages. You can set rules, like they must check in once or twice a day.

This tool already comes with your switch, but you have to make sure that there are enough ports to accommodate all employees—at-premises, at-home, and mobile. If so, there's no major cost there except for the long distance calls made by employees checking in for messages.

This indirect interaction tool is great for those workers who are contacted only on occasion by people outside of the premises; they simply leave messages. It is also an option to offer to people waiting on hold to speak to a live person. Many companies also use voice mail to take messages after hours.

I have a voice mailbox at CMP's main office in Manhasset, NY that I check regularly. It is a great screen for calls, like from PR agencies. I only give my home office number to those firms and individuals that I am in regular contact with; naturally anyone within CMP can reach me via my home office number.

Another, more expensive option, is unified or integrated messaging: where voice mails, faxes, videos, etc. become what they really are, data. With some unified/integrated messaging systems you can hear your e-mail from your phone, says *Newton's Telecom Dictionary* (more about this in Chapter 4).

Preview or predictive dialing

If your organization uses preview dialers that pull up phone numbers from databases and present them to employees to call, or predictive dialers that ring the numbers and connects the called parties to the employees, home workers can "make" those calls from their home. There are several service bureaus that hire people, employees, and/or contractors (see Chapter 11), who work from home and make outbound calls with preview or predictive dialers.

The technology is simple and straightforward. The home workers log into the servers just as they would do if they were in a premises office. The dialers then post the number to dial (preview) or connect the call (predictive). They can use any phone type to make those calls.

Premium features recommendations

When examining premium features and deciding which ones to select take a close look at which of your employees need what features for home working. Which of your employees have or need direct interaction with those outside of their immediate work group or your firm? Which of them do not, or, at best, have indirect interactions?

There are many employees who have direct interaction, e.g. call center staff, internal IT technicians, nurses, and, yes, editors and journalists. For customers or clients that have your number and call in and want to be connected to a home worker, call transferring may suffice.

Your employees may need data, such as customer records delivered via a screen-pop at the same time as they get the calls, saving time on each interaction. To ensure quality, as the familiar recording goes, you may need to record and monitor your employees' performance. To provide these features, you will need network, Centrex, PSTN, or VoIP-enabled routing.

But there are also many employees that have indirect interaction, mainly internal staff such as accountants, bookkeepers, HR personnel, internal IT help desk. Computer programmers and analysts rarely have or need outside customer interaction. Residential or business line voice mail will suffice. That also goes for those employees with established relations with their customers or clients.

Listed or Unlisted?

When looking at and deciding on providing employees with *wireline* (PSTN or VoIP) phone service you will need to decide if the numbers should be listed in telephone directories, or unlisted (wireless numbers are not listed). There is a slight charge for unlisted numbers.

If your employees supply the phone service and you will be using their home lines and they have it unlisted, respect their wishes and do not insist they list it. Get permission from them to see if they want it published anywhere, like on your website or for reception to give out.

Unless your organization is mass marketing itself to the public and your home employees take those inquiries directly there is no reason why their home office numbers should be listed. It is not the general public's business to know where your employees live, whether they work there or work on the road or in premises offices. In some places, like New York City, nearly half the home phone numbers are unlisted.

Unlisted numbers give your home office employees the same degree of privacy from callers that have no business contacting your organization (like telemarketers) as afforded to your premises office employees. Yes, some creep will know where your premises are located. But that's different from knowing where employees live; their home is their castle.

Unlisted does not mean unavailable, however. You can publish the home workers' number on say magazine mastheads or websites.

Arranging/Paying for Service

There are several options for arranging and paying for service: wireline and wireless. The option you select will depend on the services needed and the scale of the home working. If an employee is exclusively home working and is a full-time employee you can justify a greater amount of investment and commitment than if the employee is an occasional home worker or is part-time.

Employer-arranged versus Shared Personal PSTN Wireline

The key choice that you face with voice is sharing your home working employees' personal line(s) or having an employer-arranged PSTN wireline. There are arguments both ways.

Yes, you save money in installation and line charges if your employees make and take work calls on shared personal lines. I had my business and personal calls handled on the same line when I worked as a freelancer. Because I represented several different US and UK clients, juggling personal calls was easy—I greeted all callers with a simple "Good (morning, afternoon, teatime, evening) this is Brendan Read!"

But with shared lines you do not have control over the local and long distance service and charges. Employees, not you, pick the long distance carrier, resulting in additional costs.

Also shared lines cause disruption for you and for your employees. Then there are the personal and junk calls. What happens to calls that they need to make or take if there are family members or roommates at home who are on the phone? Asking your employees to keep the line clear may be next to impossible to enforce if it means for them to cut short a call from Daughter Dearest who dialed in because she met the creature from the multicolored lagoon at a dance last night who just text-messaged her an invite to a midnight swamp party.

You may want the employees to pay for a separate second line themselves and write it off their taxes. But that practice is not only unfair—especially if you provide lines to their premises—it will cause resentment over paying three-figure bills every month. Like the so-called "bonus" checks that the tax people swipe 55% or so of. Yes, the employees will get it back but that doesn't lessen the pain of the first bite.

Long distance shared line options

There are several ways to treat long distance charges on shared lines. The first is having the employee expense it. The second is giving them a calling card.

Expensing it is the simplest method. An employee gets their long distance bill, circles the calls attributable to you, totals them up, and expenses you. You review the charges, approve or question them, and get the matter sorted out and eventually the bill goes to Accounting for payment.

But expensing it may cost you more—your charges are based on the employees' own plan, and they lack your clout to get good deals. Employees selected their carriers based on what is best for their personal needs.

Reimbursing for long distance charges is a productivity-killing exercise for your employees, you, and your staff. I'm sure you have better, more pressing tasks than letting your eyes zone out on circled or underlined listings like the 46-second call to Curtiba, Brazil at 2am or checking the figures to see if they passed Arithmetic 101. Besides, mailing or couriering the monthly bill to you also causes costs and expense-sorting headaches.

Furthermore, with expensing you're asking the employees to carry the bill—and depending on the pace of reimbursement they may have to fork out huge sums—until they get reimbursed. If your Accounting division takes its sweet time with cutting checks that might cause some serious problems which you don't have time to deal with.

Expensing also could cause problems with employees at home if they are living with lovers, roommates, friends, and family members. Any of whom may have a coronary when they see the phone bills.

Calling cards avoid many of these issues. You own the card, you select the carrier, and chances are you will get preferred rates; also you get billed directly. When employees leave your organization they turn in the cards. Calling cards are great for mobile work; they avoid payphone and outrageous hotel wireline charges.

But calling cards, without separate stripe or chip readers that are not seen on home phones are pains in the tushes. People need one more sets of numbers to memorize like holes in their heads. On top of ATM, car and house alarm, computer and credit card passwords, birthdays, social security numbers, drivers' licenses, mothers' maiden names and months of birth, dogs' maiden names and dates of birth, pi to Nth power...

Calling cards also tie up personal lines that other householders may need. Where there is a choice of local carriers that decision is still the employees' not yours, made in the employees' own best interest.

Wireline Choices Recommendations

If your employees are exclusive home workers, supply the line. If they are occasional home workers but have family members at home at some point during work hours, supply the line. If they are part-time employees and living alone, or are working at-home occasionally or in emergencies, see about sharing the line and reimbursing them.

If your employees are primarily wireless users, considering sharing the wireline. Supply the line if they are not the only users during business hours, and hours that you need for them to be available.

In emergencies give preference to employees' private calls to family members. The safety and security of loved ones comes first.

Single-line versus Multi-line?

As noted earlier, the objective of a successful home working program is to enable home workers to deliver **equivalent if not superior performance than at premises offices**. Many premises have multi-line phones. They are convenient: putting someone on hold while taking another call and then flipping back and forth.

But look to see if your employees actually need and use that feature before you specify (and pay for) multiple lines and special phones to handle them, at home.

For example, I used to have a multi-line phone at my New York premises office desk: a Lucent (Avaya) *Definity* G3. I now have a single line phone—a Panasonic cordless speakerphone with a hold button, Call Waiting, Call Waiting ID and Call Waiting Alert.

Has going to single line degraded my productivity? Rarely. Has it saved CMP (and me) a bundle comparatively? You bet. CMP's Home Worker policy allows exclusive home workers to have just two lines, voice and data.

Few exclusive home workers, including myself, are about to shell out an additional $39 or so every month plus the $200 or so for a multi-line phone, even though these costs can be tax-deductible. That's up front money that could be better spent elsewhere, like on a proper workstation and chair.

But if employees need two PSTN lines, like for voice and for data, and the data line is used seldom, such as for fax and as backup for cable, DSL, cellular, or satellite broadband, employees can obtaining multi-lining by buying a multi-line phone. That way you get the most out of your investment.

Even more importantly there is the issue of courtesy. You can only speak to one person at one time. For that reason, many people who have multi-line phones ignore the second or third callers. With my Call Waiting Alert I will jot down the number (if the Caller ID can get it) of the second party. Rarely will I interrupt the call I am on to pick up the second call.

In my job I schedule interviews. I will, as a courtesy, block 10 to 15 minutes where I will not talk to anyone other than the scheduled party. But if they are running late, after 15 minutes they lose the slot. They then must get in line like everyone else.

On the other hand, if an important call does come in while an employee is on the line with a less-important call, multi-lining enables employees to switch between the calls. I've done that when I worked on premises, and there are still a few times I wished I had that feature at home.

Residential Wireline Maintenance Plans

If you do go for residential, whether employee-supplied or shared personal, take a close look at paying for or asking your employees to pay for wireline maintenance plans. Preferably plans that cover your home workers from the telco's service entrance to the phone.

These plans add to residential costs; some argue that they are like "extended warranty" packages that generate lots of revenues but very little expenses because how often is the wiring going to fall apart? On the other hand it has happened to me, in older houses I've lived in.

If your employees share their living quarters with unwanted and/or nondomesticated creatures (boyfriends/girlfriends/spouses/young children don't count) that may chew up or tear apart the lines, consider having or paying for maintenance plans. Because I'm dependent on the wireline for my job I like the peace of mind a maintenance plan gives me. But others may feel differently.

Wireless Choices Recommendation

Wireless arrangement options: both employer-arranged and supplied wireless, and shared personal wireless share many of the same issues as shared-personal over wireline. And there are other options unique to wireless, especially cellular wireless.

Cellular Wireless

Shared personal cellular wireless avoids the employer paying for the connection and the phone, but the employer pays for and must process work-related charges, inbound as well as outbound, local as well as long distance calls. Because employees selected the carrier, you may not be getting the best deals. As with PSTN, work-related calls may be in queue with personal inbound calls.

If employer-supplied you have two choices with cellular wireless: pay-per-use, in which case you buy the phone and you or the employee buys time cards. Or you sign contracts that lock you in for a year or more.

With contracts the equipment and the service plan are yours. If you fire or lay off the employees you take back the phones and/or the contracts. If you have new employees to take on the contract plan option works great—the contract is in your name. But if you've downsized, you have to buy out the contracts from the provider, which can be quite steep.

Pay-per-use avoids some hassles, but presents others. Typically the phones are not as good. Then there's the bother of keying in access numbers and tracking minutes used.

Can it get more embarrassing for you, your employee, and to the party at the other end to have the phone die in the middle of an important call? I would hate to have a story deadline, a problem needing to be solved immediately, or an important deal riding when that happens!

When I lived in the UK in the early 1990s, payphones used prepaid cards. I used to track the minutes left when I was on the calls. I much preferred those cards over coins, but coin or card counting down the time can be nightmarish especially when you are trying to get someone to wrap up what they are saying.

Compare that to wireline where the agreement is straightforward month-to-month. If you supply the phones, you take them back. If the employee supplies the phones they keep them.

If your home workers spend a majority of their working time in fixed locations (their home, your premises, either at their desks or hot-desks) share the wireless and reimburse them. If they are mobile, including hot-desking, consider supplying them with phones.

VoIP

If you go the voice over Internet protocol (VoIP) route take a careful look at the providers before selecting a carrier. For the lifespan of this book expect a jumble of firms getting into and out of the market: VoIP-only carriers, telcos, resellers and the cable companies. In this tumult pricing, services offered, and quality may be all over the map. For example, carriers such as Vonage have been using the wireless model of including costly equipment in their rate plans by throwing in VoIP boxes into their deals.

In the end I expect the winners will be the usual suspects: the "cable guys" and "the linemen." They have pockets deep enough to withstand price wars, to make technology investments, and to buy out weaker competitors.

Cable companies may provide a good-value option for home workers because the phone service comes as part of a bundle. On the other hand, the telcos and the resellers may enable easier PSTN failover: with a call to your carrier. Also, long distance firms such as AT&T are now using VoIP to offer complete local and long distance service in bundles without relying on competing local PSTN carriers.

My cynical prediction is that the telcos will move so heavily into VoIP that it will become the standard service, while PSTN will become a premium service. Why? Cost. It is less expensive to send packets down the Internet albeit with slightly lower quality than it is in tying up a line for a higher-quality voice conversation. In other words if you want 99% reliability and crystal-clear signals you will pay more for them.

Chapter 3: Home Working Equipment

You will need hardware: for your communications and to do the work. Chances are the latter set of tools will be computers and peripherals. As Chapter 1 explains, tasks that involve information processing, which today means computers, are the easiest and the most likely to be performed at home. Even call center work relies on computers to pull up customers' files, enter data, present offers, and record results.

Chances are that your employees work with data sometime in the course of their tasks. Chances are that the data will be in electronic form. If they work with data at their premises offices: from taking down customer information to writing the software that record the information, and/or write about the software for a magazine or website, then they will need the same tools and connection means at their home offices.

This chapter first looks at the hardware required to create, handle and process the data. (The next chapter, Chapter 4 explores home data connections; Chapter 7 looks at premises data connections.) The same hardware also communicates the data: mostly in non-voice form, with the exception of softphones. The first part of this chapter covers communications equipment; the previous chapter, Chapter 2, covers home voice communications.

✪ VOICE COMMUNICATIONS HARDWARE

Arguably the most important piece of hardware is your "voice communications hardware" i.e. "the phone." No matter what type of work your employees do at home: be it banging on copper to make jewelry, or banging out a story on home jewelry makers, they need the tools to communicate with co-workers, customers, journalists, and editors.

Today "the phone" is anything but the jumble of wires tied into that electronic fungus known as "the phone system." As explained in depth in Chapter 2, there are a confusing variety of communications choices out there.

The hardware your home workers need to have depends on what your needs are and the type of voice connection used. Do your employees need to transfer out calls? Do they require Caller ID, Call Waiting, and Voice Mail? How will they be connected—PSTN, cellular or VoIP? If wireline (PSTN or VoIP), one line or two?

Residential phones can be as simple as the ubiquitous $10 assemblages of plastic and cord found in the bins in your local "dollar store." But there also are multi-line residential phones you can buy that cost $100-plus that are cordless and offer features like Caller ID and built-in digital answering machines.

The cost for the average single-line business phone ranges from the old and still-popular AT&T standard "2500" single line set for $18-$20, to newer single-line models at around $40, to the same fancy consoles you have at-premises for $200-plus.

IP phones are not inexpensive, at least when compared to PSTN sets: the average cost is somewhere around $200; VoIP boxes cost under $100. But the prices for both items are dropping rapidly. Note that there are a limited number of stores carrying these products, although you should find that more outlets are stocking these units as VoIP popularity and the number of VoIP carriers e.g. cable companies grows. VoIP carriers like Vonage are including the VoIP boxes in their rate plans.

As a general rule, employees supply their own phones. But if you have premium features such as call routing you may need to equip them with the same model of phone set as on your premises.

Here are some features and options to consider:

Softphones

Softphones are popular with some call centers that use at-home employees. Soft phones are software that emulates a phone; instead of hitting keys on a phone, users click on screen icons to take or make calls. That enables you to buy cheap $10 phones rather than the $100 business telephones.

Todd Tanner, The Tanner Group, recommends staying away from softphones because they do not have the quality or the reliability of the other devices. Instead, you should buy IP phones or VoIP boxes that hook into your analog phones.

"VOIP boxes or IP phones have digital signal processing chips built in that are optimized for VOIP to voice," explains Tanner. "But in a typical computer they are not: the audio cards are SoundBlaster or their ilk that are not specifically made for VOIP to voice. Also the microphones are not designed specifically for voice as are the analog or VOIP phone handsets."

Speakerphones

Speakerphones are popular, handy, and potentially annoying. They enable hands-free multitasking: answering e-mails and messages, Web surfing, grabbing a book or magazine, signing a parcel, petting the family dog or cat that have come by to say hello, and the person on the other end is none the wiser.

Like many premises phone sets, most instruments designed for the home office also have the speakerphone feature. And it is far more useful in home offices than premises offices. Bad enough the colleague in the next cube has to hear one side of the conversation. But listening to two or more sides of the same yap session? Forget it! And what happens if the occupants of both cubes are doing speakerphone chats? The cacophony and confusion by all parties would make the floor of the New York Stock Exchange seem placid and peaceful by comparison.

But at home it's another story, and especially if the employee is the only one at home at the time and/or out of earshot of anyone else then it's handy to just click goes the button! That's the caveat. If there are people around, including above and under you, shut the door or avoid using the feature.

Yes, I am a speakerphone addict. It drives my wife nuts—she hears enough about my work as it is—so I try to remember to shut my office door when I'm on a long interview.

But the benefits of the speakerphone feature are worth bearing even her knitted brows. No need to fumble around with plugging and getting tangled in the other hands-free option, the headset (which I also have).

Cordless phones

Cordless phones permit users to walk around and perform tasks without being tethered. They are especially handy for home workers because they permit them to take care of duties, like signing for packages, and go on breaks, without losing touch.

Cordless sets have their downsides. They are more expensive and less rugged than corded, and their batteries wear out. Also, they depend on AC power to their base station, which means if the lights go out so do the phones. Lastly, some of them are too good—in picking up background noise. Which means you/your employees have to be make sure the environment is noise free since such sounds could interfere with the work.

Even so the benefits are worth it. My cordless, that I paid $149 CDN out of my pocket, allows me to stay in touch wherever I am in the house. I sometimes hook the headset into it. It saves me from running back to my desk or missing calls. So when you call I could be on our deck having a cup of coffee with my wife. Or I could be taking out the garbage.

If your employees do get a cordless make sure, or better yet, *insist* on three things.

That the phone is spread-spectrum so that some scammer can't drive by and illegally tap into the conversation.

That the employee has a $10 corded handset on hand if (and when) the batteries and the power grid crap out.

That they cut the background noise, i.e. few yappy visitors, less music and TV.

Headsets

Every phone should be headset-enabled. Mine is. Headsets (combination microphones and earphones) enable you to multitask: work on the computer without crimping your neck and without annoying the people around you. That reduces injuries (more about that in Chapter 6) that could put your employees out of action, which hikes costs and lowers performance. Headset manufacturer Plantronics cites a Santa Clara Valley Medical Center study that shows that headsets reduce neck, upper back, and shoulder tension by as much as 41%

You can buy phones with headset jacks or buy appliances that hook into the phone line. You can buy one-line or two-line headsets.

Headsets go from $60 and up. Many in the $100-plus range come with amplifiers.

I will use a headset on occasion on my cordless. That way I can walk and talk, put on the coffee, grab a slice of bread, sit on my deck, watch the sunlight bounce off the local mountains, wave hello to my commuting neighbors—the person at the other end is none the wiser.

Off switch

There is one MUST-HAVE feature for any wireline, corded or cordless, phone and that is an OFF switch, either singly or part of a multifunction button. Nothing is more annoying than hearing the !#%$^&*() phone go off in the middle of the night.

Answering machines/services versus voice mail

Chances are you will need voice mail. You can choose carrier-provided voice mail, which means you're dependent on the carrier, face having it down when it upgrades the system, and you pay them monthly fees that they make money off. Or you can go high-end and hire an answering service. Or you can buy an answering machine.

With an answering machine, you own it; you can do what you want with it, like putting it on different lines. If you want an upgrade, you just buy another machine (they are relatively inexpensive). You also can buy one that is built into a corded or cordless phone, which is even better because you save money and reduce desk clutter.

If you're buying a machine there are two key choices: digital and tape. Digital is cheap—under $30 or so—but you get very little record time, like 15 minutes. Most built-in answering machines/phones are digital.

That's fine if you're out of the home office for a quick errand, but not for a day or multi-day business or personal trip. You will have to pay more, and search around to find digital machines that do more. Or buy a separate tape answering machine. Or better yet, go for carrier-provided answering service.

Recommendations

To enable equivalent, if not better, performance at home as at premises, you should specify the phone hardware features. You should also insist that if employees have cordless sets that they also have a corded one. Additionally, you should require that their phones be headset-enabled.

But unless there are any special reasons why you must supply your employees with phone sets, such as for handling high volumes of calls don't go out of your way to buy the equipment. Phones do get broken and they are a matter of personal convenience. Tell the employees to buy them, and either write it off their taxes or expense it.

Even there you can make some cost savings. If your home workers are exclusive home employees you can ask them to take their desk phones with them.

Headsets are another matter. If your firm uses them then you should supply them to all employees regardless of where they work; you purchased the particular make and model for a reason, such as ergonomics and comfort. Or lay out the minimum features you need and let them expense it. Remember your home workers need to provide **equivalent if not superior performance than at premises offices**.

✪ DATA CREATION, PROCESSING AND COMMUNICATIONS

For most home workers, their tasks involve creating data: images and words to convey information. The most basic component for data handling is the means by which your employees create and process text and images. That can range from chalkboards to computers, to clunky manual typewriters, to lumps of soft coal on light-colored stone.

Your employees can use multiple data creation and processing means. I use a pen and notepad, e-mail and messages, faxes, and sometimes tape recordings to conduct interviews. I sometimes plan stories on paper or on my computer. When I get an idea late at night and I don't feel like tripping off the alarm and/over the cats, I jot it down on a scrap of paper.

Your employees may need to print data. That's "automatic" with handwriting and typewriters, but not with computers, chalkboards, or coal-on-stone.

Sometimes employees need to copy or duplicate data. That can range from on the machines, by copying text and graphic files, to copying printed "hardcopy."

Copying data is far easier on computers than the other methods. I'm old enough to remember (including cursing profusely at) carbons, and duplicating and mimeograph machines. When I describe the "fun" of duplication: typing, scratching out mistakes with razor blades, clamping the masters on machines and pouring in the alcohol, younger people look at me as if I was describing how I painted spearing animals on cave walls.

Employees also need to communicate data: by semaphore, "snail mail", courier, telegraph, Telex, faxes, e-mail and IM. Today the equipment used to communicate data, and sometimes voice (e.g. PC phones/softphones) is the same equipment (the computer) used to create the data.

Whatever you have your employees work on at premises they should have the equivalent to work at home. Remember the mantra: the employees should be enabled to deliver *equivalent if not superior performance than at premises offices*.

If employees scrawl coal on stone at premises they need lumps of both at home. If they perform their tasks on computers at-premises and use printers, copiers, and fax machines, they should work on computers at home and have, or have access to, copiers and fax machines. If those home workers need to mobile work or to come into the premises on occasion ideally you should consider providing or specifying that the data creation means be portable.

Computers, and Computer Choices

Yes, there are a few of us who scrawl or punch inked impressions with manual or electrical energy on paper. But when we talk about virtual work, information work, or knowledge work we're for the most part discussing computers: multi-role desktop and laptop personal machines.

Computers have become highly powerful, affordable and flexible. They have adapted with the times: with new types of memory drives, wireless connections and security devices like smartcard readers, USB tokens, fingerprint and iris scanners (more about that in Chapter 7) for added security.

So let's look at the computer options for your at-home employees. Printers, copiers,

CHECK OUT YOUR OFFICE SUPPLY PLACE

The best source of equipment for phones, answering devices, computers, software, peripherals, networking gear, plus supplies and ergonomically-sound furniture is your friendly neighborhood office supply outlet, like Staples, Office Depot, OfficeMax and, in Canada, many of the Basics stores. They have everything under one roof, at hours that are convenient to you. That includes printing, copying and faxing, which saves you the cost of buying fax machines/fax lines and copiers when you rarely need them.

The bigger chains, especially in rural and small cities, also offer third-party tech support. But more importantly, they have very knowledgeable sales staff whose answers can avoid a lot of grief and wasted money on hardware and software you don't need.

I was in my local Staples one weekend morning deciding whether or not to buy a USB hub to link two of my computers with a printer. The salesperson told me how to link through my cable network hub instead, which saved me over $50 in hubs and cabling.

Janice Reynolds, this book's editor, relates a case study in a book that she is currently writing for the home office crowd. The case study pertains to a wireless network installed by the author's next-door neighbors, Kristine Marino and Kevin Chrusciel. The recent availability of cable high-speed Internet service was the driving force in Kristine and Kevin establishing a wireless home-based network.

"Like many consumers, this couple knew about wireless networks but had no idea about how to go about installing one," says Reynolds." Their cable provider offered no support for this move to Wi-Fi. So they decided to make a visit to their friendly Staples store. They hit pay dirt. A knowledgeable Staples employee not only recommended specific Wi-Fi products for their networking needs, he took time to educate them on the ins and outs of both wired and wireless networking. He even went so far as to draw them a diagram to which they referred throughout the project.

"But before jumping off the deep end Kristine (the SOHO) and Kevin (the home worker) got a second opinion from Radio Shack. Again they found a knowledgeable sales staff; Radio Shack personnel agreed with all that the Staples employee had told them.

"Today, both Kristine and Kevin are pleased with their decision to go Wi-Fi. Kevin says that he enjoys being unleashed from his basement office and likes the freedom of being able to access the network throughout the house and the backyard. Kristine is happy with the speed the network provides, although she did have a few difficulties accessing a familiar FTP site; but once she discovered that all she needed to do was reconfigure the Linksys's firewall, all was copasetic."

faxes, and "peripherals" will be looked at later. So will computer connection options: hubs and wireless local area networks (WLANs).

For this book, I will look at computers not so much from their capability but as to the cost and security. It is up to you decide how much power and which programs your employees' computers have. But whatever you provide them, remember the mantra...

Desktops

Desktop computers are those computers with the heavy-duty components: CPU, full-size (15" but usually 17" or larger, monitor, and keyboard, which are located on or around an employee's desk. Desktops typically have data storage, with internal hard drives and user-accessible drives like CDs, floppies, Zips, or Flash memory (like MemorySticks™) that plug into USB slots. But desktops can also be storageless (more later on that option): without any means for users to install, retrieve or remove data, like computer programs.

Desktops require AC from power lines. But you can run them off large storage, backup, or car batteries through converters, though I wouldn't recommend it except in emergencies.

With desktops you can specify any size of screen, like the larger screen for your art director or programmer. You can move all of these components around; the days of the all-in-one-monitor and terminal died in the 1980s.

Laptops

Laptop is a generic term for any portable fully-functioned computer with the same capabilities as desktops: data processing, display and transmission, and/or data storage. Unlike desktops, laptops can be moved and worked on anywhere; they run on batteries with AC.

Laptops often feature interchangeable floppy or CD/DVD drives; the trend has been away from floppies to CD burners because CDs have much more data capacity (650 MB versus 1.44 MB) and have faster recording speeds. Flash memory also may replace floppies, predicts *Ultimate Mobility* magazine because it cuts costs and is lighter as well as more powerful.

While laptops' screens are smaller, you can order or hook up monitors to them, though at added cost. The same goes for full-sized keyboards and mice.

Laptops, like my ThinkPad (the less reverent call them "StinkPads") are typically equipped with docking stations: heavy-duty encased ports that you slide the machines into and from to allow you or your employees to run monitors, mice, printers and other peripherals. Many of my colleagues have long used laptops as their primary computers, hooking them into docking stations.

In emergencies, your premises offices employees can (and have) become near-instant occasional home workers by taking their laptops and running. You can't do that with desktops.

Laptops enabled many an employee to stay in touch and their company to stay in business on 9/11/01. I had returned my laptop just before CMP officially decided to close our premises after the terrorists had slammed the airliners into the World Trade Center.

I then called our IT guy, Freddie Golino who had the laptop still ready. I grabbed that, my CMP-issued emergency pack (our former overall division, Miller Freeman, was San Francisco-based), briefcase, and my uV-protecting Canadian-made *Tilley* hat, left the building, walked 15 blocks in the heat to the New York Waterway ferry across the Hudson River to Weehawken, NJ. Then I crammed into a shuttle bus to Hoboken—the only open train station near Manhattan and rode the all-stops to Woodbridge NJ, the smoke billowing out of what was the "twin towers."

Because the authorities had blocked the bridges (we were living on Staten Island) I called up a friend of my wife who lived nearby. That evening I hooked my laptop into her husband's dial-up, accessed our network, and read the first (of many) "Are You OK?" e-mails.

That leads to one other advantage of laptops: they have built-in power protection. If the lights go out, and you've made sure you've charged your batteries, and the laptop's AC is connected into a surge-protected board (like mine is) or cord you can survive outages until you lose battery power.

Desktops versus Laptops

Desktops are still less expensive than laptops—laptops cost 50% to 100% more when compared with similar-featured desktops reports *Handheld Computing* magazine. But the differential is narrowing: fast. In 2003, for the first time ever, laptops surpassed desktops in retail sales, reported market research firm NPD Group.

The same goes for technical capabilities. While desktops are still more powerful, laptops are coming on strong, fast. My wife used to do heavy-duty back-end Web programming from her company-supplied laptop.

The screen sizes have become comparable. Flat-panel-monitors (FPMs) (see box) for desktops are amazingly similar in appearance to laptops.

Desktops are harder machines to lose, steal, and fence than laptops; they also are less valuable. You or your employees are less likely to drop, kick, and punt return desktops than laptops.

Laptops don't walk out of buildings or get "recycled" by others. If your employees use laptops at premises require that they lock them up at nights, or that they take them home.

Desktops last longer than laptops. Many firms budget 3 years for desktops, 2 years for laptops.

Desktops can be made more rugged and versatile than laptops. There is room to over-engineer parts for ruggedness on a desktop that there isn't on a laptop; there is also space for CD burners and other goodies for which laptops don't have the room or the electrical capacity. The more you stick into a laptop the more it feels like a cinder block with wires when you lug it around.

More importantly, computer makers can install big powerful fans in desktops to suck out heat. Electrical resistance causes heat, which screws up components. Dumping heat has long been an issue with laptops because there is so little room inside those cases. Often the vents are on the bottom, which are blocked by desks (and laps), though some devices have the vents located on the side.

But laptops use less energy than desktops, which save employees money. They also consume less desk space, permitting smaller workstations. If you have occasional home workers you get those benefits; you can fit more workstations into a given premises.

As components improve and applications become less bulky, however, the fewer problems heat will cause. There are fewer laptops made with floppy drives, which take up space, consume power (and generate heat), and thus foul up.

CONSIDER FLAT PANEL MONITORS

If you or your home working employees prefer working with desktop computers, buy or suggest that the employees purchase flat panel monitors (FPMs) instead of traditional bulky cathode-ray tube (CRT) monitors. If they have laptops and need bigger and better screens buy or suggest FPMs for that purpose.

FPMs are cool stylistically and physically: they consume less energy and pour out less heat, which means lower energy bills directly and through lower air conditioning costs. They also take up much less room on workstations, enabling smaller, more compact tables and work areas. That can make fitting in home offices areas easier.

The prices of FPMs have been plummeting to within $300 of CRTs. Not surprisingly, 2003 FPM sales exceeded CRT sales for the first time reports market research firm NPD Group.

Laptops, however, do have one overcomeable disadvantage: they are hard to adjust for individual preferences compared with desktops. The screens, unless they can be detached, move up and down, not sideways and the keyboards are fixed. But employees can work around them by plugging keyboards, mice, and monitors into them either directly or through docking stations, although such solutions can drive up the cost of using a laptop.

To aid laptop users, there are cool gadgets like LapWorks *Laptop Desk*. This appliance lifts the laptop off the work surface, providing better ventilation, reducing the frequency the battery-draining fans will come on, and, since it is fully adjustable, providing better ergonomics—less back, neck and shoulder strain, greater comfort and ease of typing. (There is more about ergonomics in Chapter 6.)

It is far easier to set up your home workers with a laptop than a desktop. If you make sure employees have a proper home office, all you need is to give them the machine, fully loaded, and tell them to hook into IT when they get home. No boxes to unload. No evil staticy Styrofoam "peanuts" to chuck out.

When your employees leave you: alive or otherwise, voluntary or otherwise, laptops can be easily shipped back to your premises. My sister-in-law who was visiting us acted as the "mule" for a laptop my wife's company issued to her. That saved a bundle over shipping.

Tablet PCs?

Tablet PCs are a new generation laptop that allows workers to electronically write on screens; many are convertible, i.e. they have detachable or foldable screens and keyboards. The machines avoid transcribing pen/pencil/feltmarker and paper-made notes and permit e-mail checking with wired or wireless Internet connections.

Depending on the model, Tablet PCs can be used as laptops. But just as laptops compromise some design elements, i.e. screens, keyboards, heat, so too Tablet PCs. For example, having detachable components like screens and keyboards increases the odds of breakage through connecting/disconnecting. Tablets are also more expensive than either laptops or desktops, they also have less processing power, and lack external storage features like CDs or floppies.

Storageless PCs/a.k.a Network Appliances

Another hardware choice is storageless PCs, desktop or laptop. These computers lack hard or removable drives and chips; they also do not have ports like USBs that could be used to upload and download data and software.

Instead the data comes strictly off the network to be processed on the storageless machine, but it doesn't stay on that computer. Storageless PCs can be purchased that way, or your IT staff can gut an existing machine. Storageless PCs also include specially designed "network appliances" that roughly look like a box that are bit larger than a cable modem. Makers include Neoware and Sun.

Either way these machines have several advantages over conventional computers that have storage capabilities: lighter weight, cooler running, and lower energy costs (no clunky and hot-running hard-drives), smaller footprint (tiny CPU) and, most importantly, better security. The last feature is especially critical if your employees see customers' and clients' data, like call center customer service, sales and support reps, and police, or other security-related positions.

Without storage or ports users can't load unauthorized programs, viruses, and data-stealing software onto your system. They can't download and steal information and programs.

Many American, Canadian, and European laws require you to protect and trace consumers' personal data. If any employee steals it, the authorities come after *you*.

Storageless PCs are also less expensive: about one-third the cost of desktops. Because there are fewer parts and less heat the IT costs are much lower, too. And with all the software on the network side, your IT department does not have to spend time and resources loading and upgrading all of your organization's computers, especially home workers' computers that are often far away.

Storageless PCs or network appliances cost less than traditional PCs: desktops or laptops; they provide better security and are less expensive to run, upgrade and maintain. As this photo of Neoware's Capio storageless PC illustrates they take up much less room on workstations. Credit: Neoware

Storageless PCs also don't walk away. There is no data anywhere on them; they can't be turned into PCs because there's no drive. If your home employee's office is broken into and the thieves make off with it, ask your worker to check the bushes or the dumpsters. The units will be there, once the crooks realize they got their hands on a useless piece of equipment—to them—after all it is just a hunk of plastic and metal.

❂ HARDWARE RECOMMENDATIONS

Take a close look at your computing needs. Compare the options: desktops, laptops, tablets, and storageless PCs. Understand and specify what is necessary, but also leave room for expansion.

I'd lean towards laptops for your home workers, especially if they frequently travel to your premises offices (main or satellite), and if they go on the road—why buy two machines? Laptops are powerful and convenient.

But if your home workers do not need to travel, do not need to upload or download data and programs, and especially if they handle confidential data that you are legally bound to protect, buy storageless PCs. They are less expensive and easier to support.

If you go the laptop route, make sure the machines have the power and the capabilities to do the same tasks as you would need your employees to carry out at premises, *but no more than that*. Otherwise you (or your employees) could end up spending hundreds of dollars per computer for stuff they don't need. This, in turn, could make laptops more affordable. Why spend $2,000 or so per machine when $800 or so will do, for the same amount of ruggedness and reliability?

A good idea, in buying *any* computer, desktop *or* laptop, is to assess the tasks the workers will be undertaking on them—currently, and in the foreseeable future. All too frequently computers have been fitted with capabilities and installed with software that most workers don't or won't ever need over the typical three to four year lifespan of the machines. For example, I had bought a Compaq for my home office in New York that had a Zip Drive that ended up doing zip all.

Buy laptops that have the necessary ports or a docking station, so your employees can attach monitors, keyboards, and mice. If your employees work at home only occasionally, reallocate their desktops to new employees, to the training room, and fit their premises desks with docking stations.

If you already have made the investment in premises offices desktops and you haven't written them off yet, determine how many employees may go exclusive home or mobile working and find out how many other people travel at the same time. Also calculate how many truly critical employees you have who work at premises, but who may need to work at home in emergencies. That should you give a rough idea of how many laptops you need. In doling out laptops, issue them first to permanent home workers and road warriors, then put aside a shared pool of machines for the others.

As the desktops begin to break down or are written off, replace them with laptops or storageless PCs. You may be able to donate old desktops for charity, which makes you look good public-wise and possibly finance-wise by getting a tax deduction; check with Accounting.

If you are planning to go with storageless PCs you could make a sound business case to getting rid of the desktops ASAP, especially if your outfit handles confidential data. The cost of selling off nearly-new machines is pennies on the dollar compared with the potential legal expenses you could be preventing by buying storageless machines.

○ PERIPHERALS

Peripherals are the add-ons to computers. They commonly include printers, fax machines, scanners, and copiers.

Within the past few years, quality printers and copiers have come down radically in price. That has made at-home working even more feasible. No longer must you wait until you can get to the premises to zap off a clean-looking document. Or brave the traffic to reach the local computer/copier place only to stand an eternity behind the student downloading or reproducing the Magna Carta.

Attached to my personal computer is a lightweight HP Deskjet color ink printer which cost me less than $100, but it prints faster and with higher quality than a clunkier Epson that I paid three times as much for five years previously. I have connected to my CMP laptop a do-everything-but-feed-the-cats HP OfficeJet printer-copier-scanner-fax.

There is an ongoing debate between inkjet and laser printers. Inkjets are less expensive but slower and the printed product is of lower quality than lasers; lasers are fast and top quality. The choice depends on whether you need to spend the additional money for your home workers—do they really need master-class laser output?

Think hard about this; with the "paperless revolution" finally happening (the amount of paper in my files have shrunk dramatically over the past couple of years) is it essential to have quality output? When was the last time you or your employees wrote, printed, and mailed (not faxed) a letter?

If your employees are exclusive home workers then you need to supply them with the same peripheral types handling the same functions that they would need at-premises. For example, your home workers might not need a door-stop of a copier machine, only a desktop one that can do the job quite nicely.

If your employees are occasional home workers talk to them to assess their needs. Can they wait until they get back to the premises to print, copy, scan, or fax?

Peripherals Alternatives

Sometimes home workers don't have copiers, fax machines, scanners, or even printers because they have no need for them. Thus, unless you are willing to provide the machines yourself, it is not fair to ask them to pay for those peripherals out of their pockets if they only use them for work. Current American and Canadian tax laws only allow permanent home workers to write those costs off.

There are a number of ways around the issue. If employees already have their own personal computers and the necessary peripherals then they can print, copy, scan or fax off those devices. Or if they are using your supplied computers, they can send e-mails or download floppies or CDs from those machines to their personal machines and perform the same tasks. I did that when my OfficeJet died and I had to send it in to get repaired. If the documents are on the Web it doesn't matter which computer downloads and prints them. Another method is to see if there is a friendly neighborhood copy or office supply store in the at-home workers' area, and expense those employees' copying, printing, scanning, and faxing costs. When I was having trouble with a personal fax machine and the magazine was in the "edit" stage, I made the three-minute "Staples shuttle."

There are full-service international chains like Staples, Office Max, and Sir Speedy, and copy/computer/fax "hotels" such as Fedex Kinko's where, if need be, you can rent time on computers, and print, scan and fax almost anything. There are also local businesses that provide the same services as the large chains; some of these businesses can be quite interesting. When I lived in New York City working in between jobs as a freelancer I had my faxes go to (and sent them out) from a gift/office supply place run by followers of guru Sri Chinmoy in Jamaica, Queens.

Peripheral Recommendations

Let's take a close look at the needs for such peripherals, if you are supplying or paying for them. For example, how often are employees faxing these days? E-mail and .pdf files are supplanting faxes. How often do they need to copy or scan?

If your employees are using functions like copying, faxing, and scanning once in a blue moon, and there is a business nearby that offers these services it may be worth the time out of the home office to source those tasks rather than buying the hardware, or supplying or having employees supply the ink and paper.

✪ COMPUTER CONNECTIONS

Most home computer connections are "straightlined": from desktop CPU to monitor and keyboard, from the PSTN or cable/DSL to the laptop. But with the advent of broadband, to get the most from their investment and setup, people are creating networks: multiple computers and peripherals connected together.

Such networks should be allowed, and if anything, encouraged, within reason. These home networks enable multitasking and provide backups.

For example, I have my personal desktop and my CMP laptop connected to a Linksys hub that is then connected into a cable modem. I have separate data line that I use for fax and backup dial-up. I also installed a switch between two fax machines: the CMP-supplied HP Officejet and to my personal Brother.

I have my e-mail up on the laptop, monitoring messages while surfing and downloading information from the Web at the same time. Having the Brother enables me to make/take faxes when I am printing off the laptop or scanning a document.

But many people are going to the next step: a wireless local area network, or wireless LAN. Wireless LANs consist of hubs with transmitters and receivers and wireless cards with antenna that click into computers. They have a range of about 300 feet; the signals will go through most walls.

Wireless LANs take the same data from broadband or dial-up and zap it through the ether, at up to 54 Mbps—faster than the input speeds. But at the edge of the range the data transmission speed can drop to as low as 1 Mbps, which is still faster than most Internet connections.

Wireless LANs enable the same connectivity, but without the spider web of wires (which can attract spiderwebs) underneath tables and filing cabinets. Hooking and unhooking that stuff gets scary; I've done it twice on long distance moves.

Wireless LANs allow the convenience of working on a laptop and a cordless or cell-phone, enabling the home worker to go anywhere in the home, e.g. a back yard or deck. I wouldn't mind getting a Wireless LAN: except I'd have to bring the printer, my library...no it is too easy to stay in my home office.

The biggest downsides of Wireless LANs are its range (around 350 feet) and security risks. There are people who cruise areas trying to crack computers via wireless. Others pig-gyback onto a Wireless LAN to obtain Internet access: but note such "Internet hitch-hiking" can occur without the interloper being able to get at your data or any other data sitting on your employees' computers.

Finally, understand that security is not as bad as many of the pundits would have you believe. Malfeasance from unauthorized access to a wireless network is greatly exag-gerated; an individual or organization is as likely to lose valuable information via a stolen or lost computing device as from someone maliciously breaking into its WLAN. Also keep in mind that:

* Wired and wireless LANs share some of the same security issues, e.g. vulnerability to hackers and eavesdroppers.

* In order for a Wireless network's *data* to be compromised, modulation techniques, *and* radio domains, *and* channels, *and* subchannels, *and* security ID, *and* passwords must be known.

* All Wi-Fi certified wireless products produced after 2003 include Wired Equivalent Privacy (WEP), a technique that encrypts all *data* that passes between an access point and a wireless card. While, not perfect, if WEP is enabled, it usually will block casual snoops from being able to read the intercepted data.

If you require your employees to take the necessary steps to protect their comput-ing devices by password protection and by encrypting files on their hard drives and gen-erally exercising caution, you've taken a big step toward decreasing the likelihood that your valuable data (whether traversing a wired or wireless network) will wind up in the wrong hands.

Connection Recommendations

Most networking tools should be your employees' responsibility. If they want them, they pay for them. You shouldn't mind if they run their personal computers off a cable or DSL line that you pay for, chances are they also will use their personal machines to your benefit—for work as well as for backup.

The one caveat: If your employees have or plan to use a Wireless LAN require them to check with your IT department *first*. Don't approve the wireless network unless IT gives its okay. You can't afford anyone breaking into your network.

Who Supplies the Hardware?

In selecting computers and the peripherals there are choices in who supplies the equipment, you or your employees. The decision can affect your operations.

With home workers that already have a PC, the first tendency is *shared personal*, i.e. asking the employees to use their machines to do work for you. This choice utilizes the fact that most homes have computers.

Shared personal is often the first choice because it saves you money buying and supporting that equipment. Your employees are also highly familiar with their machines. And if you have to get rid of those employees, you don't have the nasty, and sometimes costly, business of trying to get the equipment back and in good shape.

But there are equally compelling arguments to supplying your employees with computers, i.e. *employer-supplied*. The big one is that you have *control*. You bought those computers and configured them with software and applications designed to meet *your* needs. Those computers have been set up to access *your* network (more in Chapter 7 on that). You, therefore, limit your network from unauthorized access and reduce the risk of your network being infected by outside computers.

If employees work on *your* machines, you can more easily monitor their work. If they communicate through *your* network you can track and manage their e-mails, messages, and Internet usage. All of which helps to ensure that employees are following your policies.

You also avoid loading your applications and software onto machines that are not your own—getting those items back can be difficult. You also may run into licensing problems, especially for occasional at-home employees.

Older software was typically group licensed; an employer would buy 100 licenses but there could be 200 users (but only 100 could access the software at any one time). But many newer software products are licensed per machine, so if you have one employee but two machines: one at-premises and the other at-home that could cost you more money.

You can (and should), as a condition of use, prohibit employees from loading unauthorized hardware and software onto any computer supplied by you—this limits the risk of hacking and data theft. You can (and should) have software that "sniffs" connected machines to check for unauthorized items.

By owning your employees' machines you are able to keep performance consistent. When you upgrade to new software everyone upgrades together; you send your at-home employees downloads or CDs.

If you have employees that are part-time workers, i.e. work for you at-home for the entire duration of part-time hours, as opposed to part-time home workers (which work at-premises a majority of the time), it is challenging to make a business case for supplying them with computers. But you need to have other access protection measures, e.g. hosted access software that can act as a bridge and buffer between your network and your employees' computers.

Employer-supplied or Shared Personal Recommendations

Avoid shared-personal, especially if your employees are working from home exclusively, unless there is a compelling financial case for using the employees' equipment. When you use employee-supplied equipment you have minimal software control and limited access to the equipment. Using employee-supplied computing equipment, thus, will make your IT department very uncomfortable—the security and configuration risks and hassles of such an arrangement are too great. (In the case of storageless PCs, these are specialty items best sourced by your IT department.)

If you have occasional home workers and you don't have enough or cannot justify laptops for their use, there are steps you can take to give them functionality without compromising your security. If your business continuity plan relies on these employees (and it should) you need to take these steps as soon as possible and test them out. For example, you can ask (but not demand) that they have installed on their computers compatible versions of the software used on your own computers. They can use an Internet-based Web mail account that you set up (they connect via a password) to access their e-mail. This method further protects your network from unauthorized snooping and unwanted viruses.

❂ PROTECTING THE EQUIPMENT

Computers and peripherals may look powerful and rugged; with the right authorization, password, and connections one can initiate thermonuclear war or order a pizza to be nuked later—the computers don't care. But deep inside they're wusses. One voltage sag caused by too many air conditioners and not enough juice, and computers fold up. One voltage spike and computers and peripherals fry.

This also goes for VoIP hardware: softphones, VoIP boxes and IP phones. If they're connected, even through DC converters, they go lights-out with the rest of your office. They, too, can get roasted on a spike if they are without protection.

Computers, because they are dumb machines (no offense, artificial intelligence, folks) they don't know for sure who has the authority to use them. So unless you program them to ask for passwords or have locks on the CPUs or laptop clamshells, or use cool biometric devices like fingerprint scanners anyone can use them to get at your employees' files and potentially into your network.

Therefore, regardless of who supplies the equipment—you or your employees—you need to ensure that the computers and peripherals are protected in more ways than one. No matter if your home worker is exclusive or occasional, if their machine craps out or someone gains access to it, they are not productive, and that costs you money.

Here are few suggestions:

Require employees to have surge protectors

Surge protectors are essential and handy devices that take the hit when there is a voltage spike, an electrical overload caused in a lightning storm, or a sudden power outage. Protectors come in the form of power strips, wall plug-ins, or power boards. Be sure when you get a surge protector that it has a light or other indicator to show that it is live. Surge protectors begin to falter after awhile from handling "events."

Surge protectors also need to be connected to cable and phone lines. Surges can route along them too.

Surge protectors are inexpensive; good ones usually cost less than $70 or so. That makes them such a great bargain when they prevent the loss of hundreds, if not thousands, of dollars of equipment. There is no reason why any home worker, exclusive or occasional, should be without them.

Consider uninterruptible power supplies (UPS) systems

UPSes are battery supplies that keep computers going during voltage sags or brief (15-30 minutes) power outages; they enable users to shut their machines in an orderly fashion if the lines are going to be down for a while. Batteries are rechargeable or replaceable: in hot-swaps. There are several firms such as American Power Conversion and Liebert that supply UPSes for home use.

UPS systems are necessary if your employees are exclusively home workers, they work on desktops, their tasks are vital for your organization, and they run heavy-duty applications that need time to shut down. But if they have laptops then they don't need UPSes, unless they absolutely must have broadband, which requires external power. Instead, they should have spare charged batteries and dial-up connections that they can hook into.

The next step above UPSes is backup portable or standby generators. If your employees live in areas where outages are frequent, they may have a generator, but they also will need a UPS, regardless of computer type, to condition the power, and prevent computer damage. Chapter 5 looks at backup generators in depth.

Locking up laptops

Laptops have a 'habit' of taking off—it's amazing how they disappear from desks. To stop such disappearances you need to suggest or require exclusive and occasional home workers (and anyone else who uses them, like mobile workers) to take preventative measures. These methods include disconnecting them and locking them up in filing cabinets and using disk drives or security cable locks. Also look at installing a tracking device on each laptop. When activated, such devices will notify your firm and the police of the laptop's location. Or if the data is extremely valuable, buy laptops with fingerprint scanners or other biometric or security devices (more about them in Chapter 7).

Make sure that you take all feasible steps to secure your laptops. Keep in mind that you and your employees may not always be able to gain access to a particular machine when needed, and the data on a specific laptop may not be accessible. Also note that recovery of both the laptop and/or the data is not always guaranteed.

Buying spare batteries

If your home workers use laptops, supplying them with a couple of spare batteries and insisting that they be kept charged, is an excellent investment. That way they can continue to work, even if by candlelight.

Restricting access

All computers, wherever they are located, at home or at premises, risk someone else getting into them. They could be the janitor, the temp, or a visiting sales rep that have been shown your employees' workstations at premises, or your employees' offspring, relatives, or friends at home.

If the data your employees keep on the computers is valuable and confidential, like customer data, you should consider having or requiring computers to have locks and/or password restrictions. That way only authorized users gain access to the machines, at least through the front door (data access is covered in Chapter 7). If the data is stored off-machine, like on floppies or CDs, consider requiring lockable cabinets.

Also, if the computers are used in a location where others can walk by, like the corner of a dining room table, consider mandating or having installed screen-wiping software. That way if no one is working on the computer, no one can see what is there.

There are many laws that require employers to take steps to restrict data access by unauthorized people. Check this with your lawyers (more about this in Chapter 8), but in general, the more you show you have met and gone beyond what the laws stipulate, the better off you are.

❂ WHO PAYS?

Here are the author's recommendations: if you supply the computers and peripherals you should supply, or allow employees to expense, the surge protection and the UPSes. If you pay for those appliances, your IT department should forward the specs. That way you get the protection your equipment needs.

If the employees supply the computers and peripherals, consider asking the employees to pay for the surge protection and the UPSes. However, if you need them to be on a UPS, then IT should specify how much backup time is required.

Chapter 4: Home Data Connections

For many of us, creating, handling and managing data *is* our work. The "information" or "knowledge economy" we are currently living in, with the nasty dirty "industrial work" performed automatically by machines governed by computers operating on and supplying, you guessed it, *data*.

I am an editor/writer for a print and online (or shall I say online and print) "content provider" that covers, in large part, how to handle customer data. My wife is a semi-retired computer programmer and analyst and database administrator who went from mainframes to Web/database integration.

So what are the methods of data communications? What connections to their homes do home workers need? Who should arrange and pay for them?

This chapter looks at the data connection components that you need to consider for at-home working employees. Home networking has been covered in the previous chapter, Chapter 3.

The data management components, including technical support is looked at in Chapter 7, which covers employer investment. It also examines data access means such as virtual private networks (VPNs), which is encrypted data communications over the Internet to your computers and hosted access, where data goes through a third party.

The descriptions and analysis provided is brief, and is as up-to-date as the print date on this book. For more details, I highly recommend discussing everything with your IT department.

To bring you up to speed, I also strongly advise perusing the library of books on this subject offered by CMP Books. To keep up to date get yourself a subscription(s) for one or more of CMP's business/IT publications and check out our websites. *CommWeb* is the best daily news source on such developments.

✪ OFFLINE OR ONLINE?

You will need to decide whether you wish to connect your home workers to either or both data-based communications and your network. And which home workers should have those connections.

This is not as easy an issue as it appears. You have to look at the costs and feasibility of providing that access, whether to use dial-up or broadband, wired or wireless. There also are security issues—data and networks are vulnerable to hacking, viruses, and data theft.

You can have your home employees, especially occasional home employees, work offline. They then deliver the work on CDs, floppies, and printouts that they courier in or drop off, or upload into your network on hot-desk PCs. Or they can fax it in.

That's how I and others worked from home, in the pre-WWW era. The sole exception for me was when I was with the Manchester (NH) *Union-Leader* in 1988 where I transmitted my stories over a modem built into a tiny state-of-the-art Tandy TRS-80/T200 laptop—arguably one of the most groundbreaking and ahead-of-its time PCs ever created. It took years before other truly personal computers could catch up to the T200— which had the now-standard flip-up screen—that first came on the market in 1985.

Many tasks that involve information analysis, creation, and editing, like accounting, bookkeeping, book writing, editing, engineering, and programming can be done effectively offline. Once the raw material is there it can be worked on.

With the proliferation of spam, aided by what can be most charitably described as selfishly stupid and destructive attempts by business lobbyists to squish effective antispam legislation, many people and employees are not bothering to use e-mail. Opening inboxes is worse than taking out the trash, they are so chocked full of perverted spew.

I can go days without answering e-mails. If my boss and colleagues need me they can call me, or reach me on instant messaging (IM), which loads when I boot up.

But these days you probably expect all of your employees to be connected, reachable by graphics and text. And that means they have to be online at home, to a data source, even to one as simple as an e-mail server. With time allotted for upchucking the spam.

The reason is that there are increasingly many tasks that rely on being connected to data. Examples include customer service, sales, management, and collaborative tasks with other employees like some engineers and programmers. Creating and updating applications and programs is less writing up code offline and more working with the application, like a web page online.

✪ DATA CONNECTION OPTIONS

There are two key components to data connections that you need to have provided for and installed at your at-home employees' offices. They are what I term:

Data communications

Data communications are the means by which employees communicate via data. That includes e-mail, instant messaging, text messaging, and chat, via the Internet, video over the Internet, and fax.

Data links

Data links are the wireline and wireless media that transport data between your network and your employees. They include dial-up on PSTN lines, ISDN, cellular data, and broadband (cable, DSL, cellular broadband, Wi-Fi and satellite).

Data Communications

If your employees use any or all of these means: e.g. e-mail, instant messaging, text messaging, and chat via the Internet, video over the Internet, and fax then they should be enabled to use them at home. There is no dispute over what was said when it is there in black-and-white.

Let's look at them:

E-mail

E-mail is the sending of electronic messages via the Internet. Despite spam, e-mail is one of the most versatile communications media ever devised. It does not matter where your employees are located, if they can get access to the Internet, they can access, transmit and receive e-mail.

E-mail, unlike voice, provides a tangible record of what was said, which means you have to be careful with what you write, but it is extremely helpful when checking what others have said. I use e-mail in my article and book fact checking.

You send and receive text, graphics, photos, movies, music: any form of data by e-mail anytime, anywhere. You or your recipients don't need to be there when it arrives; you and they can respond whenever they feel they should, if they want or need to respond at all or they can ignore it and trash it instead. They also can pass e-mails to others.

E-mail is sensitive to the capacity of your data links. A 4 megabyte file with graphics is not a problem for a cable modem that can typically handle 1 mbps (1 million bits of data per second). But that same document will cause hiccups for a dial-up line with 56 kbps (56,000 bits of data/second) bandwidth. And it will trigger electronic gastroenteritis for conventional cellular wireless modems that max out around 33 kbps (although some occasionally can top up to 140 kbps).

For that reason data-heavy (like graphics) e-mail is not really practical for low-capacity conventional dial-up or cellular wireless. But your employees easily can upload or download e-mail on higher capacity cellular broadband, Wi-Fi, and satellite.

Yes, e-mail has become larded with spam: spread by characters on behalf of firms both of whom should be barred for life from using communications media other than coal on stone. Typical human nature: come out with a useful tool and some lowlife soon finds ways to abuse it.

But such consequences do not eliminate the utility of e-mail any more than telemarketers ruin voice, and junk mailers trash the Post Office. You have to make sure that you have up-to-date spam filters, virus protections on all computers, policies (like previewing documents, and not opening attachments unless your employees are assured of the source), and that you have provided enough access capacity for your at-home workers.

Instant Messaging

A newer and much recommended communications is instant messaging (IM). IM is software that when installed on all computers permits messages to be sent and received over the Internet. IM enables you, your colleagues, your employees, and others in the loop to send and receive messages in real-time, unlike e-mail, which can take awhile. IM users must all have the same software (there are no universal standards—yet) and must log in and out.

If you have at-home employees—permanent, part-time or occasional—equip all computers with IM. IM is the quickest text communications method available today. We use IM at *Call Center Magazine.*

You and your employees can IM each other while doing e-mails and being on the phone as well as working. For example, if your employee is on the phone with someone outside your firm and has a question, i.e. in which issue will their ad run, they can IM you or one of your other employees with that information. Then you or they can IM them back with an answer—immediately.

Text Messaging a.k.a. SMS

Text messaging or short messaging service (SMS) is "mini-e-mail." Instead of sending that message to an IP address you shoot it to a special address or to a phone number. Text messages include specially abbreviated e-mail, account balances, and stock quotes.

Text messaging increasingly is being used by mobile and "fixed location" workers to reach mobile workers on cellphones and other wireless devices. It uses very little capacity or *bandwidth* compared with e-mail, and the words can be glanced at without heavy screen scrolling.

If your home working employees are mobile, and/or if they communicate with other mobile workers who use text messaging fit their computers and wireless phones with text messaging. Also, if your organization uses wireless as an adjunct to wired, install text messaging and use it like IM. It is a great tool for employees and supervisors to get in touch with each other if there is a problem or a question and there is someone else on the line.

Chat

Chat is a type of text messaging usually used over wireline or broadband. Chat can be one-on-one or in conference. Your home employees can chat with co-workers, customers, clients; and others can chat with your at-home employees if you have chat software on your server. They send a conference invite (the ability to set up conferences is built into the IM software)—the users just log on and key away.

Fax

Fax is a funny type of data communications: it is electronic mail in the broadest definition because it is the sending of data: electronic or print scanned to electronic—but whose reception is in the form of a tangible printout. Faxes use PSTN rather than data lines. There is also *eFax* where someone sends a fax that is then converted to e-mail.

Faxes are handy for transmitting printouts. These include contracts, articles: any printed matter that has been copied. I get a fax of my laid out articles every month to proof it and to make cuts.

But fewer people it seems have need for faxes, as more documents go completely data. During the lifespan of this book I will probably get the article proofs as .pdf (portable document format) files. Electronic signatures will become the norm.

Faxes can be a pain in the tail; there are still skells who send junk faxes; they and their conjob customers should also be condemned to write with coal on stone (no copies!) like their brother spammers. Only with faxes the !#%$%^&*() machines make Almighty-awful noises.

I have a fax line into my home office. But it is there more as a dialup backup for my cable modem than as a functional fax line. For that reason alone having that line is worth it. When faxes arrive for me in New York the mailroom person forwards them to me by courier.

Video

Video as data comes in two forms: one-way or two-way moving images. In the first instance, the person contacting you wants you to see them, but they don't have to see you, or vice-versa.

Call center outsourcer Convergys experimented with one-way video by having callers see the call center agent but the agent couldn't see them; VideoGate hosted the application. But callers needed broadband connections to link in.

In the second instance, both parties could see each other. In the fall of 2001, many telcos pushed DSL and many cable companies hyped cable broadband by offering free cameras or Webcams with their deals. Most large office supply stores carry both devices. Some people even received such devices for Christmas so they could see their loved ones (in the wake of the 9/11/01 terrorist attacks many people were understandably leary about flying).

In both cases the transmission is the same: video is sent as packets (as data) over the Internet. It works very similar to voice-over-IP, in more ways than one.

Video has the promise of coming as close as possible to in-person in-your face interaction while remaining virtual. You can see—hypothetically—the facial expressions (which can convey vital information) and the other party can see yours if you rig the video for two-way communication.

You or your customers see details like color shades and fabric swatches, for example. You get to literally "see" your employees and they get to "see" you.

But video to the home is one of the greatest innovations to have flopped. Whether the appliance is AT&T's PicturePhone displayed at the 1964 World's Fair, relying on good old copper wires, video-over-broadband experimented by Convergys, or the freebie Webcams—the end result has been the same, no thanks.

One of the key reasons is that quality of the video leaves something to be desired. Even at broadband speeds the images still move so slowly that if someone twitched a finger it came out as a blur.

To obtain good picture quality, video needs to run at 12 to 30 frames per second (fps). The standard that people are used to is 30 fps, which is TV quality. The typical Internet-based video has from 12 to 24 fps. You must have some type of broadband service for video conferencing to be doable—dial-up is too slow.

But broadband connections still occur over the Internet, which means no one has any control over how and when each packet arrives. In other words, video is more of a victim of latency—packets arriving late and jitter—packets not arriving in sequence—than VoIP; the results are there for all to see.

My wife and I bought Webcams in December 2001; our son, my sister-in-law and most of our friends live in New York City. We tried them on DSL and cable and with VoIP through the computers and with PSTN over speakerphones for voice.

But no matter how we configured the cameras and connections and played with the lighting the picture blurs were terrible. Now the only ones who see the cameras are the dust bunnies.

I also suggested using my Webcam for office conference calls with my colleagues. No response. I guess they don't miss my ugly face after all.

Why bother with video? What is the value-add of video? So I can count the zits on the face of a call center agent? So my supervisor can see if I shaved that morning?

Unless the resolution is TV quality—the standard for video—you cannot easily read the other party's facial expressions. If you can't see those expressions then why have video?

Such benefits like showing fabric swatches and color of a product are lost in the resolution. It is difficult to see what the color shade is of an item in-person in a store let alone through a tiny screen.

My supervisor is likewise more fascinated in what I'm putting into the story than staring at me mangling the keyboard. What I do is more important than the boring mechanics of me doing it.

Video has been used more successfully in conferencing (see Chapter 11) between premises meeting rooms. There the bandwidth exists to support high quality resolution.

Even so, there are questions about the need for video and how much, even when you have 1.54 Mbps T-1 lines links to every computer. Even if my boss can see my ugly face clear as bell even as I move around, why would he want to?

We text journalists and writers do much of our work virtually. We listen for the emotions and convey ours with our voices or with written words. We communicate what we hear and have learned with text; we paint virtual images with words.

But our skills are not that special, unique, or unusual. Anyone can do it. Considering that it costs much less to communicate via phone or e-mail than by video our skills are very cost effective to deploy and practice.

Data Communications Recommendations

Take a close look at your data communications needs. Some means, like e-mail and possibly SMS are essentials, despite the consequences (like spam). But others are more problematic.

I would look long and hard before specifying fax connections for your home workers. Examine your firm's fax usage and trend it.

Do your employees, at-premises and at-home, really need faxes? If these employees work at home occasionally, they can pick them up at-premises—unless the faxes are urgent.

Could you live with and is it cost-effective having your employees spend 20 minutes or so going to local office supply store once a week for a fax? Or having the faxes that are received at your premises shipped to those permanent home workers? Is having a fax line as a backup dial-up data line worth it to you?

I would also avoid video unless you have a compelling business case for it and are willing to invest in high-end near-T-1 connections to enable consistent broadcast quality to home workers. Video will cost a lot and may pose numerous installation and setup headaches at both ends; not every office is set up like a studio. But if you have a provable business need that can't be met any other way—and the resources—then go for it.

Data Links

The data must get from your servers to your home working employees' computers, and vice-versa. Other than couriering or snail-mailing in "storage media," e.g. music cassettes (which the first PCs used), floppies, diskettes, CDs, Zip drive disks, etc. you will need to set up a means to connect into their machines.

The key choices are *dial-up, ISDN, cellular data* and *broadband* (cable, cellular broadband, DSL, Wi-Fi, satellite). Each has their benefits and costs: measured by capabilities, flaws, and prices.

Dial-up

Dial-up connections are where data arrives at your at-home employees' computers over PSTN lines; they transmit over modems back over the same phone line. The modem "squeal" is the device establishing a connection with your server over PSTN.

Dial-up has a data transfer speed that maxes out at 56 kbps data transfer speed. But sometimes it is half that, especially in hotels and rural areas. Even so, dial-up can support, though at slow speeds, common bandwidth-intensive applications like virtual private networks (VPNs) (more about them in Chapter 7) and e-mail applications like Lotus Notes.

Dial-up can be used anywhere there is phone service, which means in most cases it can be used where your at-home employees live and work. Dial-up is less expensive than broadband and it is highly reliable because it uses PSTN. Because there is no fixed IP address at the home end with dial-up, hackers find cracking such computers more of a hassle—though not impossible.

Currently there is no other connection mode, which offers dial-up's anytime/anywhere access and data transfer speed. Also no other connection matches dial-up for price and security.

On the other hand dial-up is slow. A 1 megabyte file, like a photo, would take over 4 minutes to download on a dial-up compared with about 25 seconds by broadband, such as DSL. With dial-up, your employees run the risk of being losing their connection, especially if the connection is left unattended while they do other tasks, like editing, programming, and writing. Relaunching critical and heavy-duty applications like VPNs and e-mail/database software like Lotus can be a nightmare.

Be careful of Internet "boost" devices. While they can speed downloading of some items, like Web pages, graphics, and e-mail, they don't enhance the download speed of other types of data like files setting on an FTP server, or streaming media; you also can't use encrypted data like through a VPN, says Netscape Online. My cynical sense says these devices are stopgaps until when, not if, broadband becomes universally available and so cheap so as to significantly negate the cost/benefit of these tools.

With dial-up, your long distance carrier's point of presence (POP), which are the numbers your employees dial the modem into, are most often, but not always, nearby. If the POP is a long-distance call away the dial-up may cost you more than the add-on cost of higher-priced cable or DSL broadband.

Even so, I have dial-up ready to be launched and plugged into my laptop as a backup in case the broadband goes down, from such causes as cable upgrades and glitches. Fortunately that is not often.

Always make sure your home working agents can access a dial-up account, anytime, anywhere. That way they can work from home in those locales where there is no broadband (and even where there is) because you never know when they will need it.

Cellular data

Cellular data is just that—receiving data via cellular wireless, which is accomplished by plugging your laptop or desktop into a cellphone or by buying a network card or "aircard" and antenna and attaching it. Check the carrier where your employees live for availability. As with cellular voice, you have to buy into rate plans for cellular data.

The key upside of cellular data is that the connection is available anytime and anywhere there is cell coverage. Your employees can pick it up if they are working in their back yard, in the back 40, in the basement of a back office somewhere or in the back seat; they are free of the physical tethers (more about that in the mobile working chapters) of wireline data connections.

But cellular has two big downsides: lower data transfer speeds and higher costs compared with wired cable or DSL broadband. The top throughput on high-speed 1X networks on carriers that use CDMA technology, such as Verizon Wireless and Telus in Canada is about 40 to 60 kbps (the average speed, however, is more like 20 kbps).

But wireless carriers eventually will be able to provide match cable and DSL high-speed services—the new 3G wireless technology promises up to 384 kbps mobile and 2 mbps stationary. Still, there is greater risk of data dropouts with cell data than with wireline service. Data is more sensitive to loss over wireless than voice because you can understand "jerky" voice, dropped data packets, on the other hand, are lost forever. There may be blind spots even in one's house for wireless signals. There are also greater security risks with all wireless broadband, whether cellular or WLAN (more about that later).

As with cellular voice, there are several pricing/service options with cellular data and you get billed for inbound and outbound data. The aircard option that comes in kits from the carriers runs about few hundred dollars, less with yearly or longer contracts. Some carriers ring in network support changes of $40 or so a year. There are plans that include unlimited access.

Keep in mind that this is a fast-changing area. You can generally count on dropping prices and improving technology.

ISDN

ISDN stands for Integrated Services Digital Network. ISDN is in essence "turbocharged PSTN" that carries voice, data, and video on the same line by using signaling and special connections at the home and at the telco's central office.

At 128 Kbps ISDN has just over twice the bandwidth as dial-up and is readily available in most regions. *Newton's Telecom Dictionary* says if you can't get broadband check out ISDN.

But the connection method has never taken off. I searched around my local telco Telus's website and couldn't find any information on it; Verizon, my old carrier when I lived in New York City says, "please contact one of our knowledgeable sales consultants."

ISDN has largely been superseded by higher-capacity broadband. "The Dictionary" says carriers did not explain ISDN well, it is not an easy service to get up and running, and cost and local availability have been issues; telcos charge a lot for ISDN.

Others say ISDN has complicated configurations, does not have dedicated connections, and in some cases it is billed by the minute. That can drive you crazy thinking about how much time your workers are online.

Broadband

Broadband has become a generic term to describe any high-capacity (greater-than-dial-up) connections. There have emerged two principal broadband services: *cable* and *DSL*, but *WLAN* and *satellite* services also offer "broadband" connections. Cable and DSL also can reliably support VoIP.

For most applications your home workers need broadband to enable them to deliver ***equivalent if not superior performance than at premises offices***—this book's mantra. Dial-up is too slow. If your employees need high-speed data connections and broadband is not available where the employee lives—unless they are so valuable so as to merit very costly satellite links—then they should not work from home.

Only with broadband has home working been able to grow. Employees receive and transmit files and use the Internet with about the same perceptible speed as at premises. No more "eternity downloads."

Due to competition, broadband providers have been forced to rapidly expand their networks. I live in a small city with 19,000 people on the east coast of Vancouver Island and we have competing cable and DSL providers. Forestry/fishing towns with less than 1,000 people on the more rugged west coast also have broadband.

There are two downsides to broadband. The first is additional cost: typically double that of PSTN, though with broadband prices plummeting expect the differential to narrow dramatically, or be eliminated completely, during this book's lifespan. The second is that it is more vulnerable to disasters.

Wired broadband modems require AC current. Your employees need either a PSTN backup (PSTN is energized by batteries in telco central offices) or a battery-powered uninterruptible power supply (UPS) backup, if it is critical that they live with broadband at all times (more about UPSes in Chapter 5). Broadband connections are typically more finicky than PSTN and go down unexpectedly (usually only for short periods of time).

Cable broadband is a data connection to and from the Internet supplied by the local cable company. The signals come from the cable company, down a coaxial cable—the same type that feeds your TV—into a building, like your at-home employees' homes, through a modem, and then into their computers. The network interface and cabling from the cable modem into the guts of the computers are the same (says *Newton's Telecom Dictionary*—which is the best single-source guide in the industry) as those in your premises' LAN.

The key advantages of cable are high data transfer speeds: downstream (to the home) 220-250 Kbps for "cable lite" to 1 to 3 mbps for "high-speed cable," and 256 Kbps upstream for both (at this writing). The other benefits are availability—wherever there is enough demand for cable TV cable broadband is sure to be there—and ease of installation and service.

Nearly everybody deals with or has dealt with the "cable guy" or "cable gal." If you have cable TV, cable Internet service is often available as a low-priced package. Two jobs: TV and computer, one visit. I have Shaw Cable where I live and I've found them to be superb.

Cable is versatile; I connect my cable modem into a hub from which I run my CMP laptop and my personal desktop. While that splits the bandwidth in half, or 500 kbps per machine, for my work I don't care. I'm not downloading the entire New York and Long Island edition of the Sunday *New York Times*.

The downsides to cable are that it is only there if it is there—in only those communities that the cable companies think is profitable to be in, which can exclude many rural areas. But as the urban areas become wired, cable is reaching more isolated areas. For example, Tahsis, British Columbia, Canada is a forestry, fishing and tourism town on the west coast of Vancouver Island, with just 900 residents. The only landside way to get in on a long gravel logging road; there is a boat and seaplane service. But in September 2003 it got cable broadband.

Another downside is that if there many users on the same local line into a neighborhood, like the carpet mice scurrying home from school or playing hooky and pulling the tails of the plastic electronic variety, the performance will degrade. That's because cable is like a water line. If a lot of people tap into it—and there is enough replacement volume, the pressure and quantity will drop.

But smart cable operators—ones that realize that have competition i.e. DSL—will add juice to ensure the bandwidth is kept at a high level.

On occasion cable modems will get finicky. *Newton's Telecom Dictionary* recommends when the modem locks up, unplug it and the router from the power, wait, then plug them back in again. I've done that, sometimes a few times and it works.

DSL or digital subscriber line provides high-speed data connections over the PSTN. It does that by using a different frequency than PSTN and connects to the Internet via special modems at the local central office and at home. DSL is supplied by telcos or by ISPs renting capacity from telcos.

DSL offers fast data transfer speeds. The common-to-the-home version of ADSL (asymmetric digital subscriber line), known as "ADSL Lite, has a typical downstream speed of 384 Kbps (although it can run as high as 1.5 Mbps) and an upstream speed (back to you) of 128 Kbps (but the speed can be as high as 640 Kbps).

DSL rides on the same line as your voice line. In most instances, your employees can install DSL themselves; no "cable guy/gal" or "lineman/woman."

In short, DSL delivers. Methinks it is because of competition from cable and because the ISPs, flexing their rights under US deregulation, were beginning to offer the service and riding the telcos about delivering it.

(For more information about DSL the best book on the market is *A Practical Guide to DSL* written by this book's editor, Janice Reynolds. It is also available from CMP).

But DSL has its downsides. The key one is that your employees must live with 18,000 feet of a telco's central office (CO) to use it; and that's 18,000 feet in wire distance, *not* as the crow or seagull flies. The performance begins to degrade at the farthest reaches.

Cable versus DSL recommendations

Which is better? In my somewhat humble opinion, it depends. Both are roughly equal in data transfer rates, and in price and installation.

But which method is better depends on the area where your employees live, their proximity to COs, how many people are on the cable loops, and how good the service provider is. DSL is leashed to the CO; cable quality by how many people are drinking from the pipe and how much fresh juice is poured into it.

Where I live I have cable and I like it. But I have a neighbor on my street that had cable and didn't like it, so he switched to DSL.

Let the employees decide while type and brand of wireline broadband they want; they will have to work with the results. But whichever broadband method they and you pick, make sure they and you have PSTN dial-up as a backup in case the lights go out or the broadband connection throws a snit.

Cellular broadband

Cellular broadband is high-speed data on the go. It has top average speeds of 300 to 500 Kbps (although realistically, the customer—your employee—will obtain only modest speeds of 40 to 75 Kbps).

Cellular broadband is slowly being rolled out in many cities; it is where wired broadband was at in mid-1990s. It is much more expensive than wired broadband or conventional cellular data.

But like conventional cellular data, cellular broadband runs the risk of data dropouts; and speeds can change dramatically. Data is more sensitive to loss over wireless than voice. There may be blind spots even in one's house for those signals. There are also greater security risks with all wireless communications.

Wireless broadband may not support high-bandwidth applications like videoconferencing. One such provider, Verizon, does not recommend it for them, citing variations in upload speeds.

Wi-Fi (WLAN)

Wireless local area networks (WLAN), as discussed in this book, fall essentially into two groups: home networking (discussed in Chapter 3) and hotspots (for mobile working), which is discussed here. Both share in common high bandwidth (11 to 54 Mbps depending on the technology) that degrades with distance; the signals go through walls, and typically have a range of 350 feet from the antenna (although that distance can be extended to a few miles with special antennae). However, hackers can easily penetrate WLANs, unless you make both ends of the transmission secure, such as through a VPN. Janice Reynolds, who edited this book, has written an excellent tome on WLANs, entitled *Going Wi-Fi,* published by CMP Books.

Wi-Fi, at this early stage of development, is accessed from "hotspots": a broadband Internet connection (typically a DSL or T-1 line) plugged into access points with antennas. Hotspots can be found in coffee shops, restaurants, airports, hotels, and train stations. Wi-Fi service vendors approach these business venues and install the hotspots; the businesses hope having them will draw business or they provide it as a convenience for their customers.

Wi-Fi is a mobile application. With the viral growth of hotspots, if your employees live in densely populated neighborhoods, like in Boston, New York City, Montreal, Toronto and downtown Seattle, Portland, or Vancouver BC they may have access to a hotspot literally next door.

But because they are dependent on another source for their Internet connection (DSL, T-1, satellite), I wouldn't recommend depending on a hotspot for daily data transmission, unless the hotspot provider offers a subscription. The hotspots can only take a limited number of connections. As noted earlier there are performance degradations and security risks to examine and plan for or handle, such as with VPNs.

Satellite broadband

For people who do not have ISDN, cable, or DSL, there a high-speed alternative, satellite broadband. Users receive data via satellite dishes and transmit back via satellite or over PSTN.

The key attribute of satellite broadband is that it connects those living in remote rural areas with the Internet and thus far-flung data files. Users can downstream data at up to 1 mbps, which is great for Web surfing and receiving graphics-laden e-mails. Satellite upstream rates can go to 512 kbps.

But the equipment is not cheap: in the high hundreds of dollars, but much less for satellite downstream only (dial-up is used for upstream transmissions). Operating costs run 3x to 4x or more than that of wired broadband (cable/DSL). The one-way downstream option is good only if there is PSTN to the employees' neck of the woods and if they don't have to upload huge files.

There are also technical issues. Atmospheric conditions: clouds, solar flares and for

Alaskans and for many Canadians, aurora borealis "northern lights" may affect trans-mission and reception. There may be signal delays depending where the satellite is; the antenna must be pointed in the right direction.

Data Links Recommendations

If you have home workers, they (and you) need broadband, preferably cable or DSL, depending on the carrier and local quality of service. Only broadband can supply data at speeds that enable employees to deliver *equivalent if not superior performance than at premises offices*.

But because even broadband dies now and then and the modems rely on AC, if it is critical for your employees to be continually connected to your data network, your employees should have a PSTN line as a backup; they also can use the PSTN for faxes and for second lines.

If your home workers are largely mobile, and have cellular voice rather than PSTN or VoIP voice, consider cellular data and potentially cellular broadband. While the down-load speeds are slower and the costs higher for cellular data than wireline broadband and cellular broadband is still in its infancy, the convenience of having one anytime/any-where provider and network may pay off for some employers.

There are fewer wires to plug and unplug, which could reduce costs and downtime in repairs and replacement. But when these employees are at home, having a PSTN line is a good idea, as backup.

Satellite broadband should be looked at only to support those extremely valuable workers who choose or whose spouses choose to live in the middle of nowhere, unless they're willing to pay for the connections themselves. Or as interim connection, until the cable or DSL wire gets hooked in, but no more than six month later.

✪ WHO PAYS AND SUPPLIES?

No matter what data communications and connection components you select, you have to decide who is going to pay for it. The choices are employer-dedicated and shared per-sonal.

Employer-dedicated include employer-supplied e-mail addresses and data-only PSTN, ISDN, and broadband lines and cell services paid for and used by employees for work purposes. Shared personal includes the employees' existing e-mail addresses and lines, and cell services reimbursed or nonreimbursed and tax deductible.

What you're willing to supply your employees should be based on how much they need, which mode (online or fax), and use. If shared personal, how much of that you're willing to pay for.

With communications means like e-mail you pay nothing for the transmission: it is "free." But someone must pay for the Internet service carrying it to your employees' computers.

Chapter 7 discusses data access and its accruements, e.g. firewalls and tools such as VPNs and hosted access that give employees entry to employers' e-mail servers. But to get to that gate the employee needs an Internet connection at home.

The same for what I call limited access, also discussed in Chapter 7, such as Webmail, that lets them see, obtain, and send e-mails via the Internet, but little else. Webmail is especially handy for mobile working.

Pays/Supplies Recommendations

If your home employees are exclusive home workers, then pay for their data connection (which ideally should be broadband). If they live outside of your area, let them choose the broadband source; cable or DSL can be hell in one place and heaven the next.

Consider also paying for a backup PSTN line if they have a fax and/or rely on being online for their work.

If you decide to deploy VoIP for your employees, then pay for the data connection. But also pay for the backup PSTN.

But if your home workers only work from home occasionally, you and your employees should consider shared personal, i.e. using their home Internet connections at home for data. When your employees work at home you and your colleagues can send them e-mails to their personal address.

Your employees can retrieve e-mails sent to your firm's address from a Webmail account. If you have (and you should) firewalls, and they have an employer-supplied computer that has software such as a VPN client enabling access to your computers, they can pull e-mail and do chats within your servers.

The downside of shared personal is that your employees will not have access to your network. That will limit what they can do. Typically Webmail accounts drop off after 30 minutes—with no notice—like in the middle of reading or writing an e-mail.

Carefully examine fax. If your use is infrequent, like every couple of months, and there is no big rush on the fax receipt and transmission consider reimbursing employees for sending and receiving faxes at local office supply stores.

Chapter 5:
The Home Office

This chapter looks at home office locations and facilities requirements. There will be some overlap with Chapter 6, but that's deliberate. The lessons are so important to the success of a home working program that reading them again serves to reinforce them.

There are unique considerations entailed in having employees work from home including location, facilities and environment. People select where they live and choose their homes based on traditional factors: affordability, security, schools, amenities, and transportation access to premises offices. They furnish them as homes. They have not necessarily picked the locales and buildings and designed them with home working in mind.

For example, employees may buy a home in the middle of nowhere that only has dial-up Internet access. But their work requires them to have broadband connections. They may size an apartment or house for living in, and thus have little or no room for an office. And if they haven't planned for an office chances are they will not have suitable computer desks and adjustable chairs.

But that is changing. More people are working or want to work home. They are choosing not only to live in communities and neighborhoods that have high-speed Internet, but houses with space for offices, and they select large rooms that have minimal glare from the sun for their offices, and fit them out with proper office furniture.

Those were our requirements. When my wife and I were househunting in British Columbia, Canada nearly every house we looked at had a home office: in the front room, a part of the living room, in a spare bedroom, or in the basement. There are so many home-based businesses where we live that there is even a home-based business association that has an annual spring show.

When developing a home working program you and your employees must ensure that wherever they work, their performance does not suffer, or (better yet) is enhanced, by working from home. That includes, for example, insisting on no background noise or interruptions.

Equally, if not more important, your employees' home offices must be *ergonomically sound*, with adjustable furniture and appliances such as headsets to avoid injury. Just as your premises offices are, or should be, equipped. An ergonomically sound workplace prevents injuries. These injuries can have stiff price tags, for your employees and for you.

Chairs are an individual fit; it is best for employees to try the chairs for their liking rather than
supplying them centrally. There are many good
reasonably priced adjustable chairs that your
employees can buy locally for their home office,
such as this Global chair carried by Staples.
Credit: Staples.

They cost money in sick time, loss of productivity, and higher medical costs.

I stress ergonomics for good reason. You
may not think that seemingly sedentary tasks
like writing programs (and writing articles
about writing programs) and phone or online-
based customer service can hurt workers. But
your workers can suffer from near-crippling
debilitations like carpal tunnel syndrome and
tendonitis that can cause so much pain that
they can't work.

There is a tendency to "forget" about ergonomics in home offices because they are out
of sight, out of mind. But if your home working employees are suffering from ailments
like tendonitis you will find out soon enough: when they call in sick or say they have to
go to the doctor. Also, while authorities such as the federal Occupational Safety and
Health Administration (OSHA) have at this writing stayed away regulation home offices
per se, they are still covered under general industry rules.

Remember the mantra. Home workers must deliver ***equivalent if not superior performance than at premises offices.***

⊘ LOCATION CONSIDERATIONS

Everyone lives somewhere, but not every somewhere is suitable for working. Here are
the key issues that arise in locations.

Note that the matters discussed in this section are intended for domestic locations.
Chapter 10 covers exceptional circumstances such as international locations, moves to
small cities/towns, and handling customers and clients in person.

Communications

Your home workers will need to communicate by voice and data (see Chapters 2 and 4)
with you, colleagues, and clients. Wherever they live, or want to live, they must have
those communications media to their homes to enable them to work.

While conventional phones (public switched telephone network a.k.a. PSTN) and
dial-up Internet access is available in most places, there are still some locales that lack such
media. More importantly, many employees require high-speed broadband connections
to work; they are still less common than PSTN/dial-up. Very costly options like satellite broadband are worthwhile only if your employees are worth it.

But if you determine that you need or allow employees to work at home occasionally or in emergencies then you (and they) can get by with PSTN/dial-up. Yet as broadband networks spread out more employees will have access to those connections.

Proximity

If your employees do not need to come onto your premises regularly and you don't need to "see" your employees than it doesn't really matter how far away they live (and work) from your premises. You can stay in touch with other through e-mail, instant messaging, fax, and phone.

For example, I live about 3,000 miles from *Call Center Magazine's* premises offices. I haven't been there since 2001. The magazine has also moved to a new location in Manhattan; my old desk and phone number no longer exist.

But my colleagues don't need to see my ugly face to know I'm at my desk working. They get enough from my e-mails, instant messages and from my occasional phone calls and they get to e-mail, IM or call me. I deliver my goods by e-mail, IM or fax, and rarely, by courier.

But if your employees are occasional home workers and they need to be at premises, say once a week or month, then they need to live within a reasonable traveling distance. But that distance can be considerably greater than the average commuting distance. People don't mind longer travel and the added costs entailed if they don't have to do it every day.

There are no real limits to these travel distances. I once worked with a senior editor who took an overnight train from his home in rural Massachusetts, stayed in Manhattan for a few days each month at deadline, then rode back. The train enabled him to sleep and arrive refreshed the next morning.

But proximity also can be an issue if your endeavor gets a lot of phone calls, like a customer service, support, or sales call center. As explained in Chapter 2, those calls are either re-routed or transferred to employees. If the phone switching takes place at the premises and the employees live outside of the local calling area those calls become long distance calls.

If the costs of those calls form a substantial part of the costs of carrying those employees, for the value they create, then you might want to limit home working to those employees who live within the local calling area. Or you research and deploy alternative technologies like network/intelligent routing and voice over IP (Chapter 2) that eliminate the long distance charges.

Proximity also can be an issue when employees are working clear across the country. Your home workers may need to be available at *your* crack of dawn or well into *your* evening—a three-hour time difference can cause problems in such instances.

Employees in my Manhattan office can work 9am to 5pm or 10am to 6pm ET. Or in my case, because I am based on the West Coast those times translates to 6am to 2pm ET or 7am to 3pm PT.

But the time factor can point to another benefit of home working. It is much easier on you and your employees, if your home workers must be on duty at 4:30am or 7am, if they don't have to pour away an hour or two to put on business clothes, warm up the car or alternatively get to the bus or train: if they run that early.

Access

If your employees need to travel on business, then they will need affordable, quality access to the outside world. That includes airports with scheduled air service, good quality highways to their destinations, or if they live on an island, to passenger or vehicle ferries.

I say *if* for good reason. Many employees need not go on business trips farther than the next island, house, or town.

Access Recommendations

If your home working employees live far from your principal premises offices, you can keep travel costs affordable by limiting the destinations they are allowed to visit. For example, if your employees work in sales or support and their area is California, Oregon, and Washington State then they should live in one of those states, or adjacent to them (like Nevada) where they can grab a quick flight or hop in a car and drive to see clients.

While you can get a chartered plane into most anywhere, the high costs compared with the net value those employees supply probably aren't worth having them living in such remote locations. But for those employees who are that valuable, then the expense might be worth it.

For that reason you may want to look hard at any employee who wants to live in Alaska but who needs to fly every month or so to the Lower 48. Such costs can quickly drain the travel budget. But if these workers generate more worth than they cost compared to others in your enterprise... well come on down.

The access issue is affected by travel frequency. If the employees travel on occasion then you could tolerate higher airfares and sometimes air or boat taxis or charter flights. But if they travel frequently, and at high costs, then you might want to limit home working to those who are closer to their likely destinations.

If your employees must travel, make sure they have the necessary tools to do the job, like laptop computers. Chapter 11 offers suggestions on mobile work, including travel modes and hotel ergonomics to enable mobile workers to better survive.

Think hard on access and travel

Think hard about access as a criterion. There are tools like audio-, data-, video- and Web conferencing (covered in Chapter 11) that can limit the need, number, and duration of such trips. These tools can save time and money and boost productivity. Why pour those scarce resources into in-person interaction when you can avoid it?

I have found that there is increasingly less need for in-person interaction. People are becoming more used to e-mailing and instant messaging than phoning, let alone in-your-face visits. Conferencing from your own seat beats spending a lot of money and undergoing much hassle traveling to another seat in another locale.

I avoid business travel at all costs. I find it to be a time-consuming, stress-inducing, literal pain in the backside: an ailment that costs my employer and me money. My travel is usually limited to events like our trade shows where I network with my colleagues and with some of my sources, or an event or a story near where I live that I can get to without flying.

Instead I do most all of my work virtually: by phone, and increasingly by e-mail. That I don't see the people face to face has not hindered my ability to do my job. There are sources that I have that I have never met. I have also found that many of the people I've met face-to-face are like replaceable doll heads: with turnover so rife in management, marketing and PR I am dealing with different people almost every time. So what is the point of building "relationships" with them?

✪ FACILITIES

Home working has unique facilities: real estate and furniture requirements and issues that you need to be aware of and work out with your home working employees. Where and how they set up their home office, handle environmental factors like noise and light, how they cope with family members, pets and guests, and the furniture they select—all will impact your employees' performance and health.

These unique needs and issues have their greatest impact on employees who work at home exclusively because their home *is* their office, and *yours.* You have some say in how in how they select, construct, and organize their premises. Remember the mantra... the home office facility should enable **equivalent if not superior performance than at premises offices.**

But for occasional home workers you will have to be more flexible. While under current tax law permanent home workers can write off workspaces, furniture, utilities, and services, they can't make the same claims if their principal work location is your premises office. Therefore asking them to provide, say separate offices and buying furniture to meet your requirements, is probably unfair.

Also for employees who work from home occasionally, especially in emergencies, you will probably need to be more lenient on issues such as background noise from children and pets, and interruptions from family members. Mom or Dad working from home is the exception, not the routine. Most people understand when an employee says "oh I'm working from home today."

When looking at home worker facilities *ensure that these employees—and you—have equivalent levels of insurance protection*: such as for accidents, fire and theft. I and many other home-based workers have, for example, riders on our home insurance—which I (and your employees) pay for—that covers us in case the delivery person slips and cracks their tail end on the ice or trips over the cat on the way into the house. Chapter 8 looks at administration and management issues such as insurance.

Work Environment

Your employees need to have a dedicated workspace at home. That space could be a corner of a house, room, or a corner bedroom, a basement, or a shed out back. The main criteria is that the only tasks that are accomplished there are work tasks.

There are psychological and legal reasons for this. When your employees sit down in a dedicated area they will feel they are at work. They will then focus on work and get the work done.

A dedicated workspace means just that. No sewing kits or tool drawers next to the worktable. No laptops on the dining room table, or in the living room.

The outside deck or patio doesn't count as a workspace, although that is a nice place to go when employees want to take a break from the dedicated space. Laptops, cell-phones, and cordless phones enable you to work anywhere in the house, especially if you or your employees have wireless LANs (see Chapter 4). I've worked many a time walking around my deck while on a call.

Also, tax laws at this writing will *only* allow working at home deductions for dedicated space. But that includes portions of rooms.

I lived at one time in a one-room apartment with shared bath and kitchen in Queens, New York. I was working as a self-employed freelancer. I was able to write off that portion of my room where I worked. I calculated the square footage of my rented space and subtracted the space occupied by my desk and filing cabinets. I then assigned a portion of my rent to that space, and put that amount on my tax forms.

Wherever the workspace is located, there should be voice/data/power connections to enable employees to work. The work area should be situated and set up to minimize glare.

If your employees work with confidential information, such as corporate records, police reports, customer data, and the like, then their workspace should be set up to protect that information from prying eyes, such as children, guests, people knocking at the door, passersby, and next door neighbors. There is an increasing array of confidentiality and privacy laws that make employers and employees liable if unauthorized parties see that data. Chapter 8 covers these issues in depth.

You can ask or require your employees to take measures such as locating their workspace so that no one can see the computer screens from outside, or from the rest of the room, i.e. no "windows on Windows™," and having the workspace in a dedicated office with lockable doors. Other techniques include installing password access-to-files and screen-wiping software, specifying storageless PCs, and installing locked filing cabinets.

David Smedley is principal with a facilities service and design firm, Wave. He and colleague Keith Werner developed a presentation on home workplace facilities, and ergonomics. David recommends that:

* Your employees' home workspace should be located where there is some visual and acoustic privacy. This will allow you to concentrate and will help maintain some separation between your job and your life.

* You should be able to dedicate space in a ventilated, temperature-controlled area of your home to an office. If there are kids or pets in the home, converting a bedroom or other separate room with a lockable door can prevent loss and damage to materials and equipment.

* Avoid extremes when decorating a home office. Red walls may look wonderful, but can be distracting in a long-term work setting. Softer colors will provide a calmer

environment and reflect light better. Similar care is needed for floor coverings. Thick padded carpets may not allow you to roll your desk chair between work surfaces as needed. Rugs on hard-surface floors can also get in the way. Dense, low-pile carpets work well.

Lighting

Lighting is an issue that is often overlooked in both home and at premises offices. Improperly placed light sources can cause glare on computer screens; that glare, in turn, leads to eyestrain and fatigue, harming employees and performance. Furthermore, while having a desk by the window offers a great view, it can play havoc with one's eyes as they shift from natural light to the computer screens, especially if the sun is shining in.

Dr. John Triano, director of the Texas Back Institute's chiropractic division identifies two types of glare: direct and indirect.

Direct glare is the light that bounces off a computer screen, obscuring the visibility of the screen. While this type of glare can cause eyestrain, more likely, it will prompt a worker to shift his or her posture to a more awkward position, increasing the chance of neck and back strain.

Indirect glare is ambient light from windows that pass around the screen, overwhelming the eyes and making it difficult to see the screen. I learned that lesson real quick when I was setting up a home office in New York City. Originally my workstation was in a corner with big windows, but I soon moved my workstation from the sunlit corner to the opposite side of the room—it is one thing doing my work on a typewriter or using handwriting and quite another when working on a computer.

My home office in Canada. Note the cordless phone which has a speakerphone that also has a headset jack. CMP supplies that phone line. Underneath the desk is a personal phone and extension. I also have two printers and fax machines. The desk design and equipment allows me to operate my CMP-supplied laptop and my home computer simultaneously, which are connected through a Linksys hub. My employer-supplied and personal communications co-exist but are separate—yet back each other up. Credit: Christine Read

There is another reason why good lighting is important: productivity. Lighting affects you emotionally as well as physically. "Low light levels can make you feel sleepy and quiet while higher levels and daylight can make you feel energized," says Smedley.

Noise (from everyone/thing)

Background noise is an issue only if it disrupts your employees' tasks. Your employees' home workplaces should be free of background noise as much as possible. High noise levels can cause injury. Also, it is not the business of your customers and clients, and colleagues to know that those employees are working from home. For that reason many employers (e.g. call centers) insist on no background noise in home offices.

Home offices are generally quieter than premises offices. But homes are generally not designed to be noise-limiting. And there are unique noises that are present in the home that are not present in premises, which can be annoying to others who hear them on the phones. For example, noise from sources such as pets (dogs barking, cats yowling, birds squawking), house occupants (like children, lovers, spouses, guests), music, TVs, yelling, loud conversations, bottles smashing, cursing, "intimate acts," neighbors, and passersby. Some of these sources your employees can do something about, others they can't.

Other sound effects come from those two air-killing suburban curses: lawnmowers and leafblowers. Garbage trucks and assorted traffic, like the 5:15 to Babylon or the "Albatross Airways Shuttle" contribute to the "symphony."

Some voice handling practices can make noise worse. Key among them: speakerphones. Few people can hear what is going on the background on a standard phone, fewer still on a headset. But on a speakerphone they can hear everything: the cat chucking up hairballs, the kids outside swearing at each other, the crows yelling, and the trucks backfiring. Lovely.

But the same holds true in premises offices. Think about the times you've talked to someone on an office phone, especially a speakerphone, and heard background conversations, tradespeople banging away, and (in the evenings) vacuum cleaners.

Some of the noise sources are obvious: airports, factories, major roads, rail lines, ferry terminals, fire stations, bars, restaurants, gas stations, convenience stores, and school yards. If there are empty alcohol bottles near your home worker's house or apartment building, you can pretty well figure that these noise produced outside the worker's premises will not be exactly that of Sunday school types.

But some noise sources are not so apparent. I live in a broad valley where at the foot of my street there is a huge farm. There is a sawmill on the far side of the farm. I used to hear logs banging into chutes at the sawmill because of the way sound travels: before the sawmill owner, responding to public complaints found ways to reduce the noise.

Also, even if your employees are located in a quiet place, there is no guarantee that it will remain that way. Your employees could suddenly find that they have musicians or music-loving neighbors who like to play real or air guitar at 10 in the morning as well as at 10 at night. Or a neighbor gets a dog that howls at those hours as well as at the moon. There could be an expansion at the local shopping center or a new retailer opening at the corner that turns their street into a speedway. Or the quiet little pub a couple of door down, suddenly begins to feature heavy metal.

Temperature

The right temperature aids in maintaining work performance at premises office—the same goes home offices. But home workers are not at the mercy of the employers and the favorite employees when it comes to air quality and temperature. They can, in theory, open or close a window, switch on a fan, plug in a heater, turn up the A/C, or turn down the heat.

If employees live alone or are alone during the workday they can set the temperature in their office to meet their needs. But if there are other occupants around during the workday (e.g. family members), they may have to negotiate the temperature with them.

If your home workers live in an apartment and the landlord controls the heat, they must depend on the landlord to ensure that heat is there. In some jurisdictions, like New York City, there are laws that say when the heat comes on or off. But some landlords have a habit of ignoring such laws. And if your employees are renting apartments or rooms in houses owned and/or occupied by landlords they may find the owners shutting off the heat during the day and in the middle of the night to save money.

I had that happen where I once lived in New York City. Fortunately for me, at that time I was working as a freelancer. Also, I was living in a well-insulated apartment with few windows.

Using space heaters may pose problems. First they drive up electricity bills, which are often paid for by landlords and included in the rent. Second, such heaters can overload electrical circuits. Many older homes are made of wood and burn up like kindling in electrical fires. Such blazes are a staple of east coast TV news as they are often fatal and spectacular.

Air conditioning units have their own issues. Most homes do not have central air—A/C is usually provided by individual window or wall units. But A/C can send electric bills into orbit. And if they, and the wiring, are not sized right A/C units can trip breakers, which shut down computers, phones and other equipment necessary for home working.

Security

Your employees' castle should be a castle, within limits. If you provide them with equipment and/or they are handling your data, then that employee should be made responsible for that equipment and data.

Employees should not have their work areas where someone can reach in and grab computers and paperwork. If the only space they have to work is a window overlooking a street or sidewalk they should have the window locked and preferably alarmed. Why make it easy for a crook?

Smoking

One of the great benefits of home working is that you transfer from your premises the biggest hazard of them all: nicotine consumption by smoking. Smoking poisons the air; non-smoking workers have died from exposure to secondhand smoke. Smoking also kills and injures through fires caused by smokers' carelessness. The additional costs in healthcare not to mention in lost productivity—from having to go outside to light up— are enormous. Then there are the costs of cleaning up smokers' litter.

Building landlords loathe smoking. They and other tenants dislike the mess caused by smokers huddled outside. Nonsmokers hate running the "gas chamber gauntlet".

But you should be mindful of home employees when they smoke. Yes, their home is their castle. They don't have to go outside to feed their addiction. But the time they take to smoke is time they are not working; ashes and flames can damage *your* computer equipment. The risk of fire increases. Employees also risk poisoning their loved ones with smoke. And smokers, wherever they are, cost you more money in healthcare costs than nonsmokers.

✪ HOME OFFICE RECOMMENDATIONS

Here are some priority recommendations for the home working facilities and environment:

Glare

You should require your employees to have dedicated spaces that are glare-free. Glare and discomfort kill productivity. That includes non-glare desk and table lighting so that light does not bother them.

Have your workers examine potential home office locations for indirect light. Basements and ground floors, with bushes in front work well for those reasons.

David Smedley recommends looking at the overall work area. An overhead light that's too bright can cause glare. Although relatively low ambient light is recommended, too little ambient light can cause the space to feel like a cave. Depending on the room shape and location, you may need to use lamps that reflect light up toward the ceiling to fill in dark areas.

Another technique and recommendation from me is considering adding dimmer switches to raise or lower light during the day to balance outside light. Also, dimmer controls help in basements if there are other users and there is no partition between you and them. If they switch on or off a light a dimmer helps you manage the impact from that source.

"You/your employees should be able to adjust the light level," advises Smedley. "That's important because different people have different requirements at different times of the day. Take the time to learn how to adjust light levels in your office and make use of this option. If you have a window, shades, curtains or blinds can also help control naturally fluctuating daylight levels."

There's also another reason for adjustable light levels: avoiding outside distractions. You or your employees may have great views from their home offices. From my home office window I look out over the town, the mountains and a glacier. Fortunately I'm overdisciplined enough and too much of a workaholic where the sights aren't an issue:

by working I can afford the views. But just in case I can shut the Venetian blinds to keep me focused.

Temperature

Keep the temperature constant and humidity reasonable to maintain productivity. David Smedley says the American Society of Heating, Refrigeration and Air Conditioning Engineers recommend a work environment temperature range of 73-79 degrees Fahrenheit in the summer and 68-74.5 degrees in the winter. Air circulation speed should be maintained at .15 to .25m.per second (about 1 mile per hour).

Average humidity should be in the 30-60% range. Falling below 30% humidity can cause eye irritation, particularly while using a computer. It can also cause electrical equipment to malfunction due to static build-up. Humidity above 60% can foster the growth of mold, mildew and mite populations that can cause allergic reactions and can affect equipment.

"These ranges were set to meet the preferences of 80% of the average population," Smedley points out. "Your preferences may differ slightly. Remember too that office equipment gives off heat. It's a good idea to keep a thermometer in your office so you can check the actual temperature and adjust the thermostat accordingly. A humidity gauge is also a good idea."

If your outfit is located in a large, older metro area like New York City it is probably a good idea to ask prospective home workers about heat. Many employees may not be aware that their landlords juggle the furnace—legal or not—unless they stay at home during the day. There is no point arranging to have employees work from home if they are freezing their tails off.

Data protection

Employers can make one of the conditions of allowing employees to work from home that they take employer-specified steps to protect data. You can back that up by having those employees agree to visits by supervisors to ensure compliance (more later on that).

Noise

If background noise is an issue for you there are steps employees can take to reduce noise. Private offices, if available, cut down on the sound considerably. Employees could see if headphones work for them. If your employees have speakerphones they should have private offices or use them when there is no one else around.

Employees could see if sound masking a.k.a. "white noise" generators will do the trick. An old CB radio or a boom box set on a dead station at the far end of the dial used to do the trick for me, especially when I had trouble sleeping thanks to a noisy neighbor. Air conditioners and fans also drown out noise.

Employees could also see if there is a quieter part of the house or apartment to move their office to. There are more extreme measures employees can take like installing soundproof glass, like what quality airport hotels use and ceiling, wall and floor paneling.

One of the best measures, but not foolproof with noise, is site selection. If employees are planning to work from home whether exclusively or occasionally, then they should see who their neighbors are, day and night, before doing the deal.

The best way to approach noise problems, if it is an issue, is for you to make no background noise as part of your policy. If there are problems: a supervisor hears it while visiting the work, or hears it while talking to them, or someone mentions it to you, then talk to the employee.

By and large noise problems can be worked around; I've survived them in New York City. But if the problems can't be resolved, and it affects an employee's performance, then reconsider having that employee working from home and bring them back to the premises.

Insurance

You should have your equipment insured. But you also can insist that employees have riders on their homeowners' insurance covering home offices. You also can ask that employees lock the equipment in a filing cabinet (another good reason for laptops), have locks on the doors, and/or have burglar alarms.

Smoking

You should stipulate in your home working policy that if employees are going to smoke that there should be smoke alarms in their work area. Remember, *your property*—equipment and data—also goes up in flames in a fire, along with productivity from that employee.

Also as part of your home working policy, make it clear that if there is any damage to equipment or there is a fire caused by smoking in home offices that you will terminate home working for that employee and consider disciplinary action. You don't need "accidental" firebugs—I am loath to call fires caused by smokers to be "accidental."

Consider making it part of your organization's policy to discourage smoking, wherever employees are located. That will lower sick days and healthcare costs.

Smoking or not, make sure you are protected, and have free-and-clear escape routes in case of fire.

"Make sure you are in compliance with local codes and insurance company recommendations for smoke detectors and fire extinguishers in or near your home office," recommends Smedley.

Power

Employees should make sure their home office circuits are not overloaded. They should take steps such as limiting the loads, installing more efficient equipment or have professional licensed electricians rewire the circuits to provide more power. Their lives depend on it.

Your home workers also should protect their equipment with surge protectors on their power and voice lines. If they are using desktops and rely on VoIP they should have UPSes. If they live in areas where power outages are common they should consider backup portable or in extreme cases standby generators.

DECORATING/MAINTAINING HOME OFFICE ADVICE

The home design/decorating boom has blasted into home offices. Programs such as TLC's *Clean Sweep, Trading Spaces,* and *While You Were Out* show, by example, how to make home working environments attractive and functional, such as with wall-mounted shelves. *Clean Sweep,* especially, is definitely worth watching, almost invariably the show focuses on someone's home office. Ask your home workers if their home office looks like the before, or the after?

The "big box" home improvement chains such as Lowe's and The Home Depot are excellent sources of carpeting, curtains, doors, fans, paint, shelving and windows, and tools to install them. They often offer installation services in case your employees are all-thumbs.

Remember: improvements and renovations to exclusive home offices may be tax deductible. Bear in mind too that the local hardware store or a general retail big-box chain like WalMart may offer the same items for less.

Whatever your home workers do, insist upon them to keep their spaces *neat*. They are much more productive when they can find items like the cordless phone and their notepad.

Also, suggest that they avoid turning their workspace into another bedroom or storage area that adds to the clutter. There are a number of reasons for keeping the workspace a "single function" space: If they add other non-office uses to their home office they may risk losing that tax deduction. Safety—papers or plastic blocking heaters or by electrical outlets could lead to fires.

If there is mandatory recycling in that area urge them to set aside bins for papers and other such reusable material. But if the papers contain sensitive data e.g. customer and confidential corporate information, demand that they shred it first.

I also recommend having a corded PSTN phone as a backup (see Chapter 2). If the electricity is cut off that doesn't mean the phones go dead also. PSTN lines use separate low-voltage battery power from cells located at central offices.

○ COPING WITH OTHER INHABITANTS

One of the most unique characteristics of home offices is that most home workers will have to deal with other creatures: family members, pets, and neighbors. They can interfere, by causing noise and interruptions, with your home employees' work. That, in turn, can lead to problems with customers, clients, and colleagues.

The success of home working depends on how well the home worker has thought out and has planned to cope with such matters. There are methods by which these "problems" can be worked around.

Occupants

One of the benefits (and hassles) of working from home is being close to others who live and visit there. Your employees don't have to be a stranger to family members, to loved ones, to roommates, and to family and friends who come to visit. The two to three hours your employees waste and get stressed out commuting each day can be spent with them instead.

But these occupants, permanent and temporary, can be demanding and noisy, even if it is unintentional. Young children many times will not understand why Mommy or Daddy has to ignore them or shut the door. Slightly older offspring may resent that the home worker can't drop everything and help them with their Mulgarian History essay; or become offended when told to keep down that !@$%&*&() music or not bounce that !@$%&*() basketball outside the office window. And without hands on supervision, there is a risk that one rugrat may come out the worse for wear when fighting with another rugrat over an MP3 player one alleged the other of stealing.

Houseguests may turn on the TV at full blast, or want to interrupt the home worker to ask if it is okay if they go out to the mall in the car. Spouses may beg and plead to run them out to XXYYMart in the middle of the day to pick up "a few things."

Sometimes prying fingers get their hands on things they shouldn't, like financial reports. Sometimes overeager aspiring computer geniuses accidentally load viruses onto the hard drive, deleting all documents and spamming the universe.

But at the same time the work has to get done. And that means your employees *should not* permit household occupants—permanent and temporary—to distract them from their tasks. Also people at the other end of the phone shouldn't hear babies crying and kids yelling. Remember the mantra...

Occupant Recommendations

If you do not wish for your employees to be interrupted at their tasks you have to lay down strict policies. That means not chasing after Junior who discovered how to pull off their diaper and not running out to the store for Grandma's Depends™, when your employees should be working.

If a home worker feels he or she might face such situations, ask the employee to arrange for outside care during work hours. Or perhaps, for those employees that have child or adult care responsibilities, the answer is to work part-time, at home, and be available at home the rest of the time.

There are exceptions. Some tasks can take disruption. Some outfits permit child-care and adult care on premises or nearby as part of their amenities and corporate culture. If these exceptions fit your organization, then you can be more lenient at home. It is easier for the employee and better for the child or the grandfolks if your employees are seconds away, than minutes away.

Also create set hours, break periods, and lengths of break periods for your home working employees, and ask them to "sign in and out" (via e-mail or instant messaging)—the same as you do for your premises employees. This helps to limit set distractions. That also helps you and your colleagues to know when they can reach your home workers.

Occupants, from ages 11 up will understand work hours. They will know and understand the boundaries. They will realize that they can't interrupt your employees; that they should wait until your employees are on their breaks.

For employees with occupants in the home, there should be private offices, preferably with lockable doors. You don't know who the other occupants might bring in—especially roommates and older kids. A private office also creates boundaries between work and non-work, and reduces the risks of unauthorized access.

Remember: You and your employees are responsible for your equipment and work, including corporate and customer data and the network.

I have a private office. So does my wife. When we were house hunting our criteria included four bedrooms: for us, for guests, and for her work, and for my work. Surprisingly, we had little problem finding that much space. Nearly every home we looked at had at least one and often two home offices.

Pets

Pets are great company. There has been considerable research indicating that they help people live longer by reducing stress; stroking a dog's or cat's fur is relaxing. Also they break you from your routine, getting you active and moving. I have two cats; in the past I've lived with and looked after dogs.

If your employees are love-addicted to their pets it may not be a bad idea to have them work at home. It is far better to let them stroke Poopsie at home than to allow them to bring Poopsie into the corporate offices (something too many brainless companies did during the dot-bomb bubble of the late 1990s). The last things employers need are employees sneezing and scratching and discovering new and costly-to-remove aromatic designs on the carpets.

Pets, however, can be a pain in the butt at times, for reasons that they are not entirely their fault. After all you are responsible for them.

Dogs need to be walked at set times. They will chew up paper, like the printout of a contract. Cats want attention any time they please. Some cats will drag in a dead bird or throw up on your carpet—just for you. They will jump on workstations and step on computer keys, consigning that program your employee worked all night into electronic oblivion. Birds will chirp and squawk.

These are background sounds and interruptions that your customers and clients need not and probably don't want to know about. But in fairness to the pets they don't know about your work no matter how hard you try and tell them.

Our cats, two middle-aged ones named Casey and Tasha and two youngsters, Piewacket and Dot like to visit me when I'm working. When the cats come into my office they ask me in "felinese" what I'm up to and why don't I stroke their fur. I will do so for a minute or so and then get back to work. I don't mind except when I'm on deadline or I am doing an interview. Then I shut the door.

Though I love those fur monsters dearly, they are distracting. Casey will mournfully meow at the door until she gets fed up and climbs back upstairs to curl on the couch. Tasha will go back upstairs to bug my wife. To them we're their big, dumb, and stupid, if cute, kittens.

Piewacket and Dot use my office like a playground. They're also fascinated with my work and they try to do some of my work for me, though they are a little clumsy with the keys. Not wishing to use the excuse "the cat deleted my story!" I put them out the door or they crash out on the futon.

Pet Recommendations

If your employees have pets and want to work from home they ideally need to have a private office with a door on it. Unless your employees can assure you that their much loved animals and other creature(s) will not be a distraction.

If your employees use the phone often and they have birds, the avians should be caged and located in a different part of their residence. The employees are working for you, not the zoo. But if you are in the bird or zoo business, then the background noise could add squawking points to conversations.

Your employees should also feed, walk and clean up after them before and after hours. In other words the routine -- and pets like routine -- should be no different than if they worked at premises.

If I am busy I look at my cats and I ask: "Do you like eating salmon and tuna Whiskas cat food, and ham slices? Instead of the no-name crap of 'eat it or starve' flavors? Do you like living in a big nice house'? Do you really want to go back to the shelter where we got you?" Then I tell them, "Now leave me alone so I can buy you the food and keep a roof over your furry heads!" I'm not sure if they understand my words. But they sense my tone of voice. And they get the message.

Visitors

Yes, people will come to the house. Exclusive home workers may get to know their courier deliveryperson on a first name basis; it may be their only social contact of the day. If the home worker is alone or their office is on the ground floor (as mine is) they have to be prepared to play doorperson. That can be a hassle if the employee is on the phone, especially a speakerphone.

Visitor Recommendations

There are no easy answers to this issue. Your employees can ignore the bell-ringer or door-knocker. But the visitor might be the deliveryperson with a work shipment; often having packages redelivered or going to the courier office can be a pain in the tail. Alternatively, home workers can hang "come back later" signs during the day and make package redelivery arrangements. Or they can get "high tech" and install cameras hooked up to microphones.

Visitors are a good reason for having a cordless phone with a headset jack. When someone comes and the employee is on the phone they can open it and usher the visitor inside or sign documents and the person at the other end of the line is none the wiser.

Co-home Working

Sometimes your employees' co-habitant is also a home worker. As more people work from home the greater the odds of this happening. My wife and I worked from home together briefly.

There are a lot of benefits of co-home working. You are each other's home support group even with different occupations: people and management issues are really no different from one outfit to another. You assist each other on fixing equipment and answering doorbells. You may even share the same office, hubbing off the same cable or DSL modem, which spreads out the expenses.

But there are some costs. Both partners run the risk of getting on each other's nerves, e.g. both on speakerphones at the same time.

Co-home Working Recommendations

If your employees are co-home working with their loved ones then they need to set out some ground rules. If they can do it, they should have completely separate work areas; ideally they should have separate offices, as my wife and I have. Such an arrangement allows them to have their own work and private space and not interfere with each other's tasks.

✪ VOICE/DATA/POWER

Wherever the employees work in their homes they need to have the voice/data/power connections to enable them to do their jobs (see Chapters 2 and 4). That's a given. PSTN and cable installations are relatively straightforward.

But keep in mind that there may be limitations to the number of PSTN circuits available to your employees' homes. That may cause problems if they are living in an apartment or a duplex, with other homeowners or tenants. The employees, or you, may have to pay more for the additional lines. Alternatively, you may consider VoIP for voice if there is DSL or cable, and the cable carrier offers VoIP; in extreme cases wireless may be the only option available.

Fortunately, with the advent of powerful lightweight laptop computers, excellent and affordable cordless phones, and new and more secure wireless LANs, it's not necessary for them to punch holes and string cable and wire all over the place. Employees could have their communications go into a broom closet and wireless it the rest of the way.

They also ideally need some way to protect the voice/data/power. The tools include surge protectors (on voice as well as power lines), backup batteries (for laptops), and uninterruptible power supplies (UPSes).

But power connections are often problematic because most homes have been wired up as homes, not offices, your employees may not have enough of those utility connections and power supply in place. Employees therefore, need to pick work areas that have these connections, or have them put in.

The consequences of not having enough wiring in place to handle the loads a home office might require: CPUs, laptops, lights, printers, cable modems, etc. can be deadly—from electric shock to fires. Your heart can stop from currents as low as 75 milliamps; 20 milliamps will paralyze the respiratory system.

To put this in perspective, an HP desktop inkjet printer/copier/fax draws 2.2 amps maximum while circuit breakers are rated at 15 to 20 amps. Most people don't realize that electrical codes, like the National Electrical Code and the Canadian Electrical Code are designed to prevent fires, not electrocutions.

Wave's Smedley points out that a typical home has one outlet every 12 feet...or four per room on one circuit. But the typical home office has a CPU, monitor, fax machine, scanner, task lights, and other equipment; so you need at least six receptacles.

The typical solution is surge-suppressor power strips to provide the extra receptacles. The critical factor is to calculate the total amperage of the office equipment and add it to amperage of other household appliances on the same circuit.

The total continuous power draw must not exceed 16 amps on a 20-amp circuit breaker; or if the house is older, 12 amps on a 15-amp circuit breaker. You can find your total load by adding up the amps listed on each appliance nameplate label.

"The problem here is that many houses have only one circuit per room and may share a circuit between two adjoining rooms," Smedley points out. "In homes over 25 years old, the entire upstairs may be wired on a single 15-amp circuit. Typically the kitchen is the only room wired with more than one circuit."

Here's his breakdown:

Typical Home Office Electrical Load

5.0 amps	CPU
2.0 amps	monitor
7.5 amps	laser printer
1.2 amps	scanner
0.8 amp	100 watt light bulb
0.6 amp	3 foot fluorescent
0.6 amp	3 foot fluorescent
17.7 amps	TOTAL office load

The National Electrical Code (NEC) in the US requires two circuits to handle this load. But that's not all. There are other potential loads.

Potential Loads

12 amps	vacuum cleaner outlet
5 amps	other lights, radio
10 amps	coffee brewer
5 amps	dorm-size refrigerator
12 amps	electric space heater
12 amps	window air conditioner

"The 17.7-amp home office example exceeds the safe 16-amp capacity of a 20-amp circuit, which violates the NEC that OSHA follows for workplaces," warns Smedley.

"When other potential household loads are added, including loads from adjoining rooms with shared-wall circuits, the likelihood of a circuit breaker "blowing" is very great. Along with loss of unsaved data, overloading circuits can cause a potential fire hazard."

The options are paying to have a licensed electrician add a circuit (because this is a workplace that the employer has possible liability for, don't do it yourself or let the employees do it themselves, unless you or you're employee is a licensed electrician). Or limit the appliances, or use those that consume less power.

"You can choose equipment specifically to stay under circuit load limits and avoid all home appliances but this may not give you the tools you need to work effectively," Smedley points out. "So the message is clear. A maxed-out home office will almost surely require a purpose-built electrical system with two or three circuits of 20 amps each. This will assure electrical overload safety and allow for plugging in a vacuum cleaner or other basic appliances."

UPSes

If using VoIP, your workers need UPSes, especially if there are no PSTN or wireless services connected into your premises or network phone switches.

One of the downsides of home working is that homes can be more at risk to outages from such incidents as a power pole "jumping" in front of a car at an intersection. On the other hand, only those employees' homes on that circuit will be affected; life goes on for the employees living on the other side of town or on the far side of the Moon.

Relatively inexpensive UPSes can give 10 to 15 minutes of power: just enough to ride out a brief outage or to give the employees time to shut down their computers. The good ones also condition the power that flows through the grid lessening the risk of damage to your machines.

UPS systems consist of rechargeable or nonrechargeable batteries, power conditioners, and fast-react switches. They sit between the outlets and your equipment. They kick in the instant the power kicks out. The units have varying runtimes and power drives, and range in price from $40 upwards. Your employees or the dealers replace the batteries in the nonrechargeable units.

With today's aging and wonky utility grids, buying a UPS system may not be a bad idea. While they will not enable your affected home employees to function through long blackouts like the one in August 2003 that threw much of the eastern US and Canada into the dark, they will enable orderly shut down of computers and backup of data.

Backup Generators

Backup generators are the ultimate power protection. Companies such as Briggs and Stratton, Generac (Guardian line), Kohler, and Onan make these units. There are two types of backup generators: portable and standby.

Portable units are just that. Typically gasoline powered, they can supply up to 10,000 watts. For example, Briggs and Stratton's Model 1893 Elite Series Portable Generator features 6,000 running watts, 8,750 starting watts, an 11 HP engine and a 7 gallon fuel tank with a 13-hour run time at half load.

The prices of portable generators start around $500 or so and climb to around $3,500. The more power you need the more you pay.

You keep the portable generators in storage and roll them—literally outside and away from openable doors and windows to prevent carbon monoxide poisoning—when you need it. You also need 2 feet of clearance space on each side to ensure ventilation.

The unit must be grounded to prevent electrocution. For that reason, and to prevent fires, electrical codes such as the National Electrical Code mandate that you ground the unit. You need to consult with a licensed electrician to determine local grounding requirements. The ability of the ground to safely conduct electricity, with low resistance to the current, depends on factors like soils that you can't readily determine by the naked eye.

But to use a portable generator for backup power you will need a power transfer system, installed by a licensed electrician. They typically cost around $300-$400, plus installation.

The system includes a load manager, wired to your main circuit breaker panel. You pre-select those circuits that you want your generator to power, sized to meet the generator's output. When ready to use you roll out and connect the generator, fire it up, and switch on the pre-selected circuits. Power transfer systems like Briggs and Stratton-owned Generac Model 1276 can handle 60 watts at 120v and 30 watts at 240V, delivers 7,200 watts maximum and connects six circuits.

Standby units are permanent and larger: to 12,000 watts up to 45,000 watts, and much more costly, starting in the mid thousands leading to the tens of thousands of dollars: financing is often available. Standby units are powered by LP or natural gas and are hardwired to employees' home electrical panels. Only licensed electricians and gas-fitters should install them. These units need, for example to be a minimum 5 feet distance from the exhaust port to doors, windows, central A/C units or other indoor air outlets. These units also need transfer switches.

Backup Generator Recommendations

Portable generators are great if you or your employees perform critical work: where you need to keep going and if your employees live in areas where outages occur longer than a few minutes. Standby generators may be justifiable if employees live in areas where the electricity often goes down at an hour or two at a shot, such as on small islands, peninsulas, and in hurricane or tornado belts. LP-fueled generators may be the best bet; natural gas lines have been known to snap. But your workers should pay for this equipment, not you.

Note: If your employees have or decide to get a generator they must have a UPS system, either at their office or hooked into the generator's automatic transfer switch, which transfers the loads between power sources. The reason is that the UPS system needs to condition the power, from whatever source. There is a gap between when the grid dies and the engines fire up. When that happens, the resulting voltage spike may affect the computer equipment.

When considering a backup generator your employees should find out from their utility company just how often outages occur. They also should consult with dealers and with licensed electrical contractors or electricians to get their professional advice on UPS versus generators and making them work together.

Power Purchases

One more set of words about electricity. That should be strictly the employees' responsibility to arrange and pay for. In most cases there is only one meter for a house and it is impossible to accurately estimate for accounting purposes how much electrical usage is attributable to the job, although rough estimates are good enough for tax purposes.

The only exception is in the unusual circumstance where the home office has its own separate meter, like a legal apartment converted back to an office. Then you can work the issue of power payment with the employee.

✪ INSPECTING HOME OFFICES

To ensure your employees are complying with your locations, facilities, and ergonomics requirements you should consider having a means of verifying compliance, especially for exclusive home workers. That compliance can be as simple as laying them out in a home office policy and having the employees sign the policy, which means if they don't follow the stipulations you will take disciplinary action and/or perform home inspections.

Of the compliance measures the last, home inspections are the most thorough, but the most expensive and legally complicated. You are taking time to go to employees' homes, at your organizations' expense, and legally, that home office is the employees' property, not yours.

Still, you should require home inspections especially if the employees are handling confidential data for which your organization is legally liable, *and* thus you have stipulated protective measures, like locked doors, shades being drawn, and lockable filing cabinets in your home working policy.

If you do require home inspections, make employee permission a condition of permitting employees to work from home.

Home working/virtual office consultant Jack Heacock offers a few recommendations for home inspections:

* You must schedule home inspections, agreed upon in advance by both the employer and employees.
* In lieu of home inspections—in cases where it is not possible—consider requiring home employees to submit photographs, accompanied by an affidavit attesting to the agreed specifications to ensure that the same standards that apply to the traditional work place may be substituted for an on-premise visitation.
* NEVER have a male supervisor visit a female employee unaccompanied—always have another female or third person present during the visit to the home office.
* *Don't* overlook ergonomics. That should be near the top of the list to protect your employees from injury and you from injury claims and healthcare costs. See if the chair or workstation is adjustable. See if the employee is sitting in the chair right. Many people don't realize they are not in proper working positions until they feel pain in the arms, hands, neck and shoulders.

RECYCLE, RECYCLE, RECYCLE!

Ask your employees to recycle wherever possible. That doesn't mean only paper. You can reuse your inkjet cartridges, and save money as well as the earth. There are firms you can take your cartridges to and they'll refill them: drop them off or come back in 30 minutes or so. For example, Island Inkjet, headquartered five minutes from my home in Courtenay, British Columbia, Canada, has convenient mallside locations in the US, Mexico as well as in Canada, including in our local WalMart.

✪ PAYING FOR FACILITIES AND FURNITURE

The great aspect of home working is that the office doesn't need to be fancy; since when is the CEO or your favored clients going to drop around. Chances are your home working employees don't need costly and heavily wired partitions. Consequently your home workers can outfit their home offices for a lot less: $700 to $900 for a chair and table compared with $3,000 or so for a premises workstation.

Your home workers can get away with lower-end workstations. With the popularity of FPMs and laptops, the tables don't need to be as strong and as sturdy as they did to support clunky top-heavy CRT monitors. Printers are becoming smaller and lighter-weight too, yet just as fast and as reliable as the older, heavier monster.

I bought my setup: a manually adjustable table plus adjustable chair for under $300 in 1998; while I have had to replace the chair (I paid $119 for it) the table is going strong. That is despite having been assembled, dissembled and assembled again three times, and withstanding two moves, one cross-country and international.

You can do it for even less. There are plenty of high quality secondhand chairs and workstations from downsized premises offices on the market that your employees can truck away with for as low as 30% under cost.

The furniture makers have glommed onto the home working market. Go into any office supply place and you will see a chunk of the floor space devoted to it.

You need to work out with your employees: who pays for the furniture and appliances, them or you. As a general rule, your employees should pay for the space out of their pockets; exclusive home working employees can deduct space costs: rent, mortgage, utilities, and maintenance from their income taxes.

By allowing employees to work from home you're giving them a benefit by enabling them to avoid commuting to your premises on a daily basis. Consultants say you could be saving them $4,000 to $5,000 a year in reduced transportation and work clothing costs.

Who pays for which furniture and appliances is up to you and your home-working employees. There are no set rules. Employees should buy their own headsets or wrist pads, or take those from the premises.

There are no real advantages in supplying chairs; employees should buy the ones that fit for them. The same goes for workstations.

You can have your exclusive home working employees buy the furniture and appliances and let write them off their expenses. That way the furniture is theirs. Remember that if your home working employees leave, getting that furniture back can get awkward, if not impossible.

If you choose to pay for the furniture, you can set a budget for your home office employees, just as you do for outfitting your premises offices employees. Base the budget on expected lifespan of the furniture, research what is available from the office supply chains—let them know that they shouldn't go to the max.

If you choose to give an allowance for or buy the home employees' furniture, take a look at where they reside/work. If your home workers live in the same metro area as your premises offices, or your office equipment or furniture vendor, you can supply the furniture directly. If your home workers live outside of local delivery range, then you will have to trust them with the allowance: as long as the gear meets your requirements and your budget.

Either way, whether the employees or you supply and pay for the furniture and appliances, you need to set the standards and specifications. That way your employees can work safely, productively, without injury, liability and cost.

SAY NO TO RENTING OFFICES!

It is tempting for home workers to consider renting offices. Such space has many advantages, like high-speed Internet connections, plenty of room for filing cabinets and equipment, and to meet clients and customers. Some offices have shared receptionists, meeting rooms and copiers and printers.

Tempting yes, but unless your organization has long had a practice of doing it, don't approve it. The reasons are manyfold:

Higher costs—The employee is paying for additional space at your expense

Legal/liability—Who is responsible for that space, the employee or you?

Red-flagging—Allowing employees to rent space and expense it will quickly kill home working programs because it quickly attracts the unfavorable attention of cost-conscious top management who are already leery of home working. They will ask, quite rightly, what is the point of allowing home working if it is going to increase real estate costs like this?

Chapter 6:
Ergonomics

Ergonomics, says my old university-days Concise Oxford Dictionary, is "the study of efficiency of persons in their working environment". That has come to mean ensuring productivity of your employees by ensuring their work environment is safe and effective, enabling them to perform efficiently, without injury that costs them and you. Ergonomics, by definition, includes every factor that impacts on that work environment: furniture, lighting, temperature and noise.

I stress ergonomics for good reason. You may not think that seemingly sedentary tasks like writing programs (and writing articles about writing programs) and phone or online-based customer service can hurt workers. But your workers can suffer from near-crippling debilitations like carpal tunnel syndrome and tendonitis that can cause some so much pain that they can't work.

There is a tendency to forget about ergonomics in home offices because they are out of sight, out of mind. But if your home working employees are suffering from ailments like tendonitis you will find out soon enough: when they call in sick or say they have to go to the doctor.

While at this writing no government agency is regulating home offices—the US federal Occupational Safety and Health Administration backed off extending its work rules to home offices several years ago—it is in *your* best interests to ensure that your employees work safely and effectively *wherever* they are located. There is a general industry clause in the OSHA regulations that give that agency the power to intervene, if there are complaints. The home workplace is technically a workplace over which you have some say, control, and responsibility.

What makes applying ergonomics challenging in a home office is that the employees, not you, ultimately control that environment because it is their home. Also, most homes: houses, apartments, etc. are not set up as home offices, though there are many newer apartments and condominiums that are designed as "live/work" spaces.

But by the same token ergonomics can be enhanced in home offices compared to premises offices because employees can do whatever they want to their space, furniture, temperature, and lighting to make it fit *them*. They adjust and choose the chairs and tables to *their* bodies, not anyone else's.

Home workers don't have to worry about someone else occupying the same space, or what anyone else thinks of their choice in décor. If the authentic Lava lamp they inherited from their stoned-age father does the trick for desktop lighting, so be it. If they work better by having heavy metal blasting in the background they can crank it up. Who cares, except maybe the neighbors...

The previous chapter, Chapter 5, looked at locations and facilities, i.e. "sites" and "shells"—in real estate parlance. This chapter looks at what goes in them.

✪ INJURIES AT HOME

At first glance, it can be difficult to realize that information or knowledge workers may get injured and sick while doing their jobs. It isn't exactly heavy lifting or working with machinery.

Yet, leading experts in ergonomics and designers warn that home and premises offices are medium- to high-risk workplaces. Employees suffer from head, neck, and to a lesser extent, back and leg injuries at their workstations. They also experience headaches, neck, arm, and mid-back pain.

These conditions are often known as musculoskeletal disorders (MSDs). Two of the most common MSDs are carpal tunnel syndrome and tendonitis.

Carpal tunnel syndrome is caused by compression of the median nerve as it passes through the carpal tunnel of the wrist that eventually results in damage to that nerve. Twisted hand posture, repetitive hand movements outside the neutral range of motion (15 degrees), sustained pressure on the underside of the wrist, forceful and twisted hand movements, lead to carpal tunnel syndrome.

Tendonitis is the irritation and inflammation of the tendon. Twisted hand posture, high repetition, and force on the tendon, cause tendonitis. Call center agents, secretaries and programmers commonly get tendonitis in the wrists, forearms, and shoulders.

Unrelieved keyboarding and "mousing," along with poor placement of these appliances, and bad posture, are the common culprits of these ailments. The more people work the keyboard and mouse, especially as e-mail and chat become common methods of communicating with customers, the greater the risks of MSDs.

Dr. John Triano, director of the Texas Back Institute, points out that such MSDs are insidious and often go unreported because one often can't tell whether someone has been injured or is in pain. Often, workers do not realize that their work activities are causing their pain.

"Unlike lower back pain, where someone who is injured on a heavy job can't do their work, upper arm, back and neck pains are such for [knowledge workers like] call center and computer workers that they can continue to work before the pain becomes severe enough to stop them," explains Triano. "This risks worsening their conditions and their health."

Other key ergonomics issues in home offices are poor lighting, causing eyestrain through glare, noise, which disrupts thinking and productivity, and if loud enough harm hearing. Many but not all of the solutions to these issues are touched on in the previous chapter, Chapter 5 on facilities, but there are steps that home workers can take at the workstation level.

❂ HOME ERGONOMICS SOLUTIONS

There is a strong tendency by employers and home working employees alike to ignore ergonomics. Employers don't like doing anything that they are not compelled to do; people working from home often think they can "improvise" with existing furniture in order to avoid purchasing ergonomically-sound chairs and workstations.

But your home working employees should not think for a second that they can "get by" on old fashioned writing desks, which are too high for computers; or kitchen and living room chairs, which were never designed for 8-hour shifts. Their bodies will let them know *real quick* that's a wrong move. Such moves are penny-wise and *definitely* pound-foolish.

There are readily available solutions to many of these problems. The ones most applicable for home offices are adjustable chairs, tables and workstations.

Adjustable Chairs

Adjustable height-and-back chairs are the single most important piece of furniture in an office. Home workers spend more time in a work chair than any other item in the home. Your employees can have their laptops on moving boxes, and it is fine, as long as they can raise or lower their chair.

There is no such thing as a one-size-fits-all in chairs. Fortunately for home office workers, they can try out the chairs they like and buy the one that fits them, unlike premises office workers who must take whatever employers supply them. Or hot-deskers, who share their workstation with others, meaning they must readjust the chairs to their liking.

Cornell University's ergonomics department offers 12 tips, arranged around this drawing, for an ergonomically correct workstation:

1. Use a good chair and sit back
2. Top of monitor 2"-3" above eyes
3. No screen glare
4. Sit at arm's length
5. Feet on floor or footrest
6. Use a document holder
7. Wrists flat and straight
8. Arms and elbows close to body
9. Center monitor and keyboard in front of you
10. Use a negative tilt keyboard tray
11. Use a stable work surface
12. Take frequent short breaks

There is some disagreement as to what else you should have for a egonomically sound chair. Dr. Triano says chairs should also have separate footrests, variable angled seat pans, adjustable back rests and elbow supports.

Not all adjustments and features on chairs may be totally necessary, say Cornell's Hedge. The most important adjustment is seat height, then backrest height. Other features like adjustable seat pans that slide forward or backward, are often unnecessary. If the backrest can recline then a seat tilt is not necessary.

He says if a workstation has a good quality negative slope keyboard tray with a built-in palm rest, chairs do not need to have armrests. If a chair has armrests, they should allow users to lower them out of the way to type and control the mouse. Your employees risk injury if they place their forearms on the rests when typing and controlling the mouse.

Wave's David Smedley says that you will want to be able to adjust the chair height so that your hands can be placed on your keyboard with your elbows at a 90- to 120-degree angle with your shoulders relaxed. Your knees should be bent at about 90 degrees or greater. Your hip angle should be 90 degrees or greater. Your thighs should be parallel to the floor. Your feet should be able to be placed flat on the floor and your lower back should be fully supported.

Smedley thinks an adjustable seat pan is a worthwhile feature, especially when more than one person uses a home office space. This lets you move the seat pan forward or back so that it stops short of the backs of your knees. This gives your upper legs maximum support while preventing direct and constant pressure behind your knees. A shorter person will require the pan to be closer to the chair back, while a taller person will need the seat pan to be slid away from the chair back.

Also, a curved or waterfall-style front edge is better for blood flow. Chair edges without the waterfall tend to place too much pressure on the backs of your legs, causing discomfort.

The chair's lumbar support should press against your lower back, allowing you to rest firmly against it. To avoid backaches, an adjustable lumbar support should be set so that it curves out to rest within and support the natural curve of your back at the waist. Your feet should be able to be placed flat on the floor or on a footrest.

"When choosing a chair, look for a feature known as 'Synchro tilt' that drops the rear of the seat pan when you lean back, allowing you to open up your hip angle and chest cavity," recommends Smedley. "Another chair option to consider would be a paddle or Flipper™ armrest. These soft arm pads rotate 360 degrees and are height adjustable. They promote better working postures, particularly while using a keyboard or mouse."

He also advises having the chair casters matched to the flooring. For example if you have carpet, hard casters will allow the chair to move more easily. Soft casters roll easier on hard surfaces.

Your employees can buy good adjustable chairs for under $400, which may seem like a lot of money, but 20 minutes in a non-adjustable chair will show them (and you) that it is money well spent.

Quality, stylish and highly functional premises offices chairs that are also marketed to home offices such as the Look, by Haworth go for $450-550, depending on the features and fabric grades. That's not a bad price considering that such chairs will last a long time.

One good option for home workers is secondhand adjustable chairs from top quality makers like Haworth and Herman Miller. They can be had for a lot less than purchasing a new chair.

With many companies and operations like call centers closing their doors and/or moving offshore, there might be some good deals to be had. I know of some call centers that have been outfitted entirely with quality secondhand furniture. But like any secondhand item, employees should check them out real carefully before buying: particularly the breakable components like seat pans, lifts, and armrests.

Adjustable Workstations

Most home workstations available on the market are non-adjustable, which will fill the bill just fine—these workstations go for $99 and up. The argument has been made that if your workers can adjust their chair height that should be sufficient for good ergonomics. Your employees can swivel monitors downward for correct positioning.

But some ergonomics experts recommend that the workstation have an adjustable keyboard tray and table heights. Variable-height workstations have manual or automatic lifts that allow agents to adjust the height of their work surface.

With these workstations employees can work while standing, making the tables especially useful, for senior customer service or support desk agents who have to reach into bookshelves to access reference material. The workstations shorten the distance from the computer surface to the books. Employees can still reach the mouse and keyboard without bending at the waist, craning their necks, and reaching too far compared to sitting.

But variable-height workstations may also lead to more injuries than they could alleviate. Dr. Alan Hedge professor of ergonomics at Cornell University says that you place more stress on your muscles when you stand up to work than when you are sitting down.

"Standing burns 20% more energy than sitting and it increase strain on the lower back compared with supported, reclined sitting," he points out. "If you stand and work with poor posture and in a static position then standing will have no benefit whatsoever. We recommend sitting to work in a neutral posture then standing intermittently and moving around doing other kinds of work."

I favor a compromise: a workstation with adjustable-height keyboard shelves, *not* trays. These are hard to find, but are well worth it. I also like a table that lets me adjust the height of the monitors. That gives me the best choice of positions especially when I working on both my employment and personal computers.

David Smedley points out that most desks and tables are set at a standard 29" height. If that's too low or high to allow you to maintain the correct seated posture described above, an adjustable table or customizable wall-mounted work surface would be a better choice.

There are a wide variety and quite affordable workstations, such as the O'Sullivan workcenter available through Staples. With home office space and rooms rarely uniform your employees will need to measure and see which workstation models can fit best before buying. Credit: Staples.

○ WORKSURFACE ARRANGEMENT

The overall size of the primary work surface should allow most tasks to be done within your reach radius (an arc of reachable space to your front and side). Depending on your size, this arc space should be from 40" to 63" wide, with most tasks being done within the first 14"-18" of the arc space. Occasional tasks would be done within the outer edge of an overall 22"-26" arc; this extended arc space would require a slight reach.

When multiple work surfaces are used, it's generally preferable to have them arranged within an L-shape, or a curve, rather than along a wall. It takes less time and effort to rotate to a perpendicular work surface than to move to another part of the room. Using a separate work surface for the printer and fax machine will free up primary workspace for writing and other tasks. Storage accessories such as file holders and note displays will also help keep primary workspaces clear.

Choose and arrange furniture so that your monitor can be placed at least 25" from your eyes. It's better to place it even further than that because close viewing causes eyestrain.

The point where the eyes can rest between convergence and accommodation falls between 45" directly, and 35" with a 30 degree downward angle. Having the monitor more than 25" away is usually better for most people and enlarging document print size is a better option than moving a monitor closer than 25". Generally, the larger the monitor, the farther away it should be.

"When viewing close objects, the eyes must both accommodate and converge," explains Smedley. "Accommodation is when the eyes change focus to look at something close. Convergence is when the eyes turn inward toward the nose to prevent double vision. The farther away the object of view, the less strain there is on both accommodation and convergence. Reducing those stresses will reduce the likelihood of eyestrain."

The monitor should also be positioned so that the vertical viewing surface is 15-50 degrees below horizontal eye level to reduce the potential for headaches and dry eyes.

The International Standards Organization recommends a viewing angle of not less than 20 degrees below horizontal eye level. Also, the top of the monitor should be tilted slightly farther away than the bottom of the monitor for optimal viewing.

The keyboard and mouse should be supported on stable surfaces that allow you to type with your elbows bent at a 90- to 120-degree angle with your upper arms hanging straight down to your sides. You should not have to bend your wrists to type. Use of a height and tilt -adjustable keyboard/mouse pad is a good way to compensate for the size differences of keyboards and adjust the keyboard height to meet your exact needs.

"Also make sure there is at least 24" of clear legroom under your desk or work surface for your legs," advises Smedley. "This area should **not** be considered storage space."

✪ WORKSTATION LAYOUT ISSUES

Unlike in premises offices, where you can call up a furniture vendor and lay down a row of workstation cubes and chairs, home offices often have layout restrictions. That forces home workers to make compromises as to what types of furniture can be where.

For example, there may be enough room lengthwise for a workstation under a window but it faces south. Either another part of the room must be found or the window must be closed with blinds.

Often the best answers are to buy corner workstations. Many homes do not have corner windows, which mean no glare from incoming light. They also make efficient use of that space.

Back in 1998 I bought a corner workstation from Staples that has a manually adjustable keyboard tray and monitor stand. I assembled and reassembled it three times on my moves. I haven't found a workstation since that matches its adjustability, table area, ruggedness, and low cost.

Screens

Fortunately for home workers (and employers) one of the most costly and hassle-riddled aspects of workstations: the cube panels, are not necessary in home offices. Those panels act as noise barriers from other workers—as noted earlier, premises offices are noisier than home offices.

If your employees work, however, in a dedicated area that is part of an existing room, i.e. basement, living/family room, studio apartment, etc., your home workers may want to consider buying inexpensive Japanese-styled screens. That gives workers a sense of privacy, it also can protect data, and better define the work area. A screen might also be important for tax purposes—the workspace tax deduction is based on the space dedicated to work.

Ergonomics Accessories

There are many secondary appliances that claim to, and may actually reduce, home and premises office injury risks. There are others that may be useless.

These useful appliances include:

Headsets

Headsets, for people who don't mind the cords, have long been the leading secondary appliance and for good reason: they prevent discomfort and tissue damage to the neck and shoulders and are quite affordable.

Headset manufacturer Plantronics (www.plantronics.com) cites a Santa Clara Valley Medical Center study that shows that headsets reduce neck, upper back, and shoulder tension by as much as 41%. An additional study by H.B. Maynard concluded that adding hands-free headsets to office telephones improved productivity by up to 43%.

Headsets such as this Plantronics model (the author has the same one) are important appliances to have if your employees are often on the phones and computers at the same time. They prevent injuries caused by cradling the phones. But with home offices tending to be much quieter than premises offices the headsets need not be as elaborate. Your home employees can get away without such features as amplifiers that add to the costs. Credit: Plantronics.

Most headsets transmit sound by placing speakers over the ears. But there are newer, if more expensive, headsets that transmit sound to the temples. These bone-induction headsets are ideal for people who have hearing problems. They also enable workers to listen to conversations around them, like their spouse yelling "dinner!" and the courier guy or gal ringing the doorbell.

Docking stations

Laptops are designed for mobile and temporary use. They have not yet been optimized for all-day working. Laptop screens are typically attached, giving limited viewing angles; the touchpads and pointsticks are a nuisance; and the keyboards typically are not comfortable.

Docking stations permit laptop users to easily connect conventional mice, keyboards, and monitors to the laptops. They also save laptop users the hassle of plugging and unplugging those individual components, instead laptop owners click their machines into the docks, which have these peripherals already connected.

Laptop lifts

Laptop lifts is my collective noun for gadgets that allow workers to raise and lower laptops on workstation surfaces. They enable laptop users to overcome the inability to adjust screens and keyboard on laptops, and unblock bottom laptop cooling vents, perhaps saving the guts of those machines.

LapWorks makes the Laptop Desk that raises the screen level by over 3 inches, which can help to reduce neck strain. Users can set the desk and their machines to five set angles.

Here are some controversial gadgets that some people swear by and others swear at:

Wristpads and keyboard trays

Wristpads are, as the name implies, pads placed on workstations that users rest their wrists on. Keyboard trays are trays that roll in and out from under the workstation.

Cornell's Dr. Hedge does not recommend wrist pads; there are sensitive nerves in the wrists that they could strain. If you position yourself properly, you have little need for wrist pads. Instead he recommends keyboard trays.

On the other hand, Laura Sikorski, a veteran space designer, likes wrist pads but dislikes keyboard trays because they place agents too far away from the screen. They also interfere with the legs. She thinks you should have the keyboard on the work surface and use wrist support.

Dr. Triano does not take a position on wrist pads or keyboards trays. Yet he says neither may be necessary with proper workstation design, equipment setup, and seating.

"All of these auxiliary ergonomic aids are useful to solve isolated problems that cannot be solved, for one reason or another, with the overall workstation set up," says Dr. Triano. "Under special circumstances they may be necessary. In many circumstances they are a crutch that can be eliminated by attention to proper work surface, seating and equipment arrangements."

From my personal experience, I'm with Dr. Hedge on wrist pads and Laura Sikorski on keyboard trays. Wrist pads are nuisances; I bump my knees into keyboard trays; my wife has one on her desk. It is individual preference.

Keyless keyboards

Keyless keyboards are pads that combine mice and keyboards. The OrbiTouch from Keybowl lets workers type without using fingers or wrists by sliding two large discs in different directions to create letters. The firm claims OrbiTouch reduces strain injuries compared with standard flat keyboards. The device's built-in mouse capability allows users to keep their hands on the OrbiTouch at all times.

There are some downsides to keyless keyboards. According to CMP cousin publication, *InformationWeek*, which reviewed OrbiTouch a few years ago, the device has a much slower typing speed and a hefty cost: $399, compared with conventional keyboards.

Lighting

For the immediate work surface area, David Smedley says that placing a task light that allows you to adjust light levels is recommended. A task light is any lighting fixture that confines light to a specific work area. Some task light styles can be installed under shelves and others are freestanding desk lamps.

Avoid home-style lamps with cloth shades that swallow the light. Use lights designed for office work that can be positioned in different ways.

Noise

The best steps to control noise are picking the right locations and spaces for your home offices, and in managing the inhabitants in your space. But there steps within the work-space you can take. They include buying/specifying storageless PCs, which lack noisy fans. Employees could also see if sound masking, a.k.a. "white noise" generators, will do the trick. An old CB radio or a boom box set on a dead station at the far end of the dial used to do the trick when I had trouble sleeping thanks to a noisy neighbor. Air conditioners and fans also drown out noise.

✪ PAYING FOR ERGONOMICS

The great aspect of home working is that they don't have to be fancy, or spend a lot of money; since when is the CEO or your favored clients going to drop around. Chances are your home working employees don't need costly and heavily wired partitions. Consequently your home workers can outfit their home offices for a lot less: $700 to $900 for a chair and table compared with $3,000 or so for a premises workstation.

Unlike in the past, you can get away with lower-end workstations. With the popularity of flat panel monitors (FPM) and laptops the tables don't need to be as strong and as sturdy since they don't need to support clunky top-heavy conventional cathode-ray-tube (CRT) monitors. Printers, too, are becoming smaller and lighter-weight, while just as fast and as reliable as the older, heavier monster.

I bought my setup: a manually adjustable table plus adjustable chair for under $300 in 1998; the table is going strong, although I've had to replace the chair. That is despite having been assembled, dissembled and assembled again 3 times, and withstanding two moves, one cross-country and international.

You can do it for even less. There are plenty of high quality secondhand chairs and workstations from downsized premises offices on the market that your employees can truck away for as low as 30% under cost.

The furniture makers have glommed onto the home working market. Go into any office supply place and you will see a chunk of the floorspace devoted to it.

Chapter 1 offers an example of a complete breakout of home office startup costs, facilities, plus equipment and voice/data including installation. To repeat:

Furniture, Equipment and Telecommunications	$5,900
One time charges, Set-up & Installation	$1,200
Total per Tele[home] worker	**$7,100**

You need to work out with your employees as to who pays for the furniture and appliances: them or you. As a general rule, your employees should pay for the space out of their pockets; exclusive home working employees can deduct space costs: rent, mortgage, utilities, and maintenance from their income taxes.

By allowing employees to work from home you're giving them a benefit—enabling them to avoid coming into your premises on a daily basis. Consultants say you could be saving them $4,000 to $5,000 a year in reduced transportation and work clothing costs.

Who pays for which furniture and appliances is up to you and your home-working

AN ERGONOMICS APPLIANCE THAT WORKS

Many experts and uses are skeptical about keyboard trays, wrist pads and other such ergonomics appliances. I was given a wrist wrest for my premises office cube that I soon dispensed with.

Some of these appliances do more harm than good, like increasing wrist strain. One of Call Center Magazine's staff developed serious carpal tunnel injuries at their premises office workstation because the attachments were located too far below their desks, reports Chief Technical Editor Joseph Fleischer.

But there is one such appliance that he had found that may be helpful for home and premises offices alike: Allied Plastics' ComfortSlope. As the name implies it is a desk attachment that inclines up, with space cut out for the worker to slide their chair into, and at the top for the monitor. There is enough room to rest the wrists at the front and place a keyboard at the top of the slope.

The appliance works by having the workers lean back in their chairs with their arms and wrists straight but somewhat elevated and relaxed, instead of the usual positions of leaning forward, tensed against the desk. The position prompted by ComfortSlope reduces stress-caused injuries to the wrists, back and neck.

At $150 ComfortSlope is seemingly pricey for home offices. But it may be worth the ticket because it can save you and your workers far more in medical costs and lost down-time. Allied Plastics also makes pedestals for Comfort-Slope; if you or your home workers are looking to buy new workstations Allied Plastics makes them too.

employees. There are no set rules. Employees should buy their own headsets or wrist pads, or take those from the premises.

There are no real advantages in supplying chairs; employees should buy the ones that fit them best. The same goes for workstations.

You can have your exclusive home-working employees buy the furniture and appliances and let them write the equipment off their expenses. That way the furniture is theirs. Also, if your employees leave getting that furniture back can be awkward, if not impossible.

If you choose to pay for the furniture, give an allowance to your home office employees, based on the budget for them in your premises offices employees. The budget should be based on expected lifespan of the furniture; research what is available from the office supply chains and let them expense it (say with a corporate credit card), making it known that they shouldn't go to the max.

Yes, chances are your employees are going to be using that furniture to work on their own personal machines. But that's fine because, studies indicate, home workers work longer, including in the evenings and weekends, than premises-based workers. Well worth the investment.

If you choose to give an allowance for or buy the home employees' furniture, take a look at where they reside/work. If your home workers live in the same metro area as your premises offices, or your office equipment or furniture vendor you can supply the furniture directly. If your home workers live outside of delivery range then you will have to trust them with the allowance: as long as the gear meets your requirements and your budget.

Either way, whether the employees or you supply and pay for the furniture and appliances, you need to set the standards and specifications. That way your employees can work safely, productively, without injury, liability, and extra costs.

DIY HOME OFFICES

You can do your home office yourself, with the help of suppliers. IKEA, best known for home furniture also has products for home offices. They include workplace furniture: computer desks, desk accessories, tabletops, chairs and storage cabinets. To help you IKEA offers a selection of online tips and ideas. They include workstation planning, on storage and chair selection. There is also a downloadable office planning tool.

Chapter 7:
Employer Investment

Enabling employees to work from home, be it exclusively or occasionally, will require employers, like yourself, to invest in tools, staff, and facilities to support them. Those factors separate home-based employees from home entrepreneurs, along with the need to provide performance that is identical to that provided by employees working at premises offices.

As Chapters 2 and 4 outlined, you probably need to have a voice and data infrastructure, such as call transferring and e-mail response. You will need, then, to have the technology at your end to enable those communications and data links. You may decide to supply your home employees with computers and peripherals (Chapter 3), which means you will need to arrange for IT support to handle the inevitable technical problems.

You will want to communicate internally with your employees, wherever they are. That requires enabling tools, like instant messaging, and procedures, like regular conference calls. You may require your permanent home working employees occasionally to come into the premises, which may require spare or *hot-desks* for them to work from on premises.

✪ VOICE ACCESS

If your home working employees take calls without going through your switchboard, and/or use their own Internet Service Providers (ISPs) for e-mail and Web access, and/or do not need access to your computer network, then you need very little investment in voice, data, and networking systems to handle their communications. In the old days, before the World Wide Web and Internet e-mail, home workers often shipped in their work by mailing floppies or they had a special and, in most cases, external modems that were a hassle to set up and took forever to transmit the necessary data.

If the above describes your home work program, you may need only to pay for employees' long distance calls, for voice and data lines (and/or faxes) as outlined in Chapters 2 and 4. For little or no cost at all you can have your employees access their voice mail on your phone switch, which enables outside customers, clients, and colleagues, who do not have those employees' phone numbers, to contact them.

Or you can have reception give the callers the home employees' phone numbers. If outside callers dial up after hours they can leave general voice mails.

But if your customers, clients, colleagues, PR agencies, media, etc. need to reach your employees in real-time, you need to look at how you connect them to your home workers directly. Chapter 2 covers the options: *network routing, Centrex routing, off-premises extensions (OPXes), call transferring* and *Voice over Internet Protocol* (VoIP).

Voice Access Cost Issues

All of those methods have their cost issues, like additional long distance charges when rerouting and transferring calls, and voice quality on VoIP. Accommodating home workers may entail investing in new hardware and software, such as OPX boxes and VoIP gateways, plus special phones for VoIP.

Such investments must be thought out carefully. Yes, the latest phone switches with built-in circuit-switched or VoIP-enabled OPXes are probably better engineered and less expensive than buying a conventional switch and then buying an aftermarket box. But many of you will already have a phone switch (which has a seven to ten year life span), which cost several hundred thousand dollars. If you're in mid lifespan with one or more of those iron monsters, and you don't have any other crying needs for them to fulfill, then either develop a routing application or add an aftermarket OPX—either will enable you to get more use out of your existing equipment. This is especially true if you don't have need for premium services like skills-based routing and CTI.

If your phone switch is a year or two away from the recyclers and/or you want it to do other cool things like skills-based routing then you might find buying a new switch worth it. Alternatively, if you are faced with a buying decision on a new switch consider the options network/intelligent routing and going totally to VoIP.

Network routing enables you to route calls to home workers without incurring long distance charges. You also minimize the amount of costly room you need at premises offices for phone equipment. By outsourcing the routing to the carriers you save on up-front capital investments, gain flexibility (if your business grows or contracts you're not stuck with costly gear that is not meeting your needs). But network routing comes at the expense of contracts and paying for plain-vanilla technology that is upgraded when the carrier sees fit.

VoIP is becoming a more viable option, especially with the ongoing technology improvements. If your customers and your colleagues have a tolerance for noise, hollow-sounding voices, and the occasional dropout—acceptance of cellphones have made all more acceptable—VoIP is doable now. And note that the payout can be significant as VoIP eliminates long distance charges and minimizes infrastructure investments, i.e. one-wire.

Or you may want to go wireless: especially if a good share of your workforce is mobile. Wireless voice quality is improving; data networks are becoming faster, more reliable and less expensive. One day soon, we may have true anywhere/anytime/anyplace mobility.

✪ DATA ACCESS

Voice handling is fairly straightforward. For the most part communications skirts the core of your operations: information management. The exceptions being premium features that link into your premises such as advanced (skills-based) routing, ANI and CTI, call transferring, and voice mail.

But data handling, including data communications, is more complex because it gets to the heart of your enterprise. Unlike with voice, you cannot be sure from the bits-and-bytes packets that the people contacting you are who they say they are. Nor can you accurately determine their intentions.

If you need your home workers to have access to your data, and to communicate via data (e-mail, messaging, chat) you need to provide data access from your premises to the outside world. The access can be wired through dial-up and IP, wireless (cellular and wireless broadband), and satellite: whatever it takes.

There are two key forms of data access, which I call *ungated* and *gated*. These forms are like drawbridges and gates in the Middle Ages and serve the same functions: to let in those you trust and keep out the vandals who can wreck havoc through thievery and viruses.

Ungated

The Internet, which is how most people off the premises communicate, including between premises, is like the mediaeval forest. Anyone could be lurking there, in this case corporate thieves, competitors, disgruntled employees, investigative journalists, hackers, virus-creators, and spammers. Computers, such your premises servers and your home employees' laptops, send files to each other through various forms of file-transfer software, the most popular being the file transfer protocol (FTP).

If you *must* have secure file transfers, check out PyroTrans, a file transfer package (client, server and batch mode) that lets you securely transfer files over the Internet. Its key benefits include secure encryption and automatic compression. As such PyroTrans enables organizations to transfer sensitive data files securely.

Ungated access permits any file to enter into the kingdom, and out into the forest to the next kingdom, without questioning who is sending it or inspecting the contents. Given the enemies that inhabit the forest and the damage they could do from stealing data to destroying your computer system, it is not wise to let just anyone into your walls.

Gated

Firewalls are software and/or hardware and software that only let in data transmitted from computers that have the authority to send such data. Or as *Newton's Telecom Dictionary* put it: "limits the exposure of a computer or group of computers to an attack from outside." But the possibilities that barbarians may attack computers—yours, your employees and others—means that gates such as firewalls and anti-virus software are needed for all computer uses—premises, home and mobile.

Gated access consists of applications within firewalls and other access-sensitive software that perform the electronic equivalent of the guards yelling "Halt! Who Goes There?!", checking prospective entrants proof of entity and access authorization, such by demanding *passwords*, and patting down the occupants and looking under the canvas of their cart, or *filtering*. If you can't get past the guards you're going nowhere.

For linking at-home and mobile-working employees' computers with at-premises computers through the gates there have emerged two main methods: *secured networks* and *hosted access*. Secured networks can be hard private, on data lines that you own or lease or virtual, over the Internet, know as *virtual private network* (VPN).

Alternatively, you can offer employees limited access, through Webmail accounts. They can check their e-mail, respond to e-mails, and send and receive attachments and links. But they can't access your servers.

Secured Networks (hard private or VPNs)

In a hard private network there are lines to which you have access. These private lines link your computers. This method is practical for high volume data and voice traffic between premises but not between premises and home offices.

Using the Middle Ages metaphor again, think of hard private networks like a walkway between towers. The only way anyone can get to one tower is by getting through the gate or scaling the walls. But few Middle Ages castles were large enough to have extensive walkways.

VPNs give castle entrants formal permission to enter the castle: virtual letters sealed with wax from the monarch, and memorized passwords so they can enter unsearched. But they can only go to those parts of the castle that they have been granted access to.

VPNs consist of client software loaded onto computers. The data is encrypted at the senders' end and decrypted at the receiver's end; as the data travels through the Internet 'forest' nobody knows who it is nor can they find out; the password, confirming that the data comes from an authorized source, is memorized.

When employees knock on the gates they then launch the application, the VPN server yells "Halt! What's the Password!" the employees give it and the gates open: to e-mail servers, databases. The employee then 'goes' to where they have authority to enter, which may be the e-mail server, or in my wife's case (she was a Web programmer) into the Web software.

CMP uses a VPN—it's the Nortel *Contivity*. I boot up my laptop, click on the appropriate icon and enter my password on a form that also has my IP address. When I see "getting configuration on..." quickly flash by I know I can walk under the gate spikes.

VPNs are reasonably priced. For example, Nortel explains there is no charge for the Contivity VPN client software that users run on their PC to establish a VPN "tunnel" connection to a Contivity gateway; the only cost is the Contivity VPN gateway itself. Contivity prices vary depending on VPN capacity. The lowest end device, the Contivity 221, supports 5 VPN tunnels, and lists for $450. The highest end device, the Contivity 5000 supports 5000 VPN tunnels and lists for $45,000. This works out to a range of $9-90 per VPN tunnel, depending on platform.

Here's a sample price range offered by Nortel for an installation with 50 home workers. The least expensive: Install Contivity VPN client software on the home worker's PCs (free). They could then connect via a VPN tunnel to a Contivity 600 ($2400 list) that supports up to 50 VPN tunnels. Total cost is $2,400. Home workers in this mode could use the VPN connection *and* soft IP phone software on their PCs to make phone calls as well as transmit data originating out of the PC.

You can also link VoIP with VPNs. Again I'll use a Nortel example. If you install a Nortel Contivity 221 in each home worker's site ($450) and connect these Contivity 221 gateways back to a Contivity 600 ($2400) in the premises office for $24,900 the home workers could use a dedicated IP phone in addition to their PC; you connect both to the Contivity 221. IP telephony and PC data could then be transmitted over the VPN connection between the Contivity 221 and Contivity 600.

There are companies such as Cisco that offers advanced security to home offices via VPNs. Its system splits the VPN tunnel to enable other residents Internet access while protecting your network. It uses identity authentication to limit access and prevents wireless intrusion.

Once inside the kingdom through a VPN your part-time and occasional at-home working employees also, if their at-premises computers are turned on, can open and work with files on their internal drives. But you need remote-access software such as Symantec's *pcAnywhere*.

Hosted

Hosted access is where the king (or queen), i.e. you, the employer, hires a third party to ascertain the identity of the castle entrants by asking for a password. The entrants need not have pre-delivered letter of authority from the monarch, but unlike with VPNs, where once admitted the employees go to where they are allowed, the third party 'escorts' them. All computer interactions between the employees and the employer are "escorted."

Hosted access, by application service providers (ASPs) provides such escorts; it can also enable at-premises PC access by at-home and mobile workers. You arrange for secured links between your server and the host's server, such as through a VPN; your at-home or mobile employees log in and a viewer plug-in launches and prompts them for access codes.

VPNs versus Hosted

For the benefits and costs of VPNs versus hosted networks ask your IT person. But I'll set out briefly a few points to keep in mind.

VPNs

Because you own the VPN and pay one-time license fees you have complete control over the network. But you also bear the total costs of building, hosting, and maintaining a private network. Yet, the software can be as current and feature-rich as you need. Chances are that data transmissions over broadband are faster between at-home employees and your servers because the communications are not going through an intermediary computer, i.e. the ASPs'.

Hosted

Hosted access is less expensive up front and can be rolled out quickly; you don't have someone (like your employees) load the software and walk your employees through it as CMP did when it set up its own VPN. There is no software installed on your servers. If your home employees use dial-upm receiving bandwidth-heavy apps like Lotus may be faster with a hosted access method; I found connecting into Lotus over a dial-up excruciating, but liveable.

Another plus in the hosted access solution is that while a VPN sometimes will not synch with and/or disconnects from a home worker's broadband service, as if they had a fight and decided not to talk to each other, there are fewer of such problems with hosted access. Probably because there is a plethora of technical support devoted to maintaining the private network in tip-top shape.

Limited Access

You can offer employees *limited access* to your kingdom: the electronic equivalent of lowering the drawbridge, letting them peer through the gate and then asking a guard to get them the person they needed to talk to. That person will need a password to identify them.

There is software that enables your employees to access their e-mail via the Internet without being given access to your network. This method is typically called *Webmail*—employees launch the browsers, key in the URL, type in usernames and passwords to login.

Such programs give near-equivalent access to e-mail as secured or hosted access. Your at-home employees can get those communications and forward, file, reply, and delete them accordingly.

Webmail is not without its disadvantages. The look is not the same as if there was direct access to the application. Webmail tends to time off after 30 minutes or so and with no notice, which can be a pain in the neck. There's been many a time I've had to redo outbound e-mails because I had timed out.

But Webmail is ideal for part-time and occasional at-home workers who have machines that you don't want to outfit with VPN clients or link up to hosted access ASPs. You *especially* don't want to put security-sensitive software like VPN clients on employees' own computers.

Webmail is a great backup for VPNs—for permanent at-home workers and for others with employer-supplied computers. When the VPN and the cable "gets into tiffs" on my setup and neither of them decide to talk for more than a few minutes I send them both to a corner and boot up the Webmail on my personal machine.

Authentication

Passwords, like for VPN access, can be copied; that's why every organization and vendor urges users to protect their passwords. Therefore, if the data your firm handles is extremely valuable, consider some form of physical authentication method that proves your users' identification.

Authentication technologies range from USB plugs to contact or contactless smart-card readers to biometric devices, like fingerprint and iris scanners. There is authentication software that verifies users to your system. I/O Software makes platforms that use any manner of authentication technology and partners with leading technology vendors.

Biometrics provides the best, most foolproof security. You can forget a password, lose a smartcard or USB plug but it is more difficult—though not impossible—to lose your fingerprints or eyeballs.

There is a range of biometric hardware available. The most common form for home workers is fingerprint scanning, some of which is built into some vendors' computers (expect that feature to become more readily available), mice like the DEFCON Fingerprint Authenticator), keyboards like the Keytronic Secure Keyboard and available as plug-in modules like Fujitsu's MBF 200 Fingerprint Reader.

Iris scanning is the most secure of biometric methods: you can't play "criminalist" and "lift" iris scans. There are iris scanners like Panasonic's Authenticam that can also be used for videoconferencing.

✪ COMMUNICATIONS

One of the biggest management issues with home workers, especially exclusive home employees is keeping in touch with them and keeping them in the loop. In turn, home workers need an easy method to reach supervisors and colleagues.

Home working employees are literally out there, in their own offices, with only their cat or dog to talk to—if they're lucky. Sometimes they have spouses who also work from home, as mine did for a while until she got a supervisor who didn't like the setup and had issues with her, and cancelled it.

Let's now consider some methods and strategies to enable communications.

Instant Messaging (IM)

IM is software that permits real-time peer-to-peer communications: text, voice, even video through the Internet. IM is different from e-mail in that it is software that directly communicates with other software. IM users have to be online to see their messages; there are no mailboxes.

IM works alongside e-mail and voice. Here are some examples:

A department manager wants to hold a quick conference call but some employees are busy on e-mails or on calls. The IM icon on the bottom toolbar "waggles" and the employee clicks on the icon and sees the message. The employee wraps up the call or e-mail.

An employee is in the middle of an exchange with a client but they have a question for their supervisor. The employee IMs the supervisor, gets a quick response, then answers the client.

We use IM at *Call Center Magazine*. I can message my colleagues while answering an e-mail at the same time I am talking to yet someone else on the phone.

IM, therefore, can act like a tap on the shoulder, which home workers by definition do not normally receive. But the tap that doesn't interfere with their work; an icon "waggle" is less likely to get them to jump than someone breathing down their neck.

Keep Home Workers in the Loop

Employers should make sure that home employees are on the same distribution lists as premises offices employees. Even though such workers can't go to an afternoon cookout or take part in a charity run they like to know what is going on and may try to find ways to participate. When I found out one of my co-workers was getting married, through a distribution list, I was able to buy her and her fiancée a gift through a bridal registry.

Unified/Integrated Messaging

Consider investing in unified or integrated messaging, by buying or leasing the technology. Unified messaging (UM) permits workers at home, mobile or in premises offices to access electronic communications such as voice mails, faxes and videoclips delivered as e-mails. Some systems permit you to hear your e-mails, says *Newton's Telecom Dictionary*.

The big advantage of unified messaging is that it is a one-stop shop for all messages, no matter the medium. It avoids tying up voice lines and provides better access to corporate voicemail. Remember: every form of communication is data; unified messaging capitalizes on that fact

But unified/integrated messaging is still relatively new and it isn't inexpensive; it can run into the tens of thousands of dollars according to *The Computer Telephony Encyclopedia*, written by Richard 'Zippy' Grigonis and published by CMP. Also if it goes down; all your messages go down. Zippy cites advice from Michael K. McGarry, who is vice president of operations and founder of Merlot Communications who urges that you check out your UM vendor for reliability and support.

Collaboration Tools

Consider deploying Web-based data collaboration tools, such as eGroupware from the Literati Group. Managed by you, these tools enable co-workers and supervisors to pass on information, such as a new software problem, and bug fixes for it.

Because you control it, you can monitor the accuracy of the information and intervene if necessary. That makes collaboration far more effective than "traditional" cube-to-cube or coffee-room chit-chats. You also avoid the possible serious consequences of inaccurate information being acted on and going out. Lastly, you have a record of what was said and who said it, which permits easily followup.

Conferencing

Employers should make sure to include home workers in all meetings through audio and, if necessary, data and/or video conferencing (see Chapter 11). You should ensure that any material for that meeting is distributed to home workers ahead of time: by fax and/or by e-mail. Most outfits now have conferencing tools and bridges, which easily enable such sessions.

Consider remote control tools, such as Carbon Copy, that enable managers to take control of home employees' computers and give the same presentations on their screens as shown to premises offices workers in a conference room, or on their computer screens. Alternatively home workers (and others) can log into hosted Web conferencing.

Readerboards

In uses where employees need to see operational statistics, like call queues and sales, there is software that supplies that same information, wherever they are. Traditionally known as "readerboards," the technology is available via PCs and gives home workers data they need to work with and makes them feel part of the same team as other employees. For example, Spectrum Corporation provides realtime and historical data to home employees' computers via thin-client and browser-based applications.

Field Meetings

Another strategy is going to the field to chat with home workers: gathering for a coffee or lunch at employer expense. That gives them a chance to socialize with other workers. This method, however, works only for those employees who live with 100 miles or so of your premises, which can limit your labor pool and business continuity.

Employee Check-in and Participation

Communications is a two-way street. Employers should encourage employees to contact them if they have any concerns and issues, or just to gossip. Employees should also take the initiative and respond to open online discussions and contribute ideas and suggestions, and call or e-mail colleagues, but no more so than if they are on premises.

Employee participation works. I may be 3,000 miles away and in another country but trust me, ask anyone at *Call Center Magazine* or in our group and they know I'm there even though they don't see my ugly mug, at least in person.

✪ ADDITIONAL SOFTWARE TOOLS

You may need, besides VPNs, hosted access, remote access, and IM, additional software tools to enable home working operations. Many of these tools are extensions of software that you may already have installed for your employees, like log-in and call, e-mail and chat monitoring.

If you need such tools, check with your vendor to see if the licenses go per-installation or per-employee. You may have to pay additional license fees for occasional home workers if the software client is to be installed on organization-supplied laptops or

PROTECT YOUR NETWORK!

It is a nasty world beyond those gates. The demons have gotten into IT as well as e-mail.

Make sure that your home working employees' machines are up-to-date on the latest anti-spam and anti-virus protections. Establish strict procedures, e.g. don't open attachments from unsolicited sources and request that senders include a text synopsis so that employees will know what is inside before opening one of those electronic Pandora's boxes.

employee-supplied machines. Consider installing equipment tracking devices that when activated will notify you and the police where the laptops are, to enable recovery.

✪ IT SUPPORT

If your home workers are using hardware and software you supply, and have access to your network through a gated means (like a VPN) you will need to support them. And chances are much of that hardware and software, like laptops, printer/copier/faxes, and VPNs are going to be unique to your home workers.

"Management's message to IT should be to treat home workers no differently than on-premises workers," says home and mobile working consultant Eddie Caine. "The IT employee's individual and collective performance will also be judged on their ability to support all technology needs regardless of location."

That means you will need to train IT staff to support home workers. If you do not have remote access control and diagnostics/repair tools you may have to install them. You also may need to arrange, directly or through your vendors, for field support.

On the other hand, the additional, if any, up-front costs in buying and servicing equipment like laptops, VPNs, and diagnostic tools may be compensated by fewer IT support calls from employees. Workers who are based at home know that they can't pick up the phone and cry "Help!" as readily as they can do when they are in premises offices. For that reason home workers try to fix some of the easier problems themselves. But they may end up making a problem much worse than it is, such as by deleting files they shouldn't thus preventing their computer from booting up.

To avoid such catastrophes you may need to remind your employees, premises or home office, what they can or cannot fix. You should set out when it is acceptable for home office employees to use their personal machines, but not *your* software.

There is increasingly less difference between IT premises office and home support. Firms have been centralizing their IT support into formal help desks. If they have more than one office in a metro area it could take as long for an IT support person to reach that premises office cubicle from the time the call or e-mail is received as for a contract field support rep to visit a home worker.

Supplying IT Support

IT support for your home employees can take the form of training all your staff on those unique-to-home worker issues. Or you could have a dedicated team. CMP has a *superb* IT Home Support Team. There is also a separate hotline for IT Home Support.

CMP, for example, has a ratio of 100 home workers per IT support person. But that number is overstated as other IT staff also assist home workers, such as if there is a problem with computer viruses that affect all employees no matter where they are, or if there is a system-wide software upgrade.

A July 2002 Gartner report entitled, *Teleworkers Settle for Less in Service and Support*, found that home and mobile workers experienced service gaps (such as long bouts of downtime) up to ten days unless they seek help from colleagues. But when they received service, the quality was high.

The report recommends that CIOs and CFOs make supporting home and mobile workers a priority by setting up IT policies and strategies and service level targets. Among the suggestions: that support desks track home and mobile worker service requests separately from those on-premises. Companies also should consider outsourcing internal help desks to fill service gaps, as during after-hour calls.

The report strongly advises against having home and mobile workers install your organization's software and network access on their personal computers (Chapter 3). That will eliminate the big problem-causing headache of configuring corporate software on home machines that companies do not control and that blurs the line between corporate and personal property.

You also should have a means by which home workers can diagnose and fix some of their own problems, without FUBARing your equipment and lines. One way is to supply your home workers with a FAQ page on your corporate Internet. The FAQ page should be easily updateable by your IT staff. Couple that with an e-mail question form and a tech support forum that is monitored by your IT staff.

That way your home (and mobile) workers get issues resolved without taking valuable IT staff time and the IT staff have a standard form that enables better problem-solving. The forum enables your home workers to help each other to solve problems, but under supervision so as to check on and stop wrong information. The IT staff may find that they can learn from the forum, find suggestions that have merit, and ideas that can be entered on the FAQ page.

Field and Onsite Service

You will need field service contracts with your hardware vendors if your home employees can't bring their gear in, cannot wait to get a new machine, do not live close to your premises, or cannot leave their home offices to fix urgent problems, like a keyboard or modem dieout. But use field service as the last resort—it is the most expensive of support options; employees should try to have it fixed by phone and online with IT support or the vendor first.

I've needed field support. I had a CMP-issued Dell laptop when I lived in Victoria, BC, Canada. When the machine broke down I contacted IT support, they contacted Dell and a technician from Dell's local subcontractor, Tecnet came to my home office: which was three minutes from Tecnet's office.

Consumer products giant Procter and Gamble, which has a home working program, will have a contractor pick up the computer and return it to the office for repairs. And if there are hardware or network problems that will result in outages of more than an hour or two, P&G will ask its home workers to come to the office. (P&G has sufficient spare room, like in its training facilities, to accommodate home office employees temporarily. But P&G has rarely been force to resort to that solution; the equipment and the networks have been very reliable.)

The other option is for you or your employees to arrange or contract for field support, either through dealers (such as large chains like Best Buy) or through specialist local or national/international third parties. Your employees can also pay for support; if they are exclusive home-worker they might be able to deduct this expense depending on the tax rules.

For example, ComputerRepair.com provides outsourced field support for individuals and organizations across the US through a network of nearly 3,200 technicians, either working for companies or self-employed. You or your employees can review the techs' profiles, verify credentials, and see how others rate them.

"Most manufacturers outsource the services they sell to third parties anyway," CEO Jeffrey Leventhal points out.

ComputerRepair.com supports all hardware: computers and peripherals and most packaged software including virtual private networks. You can arrange for it to support your proprietary applications.

Payment is straightforward. You make a deposit, say $1,000, and when the problem has been resolved to your satisfaction ComputerRepair.com draws the cost from the sum. The automatic clearinghouse system pays the techs.

And unlike support provided by warranty and through manufacturers, Computer-Repair.com techs can multitask across vendors. Sometimes a hardware problem turns out to be a peripheral or software problem.

"If a tech comes to fix a computer that is under warranty they can't fix the printer or debug software problem if they're made by different vendors," Leventhal points out. "That means more support calls, much higher costs and more lost downtime. We charge the additional time and for any parts and bug fixes. The tech is already there."

✪ HOT-DESKING

If you have your exclusive home working employees come to your premises offices regularly, say once a week to work, then you need to have the space, voice/data, and if need be, computers available for them to work on: The use of the space, connections, and machines that are shared with others is known as *hot-desking*.

Hot-desking enables you to accommodate workers at premises without spending money on dedicated work areas or cubicles for each employee. Employees take turns on the desks and if provided, also computers and peripherals. Consultants estimate that each premises office workstations costs $35,000 in rent, installation and upkeep. Of that, $28,000 goes the furniture, equipment and cabling: only $7,000 to the real estate.

The average home worker: administrative/clerical, call center, IT/programming, journalists, etc. require about 100-150 square feet per workstation, says King White, senior vice president of real estate/site selection firm Trammell Crow. That includes aisle and common space like break rooms and reception areas.

Satellite offices, which are premises offices that are located near to where your workers live, are related to hot-desking (Chapter 11). Satellite offices enter this home working discussion because they minimize commutes and employees use shared resources. Employees working at satellite offices share peripherals like copiers and printers. You can also have your employees hot-desk at satellite offices.

Enabling Hot Desking

Hot desking can be informal: a spare cube, desk and/or terminal. Or it can be formal: areas, links and equipment specifically set aide for that purpose. Use hot desks in a cube formation to give hot-deskers privacy; or design hot desks so they can be shared (e.g. two or three people to a cube or to a countertop), which takes up less space).

Note: Hot-deskers are like campers; they take out what they bring in. They don't need file cabinets or storage space.

Hot desking is not just for your home workers. It also can be used for mobile workers and for employees from other offices.

To make the most effective use of hot-desks, and to gain the most real estate and furnishing savings from home and mobile working, you need to plan beforehand when you need your home working (and your mobile working) staff to come in. If these workers are in different teams e.g. a Gold Card team, the Web group, the Web magazine staff, and the home or mobile workers come to the premises once a week, stagger those weekly hot-deskings.

There are hoteling services (see Chapter 11) that you can use to outsource such arrangements. Or you can do itself with a browser-based application.

Hot-desking will require the installation of voice and data ports and phones. You can avoid re-transferring calls to the hot-desk extensions—if your home workers have calls transferred to their home office phones, they can either call in to retrieve messages or have the calls onward transferred to their cellphones (you will need to pay for those cell calls).

If you have home workers occasionally work from premises offices you will need to supply them with hot-desks that they share with others. Hot desks are small, utilitarian desks with phones and data connections, with just enough room for employees to work with their computers and any paperwork.
Credit: Brendan Read

Data pickup is easier. If your home workers bring in laptops they can plug into your ports; if they use terminals that are already in place, you give them special passwords in order for them to access your network. Alternatively, they access their e-mail through Webmail on the provided computers.

If your office has gone to wireless LANS (WLANS) then data installation is easier; if you have VoIP over the WLAN, you avoid the need for wires altogether. But the computers, either your home workers' or the hot-desks' computers must be WLAN-enabled.

You will need to have sign-in and ID tag procedures for hot-deskers. You can treat them like visitors or you can issue them ID tags that they wear when they come in.

✪ STOREFRONT OFFICES

If you intend to hire a fair number of new, exclusive home workers who live out of commuting or acceptable travel distance (which could be a half day's drive, flight, or train ride) from your premises offices you may need to consider opening regional storefront premises offices.

The reason is at this writing, you as an employer need to verify in-person that employee's ability to work in your country. Employees will need to present documents such as birth certificates, passports, alien/naturalization/permanent residency cards to prove this.

These offices could be as small as a single room in the back of a commercial building. The storefronts also could be used as sales offices, satellite premises with shared printers or fax machines, or as networking spots for your home workers.

The expense of providing the storefronts could be justified by enabling you to tap more quality workers. They may not want or could afford to fly or drive the entire day to handle paperwork that only takes a few minutes; which could discourage them from working for you.

✪ EQUIPMENT

If you choose to supply your home workers with equipment such as computers, phones, and peripherals (see Chapter 3) then you will have to source and budget for them. Gear like laptops and all-in-one printer/copier/faxes are generally unique to home-based offices and/or, at least on the computer side, mobile workers.

Good quality desktop computers tend to go about $1,200-$1,500, laptops $1,700-$2,400 retail, plus any additional software licenses. Printer/copier/faxes cost about $300-$400. Wireline phones cost as little as $10 for PSTN corded sets to $200 for IP phones; wireless phones vary depending on the rate plan that works best for your needs. If you buy in bulk you can get some good deals. When you buy computer and peripherals check the service plans; many vendors charge for support after the initial warranty.

Computers and peripherals have life spans of three to four years, if used intensively everyday. Laptops, however, have an average life span of only two to three years because laptops are less rugged and more prone to abuse, and because they are portable. The life spans grow to four to five years for equipment used occasionally, like fax machines and laptops used by occasional home workers who use desktop machines at premises. Obsolescence and the inability to obtain tech support, replaceable components like ink cartridges and parts, make five years the limit for any piece of equipment.

You will find that equipment may last longer—thus lowering your total cost of ownership—with home workers than with premises workers. Home workers tend to take better care of their gear. They know there isn't a replacement machine that could be shifted over from another cubicle, not when the nearest cube is a 100 miles or more away. Home workers also know that home working is a privilege and are less likely than premises workers to take their employers for granted.

Check with your IT department for the nitty gritty. With their information you can establish total cost of ownership of each piece of equipment.

✪ SUPPLIES

Home workers consume supplies too: e.g. ink, paper, toner and pens. You will need to decide how you are going to handle that investment.

Exclusive home workers should be allowed to expense them; so should occasional home workers—but only if they are using equipment and peripherals you supply them. You don't want to spend your organization's money on disposable goods that are going to be used by your employees to print their daughters' coming-out party invitations (though chances are they use your printer at your office for that very purpose).

If you have existing relationships with office supply vendors you may or may not find it feasible to have them supply your home workers. If those employees live in the same market radius as your premises offices then it is doable. But if they do not then it is not, like me—I live in a different country from CMP's nearest office.

Also some items (like ink cartridges) are "I need it NOW!!!" items and can't wait for delivery. Your home workers need to get them, install them, and expense them.

Fortunately there are low-cost office supply and furniture chains such as Office Depot, OfficeMax, and Staples in a growing number of communities. There also are independently-owned dealership groups such as Basics in Canada that offer excellent prices and service too. In my small city of 19,000 there is a Basics retailer (Monk Office Supply) and a Staples.

✪ STAFF TIME

Just as buildings need facilities staff, and staff time to attend to related matters (like receptionists and security) you may need to invest in staff and staff-time to enable and monitor the program. This is especially true if the program covers a fairly large (25 employee-and-up) program. This is on top of the IT staff investments.

As Chapter 12 will explore in depth, you will need people, and people-hours and possibly outside consultants to do research on home working feasibility and put the proposal together. The staff time varies greatly by program. An informal program, where you are enabling occasional home working or exclusive home working to support a few valued employees may take as short as a few weeks. A large scale formal program involving tens of employees may require eight to ten months of planning and implementation, shared between staff and outside consultants.

If your executives approve the home working program, then you will need to train supervisors how to manage home workers, screen employees for home worker suitability (including, if desired, visiting their home offices), measure and account for program benefits and additional costs, and perform cost/benefit assessments. You will also need staff-time to single out the home working variable in performance and troubleshoot any problems arising from the home working arrangements. A large scale program may require you to select, or hire a Home Working Program manager who is responsible for overseeing implementation and management.

As this book illustrates, there is a lot that goes into home working, especially exclusive home working. But as you look around and then talk to site selectors, the real estate and facilities departments and managers, and IT staff, there is an awful lot that goes into conventional offices that most people are not aware of either.

Chapter 8:
Administrative

Setting up and running a home working program poses unique and sometimes challenging administrative issues. The key administrative matters you will run into include selecting and training home workers, legal matters like data law compliance, and insurance. Also key: how do you get employees who are out of sight to comply with your policies?

Finally you need to wrap these needs, plus your organization's voice (Chapter 2), data (Chapter 4), locations, facilities (Chapter 5), and ergonomics (Chapter 6) requirements into formalized policies that lay out what you expect from home offices and home workers including what you will pay for. Your policies also should be flexible enough to examine exceptions, and exceptional circumstances. Some of those circumstances are laid out in Chapter 10.

✪ HOME WORKER QUALIFICATIONS

Good employees are good employees. But not every good premises working or mobile working employee makes a good home working employee, except in emergencies. Some need the social contact and/or nearby supervision to perform well. Or perhaps they have tasks that require in-person human interaction. Or they might not have the room for a proper home office or an environment (i.e. noise and people/pets-distraction-free) that is conducive to home working.

But at the same time, not every great existing or potential employee makes a good fit for the social life in premises offices, but they have the skills and qualifications to succeed in the home office. They may not be into unnecessary chit-chats or hanging about with co-workers, yet communicate well professionally. Or they have family/personal responsibilities that require them to be nearby or easily available in case of emergencies.

Potential or existing employees may have disabilities or have become disabled making commuting nightmarish though technically possible. While the Americans with Disabilities Act (ADA) requires, for example, that new or substantially upgraded mass transit facilities be accessible, anyone who has been on a bus or railcar wheelchair lift or watched someone in a wheelchair on them can't help but feel vulnerable, suspended and helpless many feet above the hard ground.

You will need to set out selection criteria to screen for home working *or* premises working for that matter —*on top of the employment task qualifications*—to succeed in those environments. Home working employees must also have suitable locations, facilities and the environment, as discussed in Chapters 5 and 6. There are pre-employment screening tools, deployed by staffing/screening firms such as FurstPerson that does such assessments by asking direct and indirect questions and doing role-playing with applicants.

✪ DETERMINING HOME WORKABLE FUNCTIONS

You will first need to determine which job tasks of particular occupations are home workable (discussed in Chapter 1), and which ones can be grouped together to permit employees to easily occasionally home work or occasionally premises work. To save you from flipping back there here are some of the same tasks again:

Editing and writing
Graphics design
Proofreading
Transcription
Translation (spoken and written)
Legal
Computer programming
Database management
Website hosting and programming
Customer service and sales
Accounting and programming
Engineering and design
Sound recording
Training
Management

At the same time you must be able to monitor or measure and evaluate employees' performance of these tasks. This is especially important as you or your managers are not "seeing" the employees. The monitoring can be real-time, like listening in on calls; or offline, like asking an employee to send an outline or synopsis of the work.

✪ DEVELOPING THE CRITERIA

In developing the home working criteria you will need to consult with your HR and IT departments to ensure the criteria complies with your organization's requirements. For example, HR will want to see punctuality stressed, afterall there is no way for anyone to "see" the home workers show up at their office, unless you invest in costly video equipment. For that reason, you should consider requiring your employees to log onto your network at a specified time.

Also, IT will want to make sure your home workers are computer-literate; they can't afford to spend time "telebabysitting." The IT department also will want to ensure that the home workers have data connections that enable fast, seamless links to your network.

You also should consult with your facilities, or better yet, your occupational safety or ergonomics specialist, if you have one, to help in drawing up minimum criteria for home office environment, such as lighting and workstations. That way your home office employees will be able to deliver ***equivalent if not superior performance than at premises offices...***this book's mantra.

If you have employees in your organization that are already home working, talk to them and to their managers. Ask them to help develop and to critique your home working criteria. There is no better resource than those who practice it.

There are also home working consultants you can hire who can help find "best practices" amongst home working organization that you can model or benchmark your selection criteria against. These "best practices" organizations are or should be similar in occupation, endeavor, function, and responsibility to your own, so you can get valid apples-to-apples comparisons and learning transferability.

We will now examine some of the key home working selection criteria.

Independent

To succeed at home working employees, exclusive and occasional need to be independent. They ***must be*** *self-starting, self-disciplined, motivated, and professional.* These workers need to have the abilities to solve problems, including equipment and network problems, by themselves with a minimum of fuss. They must know how to manage time and complete projects.

These employees *must be able* to work without face-to-face contact with others. Yet they *cannot* be bashful: they must be willing to go out of their way to stay in the loop; they need to contact supervisors and colleagues if they have a problem, instead of waiting for the problems to come to them.

These workers also must be able to focus on the job tasks, *regardless* of the distractions. Just as premises offices have many diversions, like the Internet on their machines, the coffee machine, the water cooler and the sexy new employee's cubicle, so too have home offices like the Internet on their home machines, the TV, family, pets and the outside patio.

I live in a part of the world which tourists the world over come and visit. I have gorgeous views from my home office. But I'm used to the scenery; I know that to keep such views I need to get my tasks done on time and well, so I turn my back on it, boot up the computer, the VPN connection, and start working.

But when you think about it, aren't these attributes that ***every*** employee must have? If so, why not screen ***all*** employment candidates for these qualifications?

Goal-oriented

Home workers must understand the goals of your organization without being told or reminded, how their work contributes to those goals, and how their performance affects the ability of your organization to meet those goals. They need to know where their contribution fits into their work team's performance and understand organizational relationships: between their team, other teams, departments and senior management.

Resourceful

Home workers must be multi-taskers and problem-solvers. They must identify, figure out and solve issues. They need to think on their feet. Home workers have little choice; there is no one "there" to back them up or to delegate tasks to.

Experienced

Equally importantly, home working employees **must have** experience and knowledge to work on their own without supervision and without annoying their co-workers. They must have the necessary skills and training to succeed at their work. Because when they're at home there's no cubicle or office to run to or people whose shoulders to tap or cry on.

Note: Home working is not for the untrained or new worker unless they are very experienced and have proven they can literally fly on their own. It is easier and more efficient for the employer to have that worker get attention now by walking over to pester the boss and the behind-deadlines co-workers to solve a problem with the computer, get the right code or the right terminology than to wait an eternity for an e-mail reply or returned call.

There is no standard for experience that you can apply when you examine your employees to see if they have what it takes to work from home. Instead you and/or your supervisors have to make that determination in consultation with those employees.

Excellent communications skills

Your home employees must be superb communicators. Because they are not "seeing" others or others are not "seeing" them your home workers must be proactive in how they use communications tools, like IM, e-mail and the phone, as well as reactive when supervisors, colleagues, clients and customers contact them via these channels.

Suitable environment

Chapter 6 goes into these issues in depth. But here are a few summary points.

Home working candidates must have the dedicated space: be it a corner of a room or separate room for their office. That space must have power, voice, and data connections to enable employees to work. That possibly could include stipulating high-speed Internet access.

The home office environment, ideally, should be free of background noise and distraction from other inhabitants (human and otherwise). If necessary, the environment should provide for privacy to protect sensitive information, like locked filing cabinets, the computers not facing windows and doors for passersby to read what is on screen, and separate offices.

Employees must have, or be willing to pay for, the proper furniture to prevent injury to them. You will need to set out the requirements (i.e. that the chairs be adjustable). Or you must be willing to finance the purchase of the furniture yourself. This precaution will prevent higher healthcare costs and lost productivity.

Access and proximity

If you require your employees to travel they need to live where there is a means of access to where you need them to be, like a commercial airport. If you need them to come into the premises on a regular basis then they need to live within a reasonable distance—that does not necessarily mean commuting distance, two or three hours one way may be adequate.

Clear employee understanding

Home working candidates should fully understand what is involved with working from home, especially exclusively. They need to know the realities of home working. That work hours *mean* work hours: not laundry, soap opera, home decorating show, or jack-up-the-monster-truck hours.

Candidates must show understanding that unless your organization's culture tolerates it, that home working is *not* day care. Distractions are not tolerated. They need to know what they are expected to pay for and supply and what you will pay for and supply.

They need to realize that because they are your employees, regardless where they are located, that they must follow your organization's policies. That includes your facilities, environmental, sign in/sign out, and data privacy policies.

Finally, candidates need to understand that working from home is a privilege, especially for occasional home workers. Screw it up and it's back to the noose (i.e. neckties), pantyhose, and four-hour round-trip commutes.

Or if your program entails not having premises offices for employees—why pay money for a building to work at when they can work at home—fouling up at home could mean out of a job. Like what happens to a field sales rep who when on the road repeatedly blows sales and annoys customers and prospects, or to a trucker that arrives late often, collects tickets, and gets into accidents. No difference.

Employee compliance

Home working candidates, exclusive *and occasional* must be willing to agree, *preferably in writing* to comply with your home working program requirements. And if you require your staff to visit your home workers to ensure compliance with the program, along with the rest of your employee policies then that *must be set out* in the agreement and signed by the employee. A person's home is their castle; employers or their agents can only enter if you have that person's permission.

✪ SELECTION

Chances are if you begin offering the opportunity to work from home you will get a fair number of applicants. Studies consistently have shown that many employees want a chance to work from home, even if it is only for one or two days a week, because of the benefits: no hellish commutes, ability to see friends, loved ones and family members before crashing for the night, and cost savings. Home working consistently ranks as the number one benefit employers can offer.

Working from home is in many ways a promotion. And you should consider it as one. The cost savings to employees—$4,000 to $5,000 a year in clothing, cosmetics and transportation—are *de facto* raises.

Thus you need to plan an assessment methodology for home workers. Here are some considerations when devising it:

Who gets to apply?

Home working requires employees to be self-disciplined, motivated, and not reliant on supervision. Those attributes usually, *but not always* are possessed by your top-performing employees.

As a *general rule* you should make home working available to your top performers, assuming they can meet the criteria. That incentivizes other employees who want to work from home to do better.

The myth of the top performer

The words "but not always" and "general rule" have been italicized for good reason. Top performers at premises offices do not necessarily make top performers at home. That is what I call it the "myth of the top performer": just because someone is excellent at one set of tasks in one type of environment that does not necessarily means they would be excellent at another set of tasks, or in another environment.

The "myth of the top performer" explains why supervision and management in many organizations are terrible. Employers promote from below without taking the time and effort to determine beforehand whether those candidates have the aptitudes and skills to succeed in their new positions.

In deciding who gets to work from home some otherwise excellent "A" level employees need in-person contact and supervision; they would goof off at home and ruin the program. But there may be good "B" and even some "C" level employees who have the independence attributes to succeed at working from home.

Avoiding discrimination

Like any promotion, employers must be very careful in deciding the criteria for home workers and in choosing candidates to avoid the appearance of favoritism or worse yet, discrimination. You don't want to leave your organization open to lawsuits.

For example, in Chapter 5 I purposely avoided in the discussion of theft and security of employer's equipment, though I mentioned data privacy. An employer could leave themselves wide open to charges of "redlining" if they restricted home workers to so-called "better neighborhoods." But security is a legitimate issue if the job requirements include having customers and clients visiting the employees at their home offices (more about that in Chapter 8).

Laws such as the Americans with Disabilities Act (ADA) forbid discrimination because employees have impairments that have nothing to do with the work. They require employers to accommodate them at workplaces.

Trialing home workers

You may want to consider trialing your home worker applicants as occasional home workers—two or three days a week for one to three months—to see how they work out. If there are no problems, then you can make them exclusive home workers.

If you trial, keep your investment and your location, environment, and facilities requirements to a minimum; ask the same of your employees.

For example, it is smart to supply your employees with laptops to use at premises as well as at home. But there is no point in arranging and paying for second or third lines only to have them ripped out again. By the same token, it is reasonable to ask employees to dedicate areas, and leave it up to their discretion where those areas should be—part of a room or in a separate room. But you should not ask them to provide separate office spaces. Similarly, you can require employees to have comfortable furniture but not specify adjustable chairs, workstations and wrist pads.

Note: *Home working is a major commitment by you, your organization and your employees. You must all work together to make it happen successfully.*

○ SCREENING PROCESS

The screening process for home workers may be a little more extensive than for premises office workers. If you require verification of whether home offices meet your standards, either in-person or by employees sending you photos and signing affidavits attesting to their accuracy (see Chapter 5), then this needs to be done before employees are approved to work from home.

The *e-Work Guide*, published by the International Telework Association and Council (ITAC) suggests that you use structured interviews and simulations to let applicants set out how they would deal with work/family conflicts and how they would plan and organize tasks to meet deadlines. The guide also advises getting recommendations from supervisors, colleagues, and clients/customers.

You also should take a careful look at how you assess, rank, and look at employees. Some people are very hard workers and make solid contributions to the team effort but are not "visible." These individuals aren't social butterflies and prefer to keep to themselves. Conversely they make ideal home workers.

When screening is to look for signs of initiative, independent working and self-motivation amongst your "A", "B', and "C" class employees. For example, did you assign them a special project that they had to conduct on their own and did they accomplish it well and on time?

You should also screen for or ask about outside activities that demonstrate that employees have the abilities to work at home successfully. Employees who have worked as freelancers, consultants, been self-employed and successful at it or in their off hours run volunteer organizations, teach, tutor and/or have hobbies like painting, sculpture, music, writing are good home working candidates. Those tasks, activities, and hobbies show initiative, self-discipline, commitment, and ability to work at the top and alone.

There are screening and selection tools that can help you. For example, FurstPerson offers a hosted assessment module in its screening and selection program that enables call centers, and other employers determine if the prospective agents have the ability to successfully work from home.

The pre-screen portion asks employees either online or through an IVR whether they have the clients' requirements. That can include anything you specify such as broadband connections, second phone lines and separate lockable offices.

The advanced screen then asks applicants indirect questions, or uses role-play to determine if they have the attributes to work from home. Those can include superior cognitive ability, problem solving and initiative, like contacting colleagues if they have a problem they cannot solve.

You can also test the prospective employees' technical skills, such as posing in role-play a computer or phone line malfunction and see how they respond. That's vital (see Chapter 7) because home workers can't easily have the help desk run over to their cube if they are located on the other side of town or in the next state or country.

✪ SKILLS/KNOWLEDGE TRAINING

If you follow the earlier advice, employees who are working from home already will have had the training and experience they need to do their jobs. But sometimes they may need or want additional training to refresh their skills or to add other skills and knowledge.

There are also many jobs, like customer service/sales, nursing, programming, and tech support where the information changes constantly and where there are often new methods of doing those jobs. In such instances, those home employees will need to be trained—sometimes fairly often.

Thus you will need to set up means by which these workers can be trained. And, yes, you can have an increasing amount of this training undertaken at home.

Again, this is more of an issue with exclusive, as opposed to, occasional home workers (who can be trained at premises).

Training is divided into two schools, literally. These are *instructor-led* and *technology-based*.

Instructor-led training (ILT) is where instructors teaches students and obtains feedback by interacting with them live. Instructor-led training is supplied to large or small numbers of students, or one to one.

Technology-based training (TBT) a.k.a. "eLearning" is my catchphrase for pre-recorded online, CD, DVD and video instruction where students listen to, follow along, to refresh their ILT or TBT lessons. Their interaction, if any, is limited to answering pre-set questions.

Both ILT and TBT can be supplied to home working employees. Instructors can teach small or large groups by audio-, data-, video- or Web conference, or one-to-one by phone or text chat. Employees can take TBT online or review CDs, DVDs, and videos at home.

A hybrid of ILT and TBT is facilitator-led training (FLT) where a staff member, like a supervisor or team leader works with the DVD or online lesson and initiates discussions and feedback on each part of the lesson. Home workers can participate via audio, data, video or Web conferencing.

FLT is less expensive than ILT because you're not paying for trained consultants and experts but the content is less customizable and fresh to your organization. Also your workers may have questions that the facilitator can't answer and that only the package's creators can reply to.

The only downside to ILT and FLT conferenced training is the lack of in-person interaction with peers and with the instructor. But as anyone who has watched and participated in chat and Web conferences sessions knows, those discussions can get quite lively too; and you can come away with learning a lot.

Alternatively, ILT/FLT and TBT training can be offered at premises offices and at off-site locations like schools, seminars, and trade shows. Both instructor-led and TBT methods are frequently offered together. For example, the University of Phoenix offers advanced degrees taught through a blend of in-person instruction at satellite locations and online TBT.

Training consultants recommend that employers deploy a mix of ILT/FLT and TBT; ILT, via conferencing, enables interaction with instructors and colleagues, the throwing of curveball questions to see if trainees are on the ball and provides motivation to trainees. TBT enables critical skills and reinforcing training. TBT also can help supervisors assess and improve employee performance.

The trends are towards ILT/FLT conferenced training and TBT training, though. The reasons are costs and productivity. Conferenced ILT/FLT avoids travel costs; conferenced ILT/FLT and TBT minimize the need for having training rooms that eat up real estate.

More importantly, conferenced ILT/FLT and at-desk TBT avoids productivity losses caused by having employees leave their work areas and, in some cases, the premises to be trained. Employees can take these lessons at their workstations, or at home.

Some employers with home workers initially conduct their training face-to-face. They have their employees come to their premises offices because they believe that in-person training is more effective. But other employers with home workers do not.

Reg Foster is chair and CEO of Alpine Access, a service bureau with all home working call center agents. His firm does all of its training remotely, without requiring agents to travel to premises offices. Alpine Access uses eLearning and instructor-led over the phone and in conference calls.

"Call center agents work virtually, with customers on the phone, so what is the point of having them being trained face-to-face when that method does not replicate how they communicate with customers?" he asks.

✪ HOME WORK TRAINING

Yes, you have to train both your employees and your supervisors on home working. Home working is "new" to most people, thus many individuals do not have all the knowledge and skills to conduct it or manage it effectively, says Bob Fortier, a telework/home working consultant who is also president of the Canadian Telework Association. They, therefore, miss out on how to maximize benefits and minimize the downsides of home working (both are explored in depth in Chapter 12).

Home Workers

Bob Fortier points out that many otherwise qualified home workers may not know how to design safe and effective home offices, establish work priorities, or develop task management skills. They may not know how to communicate and work effectively with managers and coworkers, keep office relationships intact, and deal with family, friends, and neighbors.

There are several ways of training home workers. You can use ILT or FLT in role-playing to teach and evaluate how they respond to work/family conflicts and how they would plan and organize tasks to meet deadlines. You will need to work with an HR professional or other consultant experienced with home working to prepare and deliver this training.

You can also use TBT: online or video to show the right way or the wrong way to set up home offices, i.e. "what's wrong with this picture" with a downloadable checklist. You can have your occupational safety person or your furniture dealers design, show, and release this training.

All employees, **regardless** *of where they work* **must** be taught proper ergonomics to prevent costly injuries. That includes how to adjust chairs and workstations, the use of appliances like headsets and laptop lifts, and how to use keyboards and hold phones. Employees also need to be taught how to take care of chairs, workstations, and appliances to ensure they last, saving you and them money. These lessons can be taught in premises offices by ILT, FLT or by TBT at home.

All employees, **regardless** *of where they work,* **must** also be taught proper phone and e-mail etiquette, which in today's ultracasual barely literate world is so lacking. What's cool amongst peers is amateurish and unprofessional amongst clients, customers and upper management.

Supervisors/Managers

Training your supervisors and managers on home working is the key to making home working a success. If they don't know to how manage it, or they are fearful of doing so because they worry that they'll lose control, they could derail the program.

Bob Fortier adds that many managers have somewhat inflexible attitudes towards home working—much of it is due to misimpression about what it is and how to do it well. Poorly informed managers have a hard time knowing whether and how to approve new home work situations, let alone understanding the characteristics of a good candidate, and how to effectively manage them once they are chosen.

This training is instructor-led, best done in-person. To train supervisors and managers on home working requires instructors who are experienced and knowledgeable about home working. They could be outside professionals or managers within your organization who have successfully implemented or run a home working program.

✪ LEGAL AND REGULATORY ISSUES

There are several legal and regulatory issues that may crop up when managing a home working program.

Potential for Home Employee Harassment

Organizations are especially sensitive in how they assess home workers' property, in person, especially a female home worker's office being checked out by a male supervisor. They *don't* want to have any complaints.

For that reason you *must not have* male supervisors visiting female workers unless accompanied by a female or third person present (Chapter 5). Having employees of both genders there also limits the risk of same-sex harassment.

To avoid the odds of harassment altogether, Jack Heacock recommends the employer furnish the home worker with a comprehensive suite of ergonomically designed home office furniture, along with formal guidelines and training for the operation of a home office. This greatly reduces the risk of injuries or potential actions downstream.

Additionally, to avoid home visits, he recommends having the employee take 4-6 photographs of their home office. The employers' home work program management team, along with risk management and the furniture manufacturers' representative then review the photos and make any recommendations for change identified. The photos become a part of the home workers profile and record—further insulating the employer from future claims.

Data Regulations

Data theft has become a huge problem. In response governments worldwide have passed many laws requiring that employers take steps to protect consumer and business privacy.

Some employees, such as call center customer care/sales staff have direct access to sensitive data. When customers, for example, call in, their files are popped to the agents' screen. Agents can install applications to automatically capture and retransmit this data to criminals without you knowing about it.

Having this data handled at-home increases the odds that the data could be stolen— you are not in control of those employees' worksites. On the other hand employees and visitors at your premises or outsourcers' premises offices also can steal data.

At least with your home workers you can more isolate responsibilities compared with a big premises office where employees don't control access to their premises, work areas, and equipment. You might assign a desk to an employee but someone else could sit in it or that employee could be working and a visitor let in by another or who gets through security walks by and sees confidential information.

If data theft is a concern, you have two choices: Do not have employees who have direct access to such data work at home. Or take measures to prevent at-home employee theft.

Preventing Data Theft

There are steps you can take to prevent data theft from workers. These include employee criminal background checks, having agents sign documents indicating their responsibility for data protection, virtual private networks (VPNs) with authentication, and in more critical cases, authentication tools like smartcards, USB tokens and biometric devices such as fingerprint and iris scanners (Chapter 7). You can also supply home workers with storageless PCs, and require locked cabinets and rooms.

The first step, before making any heavy investment in hardware and software, is to stipulate in your policies and procedures (more about home working policies later) that no other people other than your authorized employees are allowed access to sensitive data, and outline what exactly is that data. You then have your employees sign statements saying that they understand this and that they will comply with the laws and with your policies.

That way if your employees are found violating your policies you have grounds to discipline and, if necessary, dismiss them. If they have been found to break the law then you can turn them in.

Ideally, you should own the equipment your employees are using, especially computers, but also cellphones and PDAs. The more sensitive the data, the more vital it becomes that you have complete control over it. That's almost impossible if the data is being transmitted to, from, and residing on equipment that others own.

John Paddock, attorney with Hale Hackstaff Friesen points out that is very difficult to delete all data from equipment. You may fire employees or they may leave. They may make a genuine effort to clean out their files, but your sensitive data and trade secrets will still linger on hard drives and in memory where anyone can find them later. Also, in the case of cellphones, if your employees own them they, not you, will have the calling records.

Paddock points out that while companies may go after employees who leave to work for a competitor, taking confidential data such as client lists with them, such cases are expensive and can be difficult to prove. These cases also can be embarrassing for your organization because they not only show your vulnerability, but also imply that you have dishonest employees and use lawsuits to keep your customers. You may be better off spending that money on prevention and for marketing

Another step Paddock recommends is to make sure that your own slate is clean; that you don't have employees joining your firm with their previous employers'/clients' secret data. A judge will quickly dismiss a "trade secret" case brought for a company that specializes in pirating competitors' property.

"One of the best means of protecting yourself is to be squeaky clean at both the start and end of the employment relationship," recommends Paddock. "Carefully screen your new employees so they don't bring someone else's trade secrets to you, and be thorough about letting departing workers know their obligation to preserve the confidential information they learned from you and your customers. You will increase your odds of getting conscientious, loyal workers who won't go to the competition with your secrets."

Do Not Call Compliance

Inform your exclusive home workers that they are not permitted to put their home office numbers on Do Not Call lists; those registries are meant for personal, not work-related calls. However, if they are constantly interrupted by unsolicited calls, recommend that they get Caller ID and screen calls.

Occasional home workers, however, can use DNC registries. This is because their home lines are primarily used for personal calls.

SHRED DISCARDED DOCUMENTS!

If your home workers print out customer or confidential corporate information insist that they shred documents after they've finished with them. That includes long distance and any other bills that have information—criminals can use it.

Identity theft is on the rise. If you/your employees don't dispose of your customers' organization's information you may be giving garbage-can-diving skells looking for credit card numbers a bonus big-time.

You can shred documents with scissors by cutting the papers into thin strips then cross-cut randomly into tiny pieces. For additional protection take the pieces and soak them in hot water and soap to mush the paper and ink.

Or you can insist your employees buy desktop or trashcan shredder. Available from office supply stores and chains (e.g. Staples, Office Depot, OfficeMax and Basics), they run from about $35 to $120.

The best bet is to limit the need for your employees, home or premises alike, to print information, period. You/they save money on paper, ink, printing, copying —and shredding. You limit the need for and size of, hence costs of printers, copiers, filing cabinets, and shredders.

Legal Jurisdiction

There may be questions under which jurisdiction—state, county, city—home workers fall under with regards to matters like taxation (individual and corporate), unemployment, and workers' compensation. Because exclusive home workers work from their homes they usually but not always come under the jurisdiction of where they live, not where they report to. The situation may be even less clear for occasional home workers; they may come under the jurisdiction where the employer's office is located. There are some legal battles underway at this writing that could affect this situation.

John Paddock points out that employers who do not provide workers with premises office space may be subject to the laws of each jurisdiction in which their home workers live and work as if each home work site was a branch office. Check with HR and legal— if that is the case, they probably come under the employer's home office jurisdiction's laws.

"I say "probably" because that depends what those laws are," said Paddock. "Thus, state income taxes depend on where an employee derives income, while other employment laws, such as workers compensation depend on where a worker lives."

He pointed to two 2003 cases that reach seemingly opposite results. In one, the court ruled that a Florida home worker could only seek unemployment benefits in Florida and could not apply for the better benefits in New York, even though that's where her employer was headquartered. In an income tax case, however, the court held that New York could tax all of a professor's income because his employer was located in New York, even though he did much of his work from home.

Another issue is presence—what constitutes presence within a jurisdiction in order for an organization to be under the laws of that jurisdiction. Paddock said if your firm, for example, simply sells magazines to subscribers in a state and has no physical operations there, your firm does not "do business" in that state, which means you are probably not subject to that state's laws, do not have to pay taxes and cannot be sued in that state. But, if the firm has a call center there, then it has presence.

"These traditional concepts get murky when there are home workers because your firm could have a presence and be doing business in each state any of your home workers live and work," said Paddock. "For instance, many states require companies with a presence in their jurisdiction to collect and pay sales taxes for purchases by its residents. So, having a single home worker in one of those states could force the magazine company in this example to pay sales tax for every sale it makes to anyone in that worker's state."

At this writing it is a gray area whether employees' home offices creates "presence" within a jurisdiction. Paddock pointed back to the example of the magazine if the call center agents work from home—and do not do anything else other than handle customer service and subscription orders (like actually mail the magazines or receive payments)—does that constitute presence?

Jurisdiction Recommendations

If jurisdiction comes up or may come up as an issue, Paddock's advice is to talk to your attorneys and see about including what he calls a "choice of law" provision in your agreements or policies for home working employees.

Employers can elect to have the laws in the jurisdiction where their headquarters are located control many issues, such as protecting the employers' trade secrets and how to interpret the contract. However, regardless of any contract provision, some disputes are likely to be governed by the laws of the locale where the employees live, such as laws giving employees' remedies for unpaid wages or prohibiting non-compete agreements.

"The key is for the employer to understand ahead of time what it can mean if it must comply with the laws in each state where its home workers live. In some situations, there may be downsides but in many other cases more favorable laws in the home workers' states will be additional benefits of home working," says Paddock.

Others, like Jack Heacock, say the best advice they've seen is to have the company issue a letter stating that working remotely is for the convenience of the company and a condition of employment, not a simple "benefit" for the convenience of the employee. Again, talk to HR, Accounting and Legal about this issue. They may say that such a letter may allow you to pass legal tests and to avoid future problems for you and for your employees. In the meantime contact your business and trade organizations and see if they will lobby to end double-taxation.

Occupational Health and Safety Laws

Occupational health and safety laws, too, can be unclear for home workers. You will need to consult with HR and legal.

At this writing, American and Canadian occupational safety authorities have taken a hands-off approach to home offices; Occupational Safety and Health Administration (OSHA) backed off extending its regulations to home offices. But they haven't altogether forgotten home workers; they can still pursue employers, in OSHA's case, under general industry clauses. OSHA also can go after you and your employees for electrical safety violations in home workplaces. OSHA's Design Safety Standards for Electrical Installations regulations, which are based on the National Electrical Code, but made more flexible for enforcement.

One of the best books written explaining this OSHA rule was edited, updated and rewritten by me in 1991 when I was assistant editor, *Electrical Code Watch*, a now-defunct newsletter on the NE Code and electrical safety. The handbook, *OSHA Electrical Design Safety Standards*, is still available from EC&M Books. It contains, for example, the text of the regulations.

OSHA and other authorities will be more inclined to take a closer look at your home working employees if they have other employees working on the home premises. Now you've set up a de facto satellite premises. The home office-owning employee is now responsible for others.

Therefore, I strongly advise that you *insist* that your exclusive home workers have safe, ergonomically sound workplaces—as you should have at your premises. Make that a *requirement* of your home working program, regardless whether home workers are covered or not; assume they are covered to be literally on the safe side. Chapter 6 provides excellent suggestions on workstations, chairs, and appliances like wristpads and headphones to enable compliance to the letter and intent of occupational safety regulations.

Also, if your employees have any electrical work done in their workplaces insist—in writing in your home working policy—they contract with a licensed electrician or electrical contractor. That way you're protected as well as your employees.

Impaired Employee Regulations

Home working enables many people who cannot commute to premises offices because they are physically impaired (sight, hearing, mobility) to be employed. They provide a pool of smart, motivated, and loyal workers.

If you hire a physically impaired individual, the Americans with Disabilities Act and similar laws often have "accommodation telecommuting agreements" that allow you to let them work from home; such employees are not required to travel to your premises. Supervisors will need, if necessary, to visit them at their home offices and deliver the supplies and tools to do their job.

Laws and Regulations the can Affect Home Workers

In managing home working programs be aware and, if necessary, encourage your employees to be cognizant of laws that affect home working.

Tax Laws

In the case of taxation laws it is the employee that is affected more so than the employer; the governments go after the employees for the additional taxes. Governmental taxation departments pay a real close eye to home office deductions—they are always on the lookout for taxpayers who attempt to claim home office expenses when they work at premises offices.

For example, American and Canadian tax laws currently allow *only* exclusive home workers—those that do not have employer-provided workspaces—to deduct expenses like share of rent, mortgage and utilities from their taxes. The Canadian government also requires employers to fill in forms stating what at-home expenses they pay for, which employees can't write off, and what expenses the employees pay for, that employees can deduct.

There are also turf issues on taxation that the previous section on jurisdiction covers. Each jurisdiction jealously protects their right to tax, even if it means double-taxing the same people, like exclusive home workers living in one state but their employer is based in another state.

Local Regulations

You and your home workers need to be aware of local bylaws and ordinances affecting businesses, such as zoning. Most of these regulations don't concern themselves with home workers as long as they are "invisible," i.e. do not have big signs, generate excessive car and delivery truck traffic, disturb others, or have products and supplies on the front lawns.

Many communities, however, do have bylaws aimed at home based-businesses. These laws may have to be complied with if you have home workers who have other employees, clients, and customers who come into the home office and engage in activities other than working on computers and phones (see Chapter 10).

If the above conditions apply to your home workers have them check out the laws and refer that information to you, and in turn forward that information to Legal. Write that into your Home Working Policy.

✪ INSURANCE, AND LIABILITY

You and your at-home workers need to take a close look at insurance and liability. At the very least you should require your home workers to carry homeowners' or renters' insurance, if they don't do so already. That protects their building and personal property.

Require your home workers to get riders to their insurance policies to cover incidental office use to protect them—and you—in case anyone who steps onto their property. e.g. a deliveryperson, computer/telco tech, or anyone else there on business, and who is injured. If your employees do not have this insurance *you* and/or your employees could be sued. Lawyers go after those with the deepest pockets and that means *you*.

If your employees are doing more than just working at their home offices, such as meeting with clients and customers (see Chapter 10), you may need to consider policies covering home-based businesses or get a commercial policy. If that is the case, have your employee contact their insurance broker. Because meeting clients and customers is over and above the usual home working arrangement you should pay for that additional coverage.

To reduce the odds of fire and theft and to ensure work continuity, I recommend suggesting that employees equip their homes with monitored fire and burglar alarms. I live in a safe, semi-rural community and I have them. Homeowners' insurance gives breaks on such systems: 10%-20% in my case.

✪ RECOMMENDATIONS

Talk to HR and Legal when any legal or regulatory issue comes to your attention. Ask them to see if it is prudent to make home working *required*, even occasionally, as a condition of employment, to pass legal tests and to avoid future problems for you and for your employees.

✪ HR MATTERS

Home employees have their own unique issues when it comes to human resources. Let's look at a few of the leading ones.

Compliance with Immigration Regulations

All employees must comply with national immigration laws; new hires must prove they are legally permitted to work in the country that they are being hired to work from. The US requires, for example, that employees show proof, such as birth or naturalization certificates, passports, and "Green Cards," *face-to-face* to employers' representatives.

This requirement can get tricky to implement if the home workers live (and work) a long distance from the premises office. The usual option is for the workers to travel to the premises office, at their expense or yours.

But if you have home workers living outside of traveling distance of your premises office you can contract with a notary public or other such official agent. They inspect the documents for compliance with the laws and sign off on the documentation as your representative.

If there are many such home workers in that faraway area and you expect many more, like for high-turnover/fluctuating volume call center work, then consider opening a small storefront office and have your HR person fly out to process applicants, say once a week or every two weeks, as needed. You may find that storefront useful for other purposes, such as satellite working or as a base for field sales reps.

Proof of Work

Home workers, exclusive or occasional, do not physically sign in and out every day. But you need to ensure that they are there to work for the period you need them to work and that they did work required for that pay period.

There are several methods to ensure this. They include having employees log on and off computers and phones and simply filling out and e-mailing in weekly time sheets (which is what I do).

You can demand that your home employees work at the same times as all other employees. Also, that they notify you if they are leaving their home office during the work period for any length of time.

On your time sheets, you should also have a box or a place for "Work at Home" for your occasional home workers' use. That way their absence from the premises is accounted for.

Hourly/Overtime

If your home workers are hourly you will need procedures to account for their hours. You also will need procedures to initiate, request, and approve employees' request for overtime. For those reasons, if you have hourly workers, you should have formal log-in/out on your phones or servers. Home workers also can call in when they come in on duty and when they go off-duty. I did that in my youth when I worked as a security guard on construction sites.

✪ CHANGING CIRCUMSTANCES

Employees, like employers, don't remain constant. They are hired, get promoted, get demoted or fired. Some are relocated while others are brought back. As individuals they move, marry or otherwise get into permanent relationships, move, have kids, kids move out or their parents move in or out.

For example, you may have decided that only those workers who live beyond X distance or X commuting time from your premises qualify to for your home working program. That makes sense if the reason why, and the justification of your program, rests on improving productivity. If an employee lives two minutes from your town it may take them longer to get their home offices than to your premises.

But what happens if a home working employee moves to within that radius? Is it fair for them, who have otherwise met your parameters and who have been successful working at-home to suddenly be required to yank on the tie and/or the pantyhose and walk away from the investment they and you have made in home working and its equipment?

Conversely, you require employees to come in once a week, or every two weeks. Then one of your workers decides, for whatever reason, to move out of visiting range, like to the other side of the country. Do you terminate that employee or do you find ways of accommodating him or her?

Here is another issue: you have a home working employee who travels occasionally as part of their job. They then decide to move, with your permission, to a smaller community that turns out to have less air service and higher fares. Do you bite the bullet on higher travel costs or do you cut back on the trips they make?

The key to coping with these circumstances is to look at each issue with home working like you do with any other management issue: on a case by case basis. You cannot write rules to cover all circumstances.

Moves (by your employees)

You need to pre-plan in your policies for home employee moves. That includes listing the criteria for home offices including voice/data access e.g. broadband, proximity to premises, including satellite offices, commercial air, ferry or rail service availability.

You should ask your home employees to give you substantial notice i.e. when they have move-in date so you can make any necessary accounting/HR and IT changes at your end. Your employees need to make arrangements with the local voice/data providers in their new locale to ensure the service is live when they move in. Ideally the lines should be installed and functional a few days beforehand. That way if there are any bugs—for example your employees can call the new lines to test them—they can be exterminated ASAP.

Employees should also have the old service switched off, preferably a day or two before they move out just to make sure, thus avoiding any undue charges like the $5,000 call from the new owner's son to Uncle Bob in Antarctica. With wireless service they can still stay in touch, and check for messages at your premises offices: a good reason for having voice mailboxes for your home workers.

You should allow some downtime for moves. But with mobile communications and laptops moving does not necessarily mean a loss of productivity, though you and your employees will need some pre-planning. When I moved from New York City to Victoria, British Columbia, Canada in October 2001 I worked for 2-3 hours every morning in the hotel rooms on my laptop, connected to our e-mail server using my cellphone or the hotel phone. I then drove on for 6 hours a day, making or taking the occasional call (with my hands-free headset or on the side of the road) enroute.

Moves (by you)

When planning your move from your premises offices don't forget your exclusive home workers. If they get calls rerouted from your switch remind IT to make the reroutes in the new premises, or tell the carrier to do so if you're switching to network or Centrex routing. If they are changing the voice mail system notify your home workers too, so they can get up to speed on it. Supply your home workers with your new address ahead of time.

If the home workers also hot-desk, you should allocate the necessary hot-desk space in the new place. Remind the exclusive workers that they may need new security tags.

✪ TERMINATION

Terminating employees—either by firing or laying them off—is challenging. But ending the work for exclusive home employees poses a few more obstacles.

Chapter 15 covers ending home working in depth, but here are a few points and suggestions that are worth repeating.

You may have angry employees, who may want to wreak havoc. To limit that possibility, terminate network access immediately and from the time of notice to the actual departure give them Webmail access only.

If you have physically impaired employees under the ADA accommodation telecommuting agreement, if you end that agreement you are terminating that employee. You can end that agreement if your organization ends the home working program, believes the accommodation is no longer fair if your needs are not being met.

Whatever you decide you will need to have decided your policies beforehand, check them with legal counsel and make employees aware of them when you agreed to have their work from home. Retroactive policies are unfair to everyone.

✪ THE HOME WORKING POLICY

You will need to have worked out all of the elements of your home working program before implementing it. You will then need to formalize this into a written and ideally online-posted *home working policy*.

With a home working policy everyone knows what to expect from home working: employees, supervisors, senior executives and support departments such as accounting, HR and IT. If you are recruiting for positions that are exclusive home office your recruiters can use this policy to stipulate some of the position's requirements and be ready to inform applicants what the position entails should it be offered to them.

Here are a few key components that should be in your home working policy:

* Voice/data and equipment requirements
* Home workplace facilities, environment and furniture requirements
* Insurance
* Confidential data protection procedures
* Monitoring, if required
* Employee assent to home visits by supervisory staff
* Sign in/out and office absences
* Termination

The home working policy, also, should differentiate between the requirements for exclusive and for occasional or emergency home workers. The requirements for occasional home workers are not as great as for exclusive workers. On the other hand, you will need to maintain premises offices for occasional workers.

Chapter 9: Management

Home working may require changes in how you manage your employees. Because employees are not "there" at your premises—where you, your supervisors, and senior executives can see them—you may need to find and set up unique ways to monitor and manage performance, and assess and develop careers.

I say "may require" for good reason. There are many organizations that have long practiced what I call "performance-based management," i.e. managing by the work and the result that employees do rather than whether or not they are warming seats.

For those outfits that require tools to help reassure management that their "out of premises" employees are doing their job, check out the use of computer and phone logins/outs and call and e-mail/chat monitoring. Organizations, especially call centers have long used these tools at premises offices. The only difference with home workers is that the wire—or wireless connection—is slightly longer, with no measurable decrease in performance.

At the same time, you will need to manage the home work program as well as the home working employees. You will need to set goals, isolate, measure, and analyze your program to see how it is working and what changes need to be made. You will need that information especially if you are trialing or gradually introducing home working into your organizational cultural.

✪ WHY "OVER-THE-SHOULDER" MANAGEMENT IS OBSOLETE

Arguably the biggest change many employers will have to make when having employees work from home, whether exclusive or occasional, is how they manage their home workers. Managers cannot as easily look over the shoulder of subordinates if they are not on the premises.

The inability to "see" whether someone is working is the biggest rationale, spoken or unspoken, that top organization executives and low-ranked supervisors give for not having home working.

Over-the-shoulder management is *obsolete* when you're supervising information and knowledge industries; it is a relic of Victorian times. Monitoring and assessing information work is not like peering over a garment worker guiding a sewing machine over a seam or welder wielding a torch on two pieces of steel. When a stitch or weld is complete that's it. They are standalone in the entire process. It is done or it is not done.

But information is different. Each word, each fragment of code is interconnected, it can be connected or disconnected, adjusted or corrected until the task is finished. Thus managers do not have a complete picture of what has been done; or needs to be done to complete the task, because information work is typically a work in progress.

You can't tell by looking at a screen or two what is the full program or story. It may be a draft you are seeing—the employee is working on another story or program and will get back to it. You can't accurately judge the quality of a customer service interaction by listening to five seconds of a phone conversation or looking at one e-mail. You need to hear and read the entire exchange.

Over-the-shoulder management of information/knowledge workers is like back-seat a driver in a big-rig truck. If you keep bugging the driver, questioning the speed, fuel consumption, passing techniques, and how they smell, *at best* you'll be left at the side of the road in one piece.

Yes, you can mount cameras behind the driver's head and either record them or send the images by cellular or satellite data. You can insist that home workers mount cameras in the corners of their work environment and have the images transmitted over their wireline or wireless. But such supervision is like Schrodinger's Cat: the famous illustration of quantum mechanics. Open the box (or story) and you may kill it, in this case by disrupting the employee who could make more mistakes had you not intervened.

Okay, so the employees know the boss can check up on them. *So what?* Does that mean they will do a better job of getting the program done, answering the call, finishing the story, or for that matter getting that load to the customer's shipping dock on time?

That mindset assumes that if employees know the overseer can't crack the whip on their back anytime, anyplace they will goof off. The cat's away so the mice will play.

Think about that for a minute. Does that mean you or your supervisors are going to have to be on their backs continuously? If so, how much time and dollars in productivity is that going to cost you? What will that do to employee productivity and turnover? If you breathe down employees' necks you are going to be disturbing them, making them less productive, and more likely to leave.

You therefore risk setting yourself up for a self-fulfilling prophecy. If you micromanage workers they *will* goof off, but what is more likely they will backstab you and look for opportunities to make you look bad every chance they get. If you or your supervisors treat these workers badly pray that you won't be dependent on them, ever.

✪ THE FUTURE OF EMPLOYEE MANAGEMENT

Supervisors, managers, and directors—anyone who has authority over other employees should *manage by performance*—no matter *where* the employees are working. Performance-based management, in a nutshell, is determining how well your workers accomplished or performed their tasks compared with your expectations and standards.

There are many types of objective performance metrics. They include:

* Customer satisfaction

* First call/contact resolution
* Sales
* Meeting deadlines
* Error rate
* Time/dollars spent correcting
* Employee satisfaction and turnover
* Meeting budgets

Managing by performance is the most effective means of assessing especially information and knowledge workers. Why? Because you cannot make a complete and fair assessment until the tasks are accomplished.

Virtual Management

There is mounting evidence that with home workers the combination of performance-based and virtual management is more effective than the "over-the-shoulder" management style used by many at premises supervisors. While this will be discussed in depth in Chapter 12, here are a few of the multitude of examples of the benefits of virtual management:

* American Express reported productivity increases of 43% plus.
* AT&T generated over $150 million in benefits in 2003 from lower real estate and other overhead costs, higher productivity and improved employee retention and recruitment, which lower staffing costs.
* Each home worker saves Merrill Lynch $10,000 annually through lower absenteeism and turnover.

Not bad company to be in. It does pay to have your workers "out of sight" but not out of mind.

The combo of performance-based and virtual management is fairer, in many respects. Unlike the premises office workers, assessment of home workers' performance is not influenced by employees' pretty faces, snappy suits, cute smiles, firm chins, and tight butts or clouded by the stinky aftershave, the spare tires over their belts, or the bad if inoffensive jokes they tell.

With lawyers seeing $$$ behind wrongful dismissal or passed-over-promotion suits, especially if the employee can prove discrimination on the basis of gender, ethnicity, sexual orientation, and/or size you need to be to-the-letter impartial. Out of sight or *virtual management* puts many of those issues out of mind.

Monitoring Ongoing Performance

Performance-based management *does not mean*, however, that your managers are giving up the ability and the right to monitor and supervise employees, especially those working from home—far from it.

You need to track performance. And you need to be in touch with employees to see how they are doing. By the same token employees need to check with you to see how they're doing.

There are many methods and tools to enable this. They include:

Call/contact monitoring and recording

The exact same call monitoring and recording, e-mail, and chat response buffering tools managers use to supervise their customer service, support, and sales agents at call centers can be and are used successfully to supervise agents (and your workers) at home. You can also monitor instructors working from home if you have them go through your voice conferencing services and Net servers.

Conferencing

Audio and data conferencing among work teams are great monitoring tools (see Chapter 11 on conferencing). With conferencing, everyone—home and premises workers—alike, feel that they are on the same team. Supervisors see, ask questions about, and offer suggestions, on everyone's work; colleagues can see and offer suggestions on each other's work.

The beauty of information is that you can distribute electronic drafts to supervisors and colleagues; noting that these are drafts—not completed tasks. You can do this via data conferencing. You also can have recordings distributed to workers—premises and home workers —to go over what a team did right and wrong and how to make improvements.

I've been conferencing with my colleagues since 2001 and it has worked out great. We will sometimes e-mail, fax, or courier ahead of time text such as story synopses or awards applications before we discuss them during a conference session.

Collaboration

Electronic collaboration, such as through groupware is an up-and-coming management tool. Arguably a subset of conferencing it enables your employees to communicate information in a manner that you see, supervise, control and correct.

Phone, e-mail and messaging

You can and should ask how employees are doing, such as with a program or story. I send in synopses of my stories to my editor. As Chapter 7 notes, there are excellent communications tools: from the traditional phone to instant messaging (IM) to enable this contact and to encourage ongoing dialogue.

IM, to reiterate, is an excellent management and supervision technology, especially when employees and/or supervisors need frequent communication. IM overlays e-mail, not interfering with it; IM takes place in real-time.

Electronic checkup and supervision is easier and less disruptive for employers and employees than looking over the shoulder. It saves time and keeps productivity high, unlike a supervisor walking over and disturbing employees while they are working.

Most premises offices are going to electronic supervision. A study for staffing firm Accountemps in Canada by International Communications Research revealed that 94% of Canadian executives say managers often send an e-mail message rather than meet

Home working employees need to be checked on now and then, either electronically or in-person. Casey does that as a volunteer; she checks in on the author once or twice a day. Credit: Christine Read

one-on-one: 67% said managers often favored e-mail over in-person contact.

I can attest to that. When working at the CMP offices in New York City, I spent literally days working where my supervisors or colleagues did not come to my cube, or I go to theirs. Instead, we e-mailed or, if urgent, we picked up the phone. What, then, is the point (I figured) of working on premises? So, I sought permission to home work, and once permission was granted, began to work from home.

Shared information

Also as noted in Chapter 7, it is vital that your home employees see the same information as your premises employees so they know what is happening, what is expected of them, and can feel that they are part of the same team.

Spectrum Corporation provides realtime reporting and historical data to home workers' PCs via thin-client or browser-based applications. Call centers are big users of this information. Home workers can see, for example, if there are potential buyers-customers that have called before. They can then let supervisors know through instant messaging that they can let these callers queue-jump.

The Spectrum system also does more—it lets call center supervisors know if they need to bring more home workers on the lines. By tracking call volumes it warns supervisors of possible service-level-exceeding spikes. The managers can take that information to see if there are any home workers are available and call those up on their days off to see if they can login. Or they can dial up their outsourcing partner to see if they can hook up their home agents.

In-person monitoring

If necessary, you can ask your home workers to come to the premises offices periodically if the individual tasks they accomplish do not take long, like answering calls and e-mails and can be easily assessed and if they live in traveling distance. You may need to set

ENSURING PRESENCE

A home worker is seldom, if ever, late to work. The great aspect about home working is that employees don't have the excuse of traffic, the cool crackup on the I-whatever, or the infamous "sick passenger" that holds up so many New York City subway trains, for showing up late for work. Stumbling over Poopsie the Pomeranian doesn't count.

However, as covered in Chapter 8 you will need to ensure the home workers are there when they said they will be there, that they haven't left early, haven't taken 3-hour lunches, or gone fishing for the rest of the day. You are paying them to work and for the time that you want them to work.

This is especially important for work such as customer service and sales; if an employee is late or hasn't shown up, additional work is thrown onto their colleagues. Also the work begins to back up, annoying customers and others dependent on the employee being at work on time.

There is a big reservation many supervisors and senior executives have about home working. If the worker are not punching in and punching out, or someone is not at Reception to take attendance (as if the employees were in first grade), how do they know the employees are available to work?

Fortunately, there are tools—ranging from trust that employees are adult and responsible enough to show up and do their jobs, to take the breaks for the periods you specify and follow procedure for extended absences or early departures to automated computer and phone log-ins/outs. Home working employees can fill and send in weekly time sheets.

If a worker—home or otherwise—is not there when they are supposed to, that will show up in their performance ratings real fast. Calls and contacts will get backed up; service levels will drop; work will be late or incomplete; Clients and customers will complain about not being able to get a hold of that employee, or that they are rushed off the phone.

By collecting such hard evidence you will have incontrovertible proof that an employee is slipping up—you or your supervisors are not bringing disciplinary action it because you or they don't like them—and you and they can deliver the message to shape up or ship out. Like back to the premises office, or out the door altogether.

them up with hot-desks (see Chapter 7) to enable them to continue working while there. Employees also find out what is happening with your outfit and check up on gossip.

Also, if your home working employees go to conferences and on business trips frequently and their behavior and performance at these conferences and on trips are key parts of their tasks it may be a good idea to go with them. That way you get to see how they perform, see how they are doing, talk over any issues that need to be done face-to-face. You also can network at these shows and meet with clients. In turn, the employees get that chance to catch up with your outfit.

If you require your staff to visit home office to check up on your home workers—if they assent as a condition of working from home—then that also is a good opportunity to meet and chat with them. Most people are proud of their homes. You can learn a lot about a person by how they live.

Deploying Performance-Based Management

Implementing performance-based management for all employees *wherever* they work requires the following pieces:

Having the right people

The key to performance is people. You need to have responsible, self-starting employees who have the skills and the ability and willingness to learn. They also must be motivated to succeed, and sufficiently mature to show up, stay, and leave on time, in accordance with your rules and with supervisor requirements.

The previous chapter, Chapter 8 looks at the special skillsets for home workers, especially exclusive home workers. They must be self-starting, self-disciplined, resourceful, motivated, and professional. These workers also need problem-solving abilities, like fixing equipment. And they must not be shy—they often can't wait for others to come to them.

Home workers also must have locations, facilities, and environment to succeed at home working, which is the focus of Chapter 5. They need the proper non-disruptive and safe workspace, ergonomically-sound furniture, minimal noise, and the right lighting to perform effectively (Chapter 6).

Equally importantly, home working employees **must have** experience and knowledge to work on their own without supervision and without annoying their co-workers. They need the necessary skills and training to succeed at their work. Because when they're at home there's no cubicle or office to run to or people whose shoulders to tap or cry on.

Having the right attitude

To make performance-based management work, especially for home office workers, you and your staff need to regard your home workers as *professionals*. You need to *trust* them, as adults, to get their tasks accomplished to your satisfaction or better. Thus you need to *empower* your employees to make decisions, including the occasional mistake. Mistakes, handled correctly, are often opportunities to learn and to do better.

Home working *will not work* if your organization and your managers feel they have to look over the shoulders of their employees and bug them every few minutes. Such micromanagement is ultimately counterproductive because it wastes everyone's time and, as such, lowers productivity.

If you treat people like adults, like professionals, then they will behave like adults, and professionals. Only when you do will you get the exceptional performance that nearly every employee, in one form or another, is capable of producing.

Setting goals and objectives for all of employees.

These goals should tie back in the efficiency and effectiveness measures for your department. For example, meeting deadlines with the work done correctly enables all of your staff, regardless of where they perform their tasks, to complete all of their work on time, allowing you to avoid higher production costs, to satisfy your customers, and to increase revenue. Resolving service and sales issue on the first call to a call center avoids costly escalations to other departments or repeat calls, while keeping those customers who might walk away (with money that could have boosted your revenues) if they believe they are not being served well.

Having employee communications programs in place

You need to figure out which methods of employee contacts that you and your colleagues will use. That can be as simple as e-mail, IM and phone and be slightly more complex like collaboration, conferencing, internal newsletters and personal visits.

Training supervisors

To make home office working and performance-based management happen, your line supervisors must be brought on board. They can make or break any program. They need to be trained on how to manage by performance, measure results, how best to communicate, how to stay in touch with home office and other remote employees and how to stay on top of problems and correct them.

Encouraging employee health and well-being

To keep your home workers performing effectively you should consider asking them to take some time to limber up. That includes allowing time to go on an exercise machine, take a quick run or brisk walk. They will feel better and more refreshed and perform better. That will help you also to save money by reducing sick time and medical costs.

Note: *A well-managed organization will make home working function well; in turn home working will make it a more successful organization. But a poorly-managed organization will make home working fail.*

Career Development

By managing by performance (by the goals set and by their colleagues' performance), rather that looking at "seats in seats," managers can fairly gauge how well their home employees are performing. If the home workers exceed the standards and show other skills then they could and should be considered for advancement.

But that's the easy part. As noted in Chapter 8, the hard part is that exclusive home workers, especially those who live outside of travel distance to your premises, risk being out of the loop on advancements and promotions that they could put in for and which you could support.

To some extent this is the tradeoff that employees who want to work from home make. And chances are those employees will know the extent of the tradeoff.

There are two categories of employees. The first is *career-builders*, employees who are still growing their skills and knowledge and that want to advance and that you recognize as having advancement potential. The second are *cresters*, employees that do not wish to advance and/or have no potential of doing so—their careers have crested in the firm—but who are good employees that you want to retain.

Of the two, cresters are the most likely to want home working. The reason: they no longer are interested in hanging around the water cooler waiting to hear about new openings, nor do they want to be in your face to impress you. They are after instead quality of life, which home working offers.

There are and will be more cresters than career-builders. With the flattening of management ranks over the past 10 years to save money, and with much higher employee-to-supervisor ratios, there are fewer opportunities to move up. Home working offers a way out.

But career builders can succeed at home working (see next section). Home working can be especially helpful for career-building parents and caregivers. The home environment is ideal, for example, for career builders who just had children. They can work part-time at home and look after their children when they are not working—until the progeny are old enough to be on their own.

But at some point, most career builders will start feeling out of the loop and want to return to your premises offices. They know that by working at home they can't easily nip out to the next department's office to apply and be interviewed for a new position. But as more organizations provide home working, and as more directors and executives manage their charges remotely, the more being at home will become less of a hindrance to ambitious employees.

You or your supervisors, if they are any good, will know which employees are career-builders and which are cresters. Ask both if they are prepared to commit at least a year to home working; ask both if they have any reservations that working from home will hinder their careers.

Home Working can be a Career-building Asset

There is growing evidence to suggest that home working *builds* rather than inhibits career growth. Home working leader AT&T reports that managers in these "virtual offices" are more likely to be promotable than managers in traditional premises offices. Nearly one in three managers reports that teleworking (including home working) has had a positive effect on their career.

The reasons are multifold. Among them: home workers are more motivated, productive, work longer hours, and take fewer sick days. They also are more self-disciplined and good problem-solvers. And they take the initiative to communicate with others.

AT&T reports that home-based managers worked *7.6 hours* per workday compared with *6.8 hours* for all managers. Home workers gained about *1 hour* in extra productive time.

Furthermore AT&T found that their home workers are happier and more loyal. Over 63% of home working managers report increased job and career satisfaction; 47% who had received competing job offers factored in working from home into their decision to stay within the company.

In short: ***successful home workers show the qualities needed for successful careers.***

To help your home workers to build a career, you must make sure they are aware of opportunities. They should be told, through e-mails and other communications, of any openings. If you have an aspiring candidate, you should bring him or her to the appropriate manager's attention.

Keep in mind many of the positions that do come up, either promotions or in other departments, may require employees to return to work at premises. Those employees may decide the tradeoff —returning to the Monday morning grind—isn't worth it. But let them make the choice.

But as more work becomes home worked, there is a growing chance that those employees may not have to make that tradeoff. That's win-win-win for you, your organization, and your home working employees.

Seeing and Coping with Problems

Because you or your supervisors do not "see" your home working employees you may not feel you can detect problems they might experience, especially difficulties with the home working arrangement. But there are some indicators of such problems. They include:

Dropping performance, such as sloppy and late work. Late log-ins and unannounced absences, like when you called and they weren't there and didn't call back for a long period of time. Complaints such as not responding to calls and e-mails and of background noise are yet another clear indication that home working is not working out for a specific employee.

When these issues arise you will need to get in touch with your employee *immediately*. Determine the cause of the problems and find out if it relates to the tasks or to the home environment. If the latter, discuss the problem with the employee and see if a change can be made to improve the situation. That can include getting them to sign in and out, respond to calls and e-mails quicker, and doing something about the noise sources. Where home working is a privilege, remind the employees that these issues need to be worked out if they are to continue to work from home.

If the matters cannot be resolved, then give the employees the options of quitting or coming back to the premises offices. The last chapter, Chapter 15, outlines suggestions how to handle terminated home employees, including patriation of equipment.

This is one of the reasons trialing home workers is a great idea. When you trial home workers first, say letting them home work only one or two days a week for a month before letting them home work exclusively, you can see fairly quickly any differences between their performance at home and at premises.

✪ MANAGING A HOME WORKING PROGRAM

The other side of home working management is running the home working program. Chapter 13 discusses planning home working, while Chapter 14 outlines rolling the program out.

Setting Goals for a Home Working Program

You will need to set goals for the program to achieve. These goals are manyfold. They can include:

Enabling work when employees have to be at home

This is one of the two most common goals of home working and goes without saying. But for a goal to be reached, there first must be a plan.

Nearly every employee, at one time or another, needs to be at home during their workday. The reasons include waiting for furniture to arrive, the rugrat is home sick, the car died and there is no mass transit, they have the electrician or plumber over, or they need to go to the doctor to get their own plumbing (or wiring) checked out and repaired. Or these workers are not feeling well enough to commute: they may have a communicable disease. But they are well enough to get some work done. Who *doesn't hate* going through the barrage of voice mails, e-mails, and faxes when they get back into the premises office?

Home working enables employees who are on maternity leave or long-term sick leave to stay in the loop, and productive. That is very important; many people who are out of the office for long periods of time feel they get behind in their careers when they are not in the office.

Attracting, meeting the needs of and retaining staff

This is the other most common goal of home working. Employees can be exclusive as well as occasional home workers.

Organizations have invested too much in their good people to let them go. With the *experienced* labor supply being what it is today, most outfits will fight to keep their trained employees.

Home working becomes an attractive benefit that costs employers comparatively little compared with the wage-and-benefit-bidding wars and pricey amenities like on-premises gyms, and day-care. Yet it saves employees to the tune of $4,000 to $5,000 a year in transportation, clothing, and cosmetics costs.

Achieving this goal requires employers to examine and decide on employee eligibility, either formally or on a case-by-case basis, especially if an employee will be working from home exclusively. Key amongst the criterion, as discussed in Chapter 8, is whether the employee has the skills and the self-discipline to carry out the required work tasks without in-person supervision.

Employees must also satisfy their supervisors (and you) that they have the facilities, environment, equipment, and if need be, the location to work successfully from home (Chapters 5 and 6). That includes being willing to make any investments on their part such as in ergonomically-sound workstations and chairs, lockable file cabinets, and shredders.

Business continuity planning

Being prepared to have your employees work from home (even those that prefer to work from your premises offices and/or whose functions require them to be there) in readiness and in response to disasters is a sound strategy. It should be part of your business continuity planning. As pointed out in Chapter 12, you can save lives and keep your operations going that way.

But you need a plan so as to minimize chaos and to respond quicker when a disaster strikes. As one of many who lugged a laptop from their offices while evacuating from Manhattan on September 11, 2001, I can testify that the strategy works. My company already had a home working plan, and support for such plan, e.g. people and technology such as virtual private networks (VPNs), in place prior to the World Trade Center attack. When I ended up staying with a friend of my wife's in New Jersey because I could not get back to my home on Staten Island, I hooked up the laptop, entered the password on the VPN client, and I was connected.

Meeting defined objectives

A common reason by organizations to have a home work program is to meet defined objectives. They include reducing tardiness, lowering sick days, improving productivity, shrinking turnover, meeting government-mandated employee trip-reduction requirements to diminish air pollution, cutting parking demand, and downsizing property costs.

To meet these defined objectives will likely require employers, like yourself, to set targets as to how many employees you want to work from home, exclusively and/or occasionally. You then have to figure out what functions can be worked from home and find out how many employees will and are able to do so.

Your analysis may indicate that not enough of your current employees are home workable to enable you to meet the defined objectives. You will then have to make some possibly difficult choices: like letting go lower-performing employees who can't work from home—just as you let go those who can't commute to a new location if you move your premises offices.

Note: *Consider modest goals, e.g. enabling work when employees have to be at home, FIRST. Many organizations want to see whether you can crawl before you can walk, walk before you can run. Otherwise you run the risk of falling flat on your face.*

Measuring a Home Work Program's Performance

You may need to set up measures of home working performance based on the goals that you want it to achieve. If you do that, then you need to set up comparable measures of premises working performance. That way you can see if it is working and justify it to your superiors.

I say "may" for a reason. You may decide that all you need home working to accomplish is to accommodate occasional home workers or those valuable employees who can no longer, or wish to no longer, commute to your premises, in which case you probably don't need exacting measurements.

As noted in the Introduction and throughout the book, and explored in depth in Chapter 12, home working has many advantages including increased productivity, decreased tardiness, less sick time, lower real estate costs, lower turnover, and reduced environmental damage.

The key indicators that are measurable and that should be considered when determining a home work program's performance include:

Productivity

There are several ways of tracking productivity, by your own internal metrics including customer satisfaction and turnover, by comparing availability, and employee contribution. Let's look at some of those methods in more detail.

Internal metrics

Nearly every organization has means of measuring employee performance. Because you know where your workers are located, you can isolate the home workers from the premises workers.

Some of the best examples come from call centers, an industry group that is fanatic about measuring monitoring performance and productivity. PHH Arval's 10-member home working Elite Action Team achieved a 97% customer satisfaction rating. That is nearly 7% higher than at-center agents. That team won *Call Center Magazine's* Call Center Leadership Award for best team in 2002.

Procter & Gamble (P&G) saw productivity grow between 5% and 10% after it implemented home working for its consumer call center in 1999. The gain allowed the company to close one of two call center floors after a three-year pilot with home working.

Turnover

Employee turnover is a very effective productivity measurement. You can put a hard cost to it. Turnover typically costs organizations anywhere from 25% to 250% the departed employees' compensation in recruiting, selecting, training, and in lost productivity as you bringing up to speed replacement employees.

As with internal metrics, you can isolate the employee location variable, e.g. premises office workers, premises office/occasional home workers, and exclusive home workers within the same job titles and responsibilities. You can track and compare tenure length. You have (or should have) established a turnover cost per employee.

Let's say the average compensation per employee in a particular job title, e.g. programming, is $50,000 per year. The turnover cost is $80,000. You have 20 programmers. If all of them left in one year (or 100%) the cost will be $1.6 million. Your annual turnover rate for premises workers who are programmers is 20%. Your total turnover cost is *$320,000*

Let's compare premises workers with exclusive home workers. You found the turnover rate is 4%. Your total turnover cost is *$64,000.*

Annual premises office employee turnover costs:	$320,000
Annual home office employee turnover costs:	$ 64,000
Annual savings with home workers:	$256,000

Employee availability

Comparison of employee availability can be achieved by recording and analyzing on-time arrival and departure, and absence for each employee group: exclusive premises, exclusive home and occasional home workers. Then match the time lost or gained against their compensation (wages/benefits). Simply put: employees who are not there when you expect them to be are not productive. They are therefore costing you money.

Employees who are there before and after you expect them to be add productivity; they are adding to your bottom line. There is increasing evidence showing that home workers do just that: they work longer on workdays, are likely to work weekends and holidays, and are more likely to come on duty when asked.

Here's a sample calculation. Assume that the average employee compensation (wages/benefits) is $40,000. In devising the plan you found home working improves productivity by five hours per employee per week (The Chicago Sun-Times reported in 1999 that AT&T's home-based employees worked that additional amount of time compared with premise-offices employees). That amounts to about $19.23 per hour per employee.

$19.23 x 5 work hours per week x 52 weeks/year = approximately $5,000/year per employee in productivity gains.

Now assume there are 100 employees eligible to exclusively home work.

Annual gains per home working employee:	$5,000
Number of employees	x 100
Annual availability benefits from home working	$500,000

Employee contribution is a different measure of productivity.

Employee contribution

Contribution is the *value* per employee. Consultant John Edwards recommends this measure because it gives a truer mark of what an employee is worth. Remember that's why you have employees in the first place: you pay them compensation to give you value.

How you set the value scale is up to your organization. One method is to base it like you do for bonuses: you scale the bonuses based on employees' compensation level or rank, e.g. managers are worth X and directors are worth 2X.

Edwards recommends that for businesses, you express the employee contribution value as *net sales contribution, after returns and allowances, per worker.* The easiest way to calculate this measurement, says Edwards, is to take the total net sales, after returns and allowances, and divide it over all employees in that business unit or department to get the net sales per person. Then you divide that per hours worked.

Never mind about those employees who do not directly contribute to sales, i.e. programmers, support, says Edwards, even though the effort of each of those workers contributes to the success of that department.

For governments and non-profits, the equivalent measure is *budget value per worker.* To get that you take the budget totals and divide that over each department's employees, advises Edwards. Then divide that over the hours each employee worked.

Environmental impact

Another key measure is environmental benefits. Burning just one gallon of gasoline produces, for example, 19.64 pounds of carbon dioxide CO_2 (a leading greenhouse gas), plus 9 pounds from upstream refining, distribution, and refueling, reports the Surface Transportation Policy Project.

You will know or can easily determine through employee surveys and parking lot checks how many people come to work on your premises by private vehicle (alone or shared), mass transit, cycling, walking, and electronically (i.e. home working). You are most interested in the number of employees that commute to work by car.

As a general rule, home working is most attractive and "cleaner" option to driving, especially where premises offices are located in edge city locales, far from mass transit, and out of cycling or walking distance of homes. But home working also is very attractive to people who have very long mass transit commutes. For example, it took me 90 minutes to get to my Manhattan premises offices from my home in the New York City borough of Staten Island because the borough lacked a direct rail link with the rest of the city. Many people also drive to mass transit park-and-rides, which incurs emissions.

Check with your local environmental protection department or with activist organizations such as the Sierra Club to get the latest data on commuting pollution impacts. Also check with AT&T. The telco and home working leader provides a Telecommuting Calculator (www.att.com/telework/calculator).

In the calculator your employees enter their vehicles' miles per gallon and miles to work round trip. For example, if an employees' vehicle gets 20 mpg and the miles to work roundtrip is 15 miles, the calculator offers a breakdown on CO_2 emissions saved per frequency of home working. Here is an example:

1 day/week	736 pounds per year
2 day/week	1472 pounds per year
3 day/week	2208 pounds per year
4 day/week	2944 pounds per year
Full time	3680 pounds per year

As more people work at home, calculate the amount of fewer pollutants that go into the air. The less pollutants, the lower the environmental impact/costs.

Because some employees carpool or drop off friends, spouses, etc. on their way to premises offices, you will need to find out how many do so, and how often, and then split up the environmental costs. For example you have an employee who drops their loved one at school or their premises offices on their way into your premises. The environmental savings attributable to you is the mileage and gas consumption saved from the drop to your parking lot. The loved one may drive themselves instead or take mass transit.

Transportation impact

You can also assess traffic impacts caused by commuting. Chances are, your employees drive part or all the way to your premises; they—and you—thus are part of the traffic problems. To check to see how much of a part your organization plays, calculate the impact of each additional vehicle attributable to your premises workers—on congested roads and additional delays, and the costs of those delays per commuter.

Check to see if there is methodology that calculates the impacts of each additional vehicle on congested roads and additional delays, and the costs of those delays per commuter. Your local transportation authority may have that data for your area.

The 2003 *Urban Mobility Study* by the Texas Transportation Institute uses constants of 1.25 persons per vehicle, average cost of time of $13.25 per person-hour.

Cost comparisons

You will need to determine, from your accounting and IT department especially, how much it costs operationally to support exclusive home workers, exclusive premises workers, and occasional home workers. Those costs include voice/data, equipment, supplies, support, and administration.

There is a trunk support cost for *all* employees, *regardless* of where they work. These expenses include administration, HR, and supervisory.

But be careful here. *Do not* assume re your trunk phones, computer equipment, support, and furniture—home workers, depending on the program, may be responsible for such expenses.

Your analysis, therefore, should branch off into work types, which can include mobile work. You attach to those "branches" the "leaves," i.e. each cost item such as hot-desks (apportioned per exclusive, mobile or visiting employee), high-speed Internet, laptops, and cellular data.

Real estate performance

Real estate is a tricky cost to measure. Only if your organization is faced with moving does real estate *generally* comes into play re a home working program. This is because you can now calculate into your real estate metrics how much you would save if a percentage of your employees worked from home.

Let's use an example from The Boyd Company, a leading and cutting-edge site selection firm that was cited in *Designing the Best Call Center for Your Business*, which I wrote for CMP in 2000. Here are the costs that are real-estate-relevant.

The analysis looks at the costs per year a 150-employee call center on 30,000 square feet (sf) in Chicago, IL. There is no assessment for allocated HR or IT costs.

Electric power costs	$50,400
Office rent:	$488,125
Heating and Air Conditioning	$36,025
Total real estate costs:	$574,550
Total annual real estate costs per employee:	**$3,830**

Those costs have to be paid regardless of how many employees are working where. But you can use this analysis to determine your *opportunity cost*.

The opportunity cost measures for every person working from home exclusively, excluding the space for hot-desks (if you have them), and how many employees you can fit in who have to work at your premises exclusively. That is the cost that you would have to pay for that employee to work on premises.

The opportunity cost comes into play *especially* if your lease terms allow you to sublet space; or if you are faced with renting another office in the same market for another department; or if you need to add adjacent space. Now you're talking *real* money.

Let's go back to the Chicago example. Let's assume that 100 of the 150 employees work at the call center's premises and none hot-desk, i.e. there needs to be one workstation for each of the 100 employee. But the other 50 employees are exclusive home workers, and there are no hot-deskers. You sublease at cost to you; no being "mini-landlord."

Total annual real estate costs per employee:	$3,830
Number of home worker employees:	x 50
Total annual real estate savings from home working	**$191,500**

Deploying Methodologies

To see if a home working program will meet your goals, you will need to decide on and set out methodologies and information collection mechanisms to tabulate results. These methodologies and mechanisms are vital, especially for pilot programs, which can help you make the case for, and help tweak your overall plan.

For example, if your goal is to reduce tardiness then you need to develop and deploy methods to track attendance: in premises offices and home offices alike. Those methods can be as simple as home workers sending in an "I'm here" e-mail and receptionists writing down arrival times; or as complex as having both premises and home employees logging into servers, and recording log-in times.

If your goal is to improve productivity then you need a means to assess and track that productivity—the methodology for which may or may not already be in place. Such productivity measures include customer satisfaction scores in customer support: how did enabling employees to work from home improve customer satisfaction? They also include fewer mistakes, such as data entry and programming; did working from home have any impact on the rate?

If your goal is to lower turnover and increase the length of employment then you must carefully track those results and accurately compare them with projections and with premises office working. You will need to conduct exit interviews with premises and home workers to determine why they left, what was the impact of home working in their decision.

If your goal is to reduce environmental damage or traffic congestion then you must set up methods to determine how many employees use which mode of transportation and how often and for what distance. That could include employee surveys and calculation methods such as AT&T's Telecommuting Calculator.

When developing these methodologies see what your peers are doings and, if applicable, deploy similar tracking and results-gathering methods. That way you can benchmark your performance against those of other organizations, including your competitors.

Senior executives want to know, as well as you, how your home working program compares with others so everyone can know if it is on track. If your program is better than the rest, they can learn from you. If it is not, then you can learn from them steps you might take to improve your program.

A home working consultant can assist you in devising the methodologies and doing the benchmarking, including selecting appropriate peer groups. That way you can accurately obtain and track the results, and compare the data "apples-to-apples" with other organizations.

Assessing Home Working ROI

When managing a home working program, especially a large scale one, you will need to assess and determine all of your costs and do a return on investment for it (ROI). Senior executives will want to see this data to justify home working.

Here's one model I devised, which I call "Infrastructure Cost per Employee" (ICE). There may well be others. I recommend working with an experienced home working consultant to help determine the model that functions best for your organization.

Determining Infrastructure Cost per Employee (ICE)

Organizations will want to know how much it costs to support home workers. That includes voice/data/equipment, communications, administration, HR, IT, management, and in the case of hot-desks and satellite offices, facilities to support home workers.

They will want a comparison with the costs to support premises workers, which *must* include those costs *plus* power, onsite equipment, furniture, maintenance, parking, transportation, security, amenities such as child care and gyms, and the big one, real estate. If organizations *really* want to get a fair and accurate assessment of premises working costs, they also will include environmental damage and traffic impacts—both of which will literally be paid out of their hides.

One means of assessment is to determine, for comparison purposes, how much does it cost to support each employee, whether in premises offices (including satellite offices), or home offices. In the case of occasional home working employees, you need to know how much more will it cost to support them in home offices, on top of what it costs you to support them in premises offices.

The basis of comparisons like ICE is calculating the cost to support the employees' means of production, which in the information/knowledge industries is the workstation: furniture and chair. Each workstation takes up space that has an assigned value and cost: rent or in the case of employer-owned buildings, allocated capital or mortgage costs.

Each workstation requires voice, data, and power. Each workstation typically has a computer and "communications device," i.e. a phone or e-mail/IM.

The components of ICE includes rent, furniture, voice/data connections, equipment, maintenance, transportation (parking and employer-subsidized mass transit), and directly or indirectly, subsidized amenities such as food service, child care, and exercise facilities. ICE can (and should) include office supplies, chairs (which get a lot of wear and tear), and per-employee-allocated IT and HR costs.

The analysis looks at and lays out which costs are higher or lower: premises office versus home office. It also examines how you propose to support your home workers.

For example, home office voice/data costs may be greater than premises offices if the assumption is made that calls will be transferred directly to home workers. That raises the ICE for those employees. But if a cost/benefit case for installing VoIP shows a net return, say in one year, that takes away much of the higher ICE for home workers.

To obtain a clearer analysis you need to break ICE at the premises side between *premises unavoidable ICE* and *premises-avoidable ICE*. You can calculate and compare ICE between premises and home working, or *Home ICE*.

Premises-unavoidable ICE

Premises-unavoidable ICE are those infrastructure costs—rent, power, equipment, heating, plus voice/data costs—that have to be paid *regardless* of how many, if any, employees are working from home (you are committed by lease to be at your premises for a given period of time). That goes for occasional home workers and—to an extent—to exclusive home workers.

To calculate premises-unavoidable ICE, first you take all those costs per period of time (per year/per month) and divide them per workstation: the means of production. If there is more than one employee per workstation, say on shifts, then you divide the number of employees by the number of workstations.

Let's do a very rough ICE calculation based on the Boyd study, with additional costs brought into play. Amortization includes workstations, phones, and computers. There is no assessment for allocated HR or IT costs.

Electric power costs	$50,400
Office rent:	$488,125
Equipment amortization:	$960,000
Heating and Air Conditioning	$36,025
Telecom	$1,578,123
Total	$3,162,673
Annual Unavoidable ICE	**$21,085**

Premises-avoidable ICE

Premises-avoidable ICE refers to those directly-property-related costs that can be avoided if the organization can or decides to reduce existing or future property expenses with exclusive home working. Those directly-property-related costs include rent, furniture, maintenance, and amenities.

Having a portion of the staff become exclusive home workers could enable employers to add staff without adding space or moving into a larger building. Or it could enable employers to move to a smaller building or consolidate two or more premises into a single building to save on property-related costs.

This scenario arises when the lease is coming up for renewal, or if the organization has sublease clause or an escape clause and decides to exercise it. It also applies if an organization owns the building and can sell it.

Unless an organization goes completely virtual, i.e. no premises offices altogether, they will need some office premises somewhere, if only for administration, training, and possibly hot-desks. They (you) will need to allocate those costs for those employees.

Let's look at the Chicago example again. This time we will assume one shift and that 100 of the 150 employees can exclusively work from home, and no hot-desks. They live within the Chicago local calling area so there are no additional long distance charges from rerouting or transferring calls. You own the computers, but the employees own the phones and furniture, cutting your equipment costs per half. The calculation assumes no layoffs, but realistically, you can probably cut an administrative position or two.

Those employees who now work at home had consumed 20,000 sf of space. You now need only 10,000 sf of premises space. You can probably get that space for less too, but your heating and air conditioning costs may be more, so it's a wash.

Electric power costs:	$33,264
Office rent:	$332,162
Equipment amortization	$480,000
Heating and Air Conditioning	$23,777
Total Avoidable Expenses	$869,203
Annual Avoidable ICE (allocated amongst all employees)	$5,795

Net premises ICE

Net premises ICE is how much infrastructure costs per employee you will have to allocate with a home working strategy. You will then add those costs to the Home ICE.

Once again, let's look at the Chicago example, with 10,000 sf of premises office space, 150 employees, 100 working at home, 50 working on premises. The telecom costs are split 2/3 home office 1/3 premises office. The premises employees will eat up 1/2 of the reduced equipment costs even though they are 1/3 of the work force because they need workstations, chairs, voice/data cabling, coffee machines, fridges etc. that exclusive home employees pay for.

Electric power costs:	$17,136
Office rent:	$155,963
Equipment amortization:	$240,000
Heat and Air Conditioning	$12,248
Telecom	$520,781
Total Net Premises Expenses	$946,128

Annual Net Premises ICE	
(allocated for all employees)	$6,308
(premises employees only)	$18,923

Home ICE

Home ICE refers to infrastructure costs incurred in supporting home offices. They can include, depending on your policies and what you work out with your employees, voice (including off-premises extensions and VoIP gateways at the premises side), data (e.g. home cable or DSL connections) and equipment.

There are additional infrastructure expenses you may wish to add for exclusive home workers. These include their share of HR and IT support.

Home ICE also comes in *unavoidable* and *avoidable* flavors. The unavoidable variety assumes premises-unavoidable costs (see above) especially including equipment amortization costs.

But let's assume in the Chicago case with 100 employees working from home, that the employer picks up the employees home working voice and data line charges, including outbound long distance. That adds 20% to telecommunications costs because there are still economies of scale in supplying phone lines to premises offices (though with advances like affordable network/intelligent routing and VoIP that may disappear).

Unavoidable Home ICE:	
Additional telecom charges	$315,625
Additional Home ICE	$3,156

Total Premises and Home ICE	
(allocated across all employees)	$24,241

The *avoidable Home ICE* looks at a scenario where the organization needs to add staff but doesn't want to add space, or wants to cut property-related costs. It examines what the ICE are for home workers and for all employees with that strategy.

Let's return to the Chicago example, assuming the above amortization cost split between net premises and home costs, with home employees incurring lower costs. Let's also assume that the telecom charges will be a wash: the 20% higher charges for home workers in unavoidable home ICE are made up by lower economies of scale in the smaller premises.

Equipment amortization:	$240,000
Telecom	$1,057,342
Total Home Premises Costs	**$1,297,342**
Annual Avoidable Home ICE	
(allocated for all employees)	$8,649
(home working employees only)	$12,973

Analyzing ICE

The way to analyze ICE is to line up the premises-unavoidable and premises-unavoidable costs to one side and the Home ICE costs to another. Let's look at the Chicago example:

Annual Unavoidable Premises ICE	$21,085
Annual Unavoidable Premises ICE	
with Unavoidable Home ICE	$24,241
Home ICE "penalty"	$3, 156
Home net "loss" (x 150 employees)	$473,400
Annual Avoidable Premises ICE and	
Home ICE	$14,957
The 'Home ICE "advantage"	$6,128
Home net "win" (x 150 employees)	$919,200
Win over loss	$445,800

Calculating Productivity Measure into ROI

Infrastructure is only part of the equation. The other key element is putting it together with productivity, employee contribution, turnover, and other performance gains. By calculating productivity measures, you may find that a home working program will pay off even if you still have to pay rent on premises offices.

Let's go back to the Chicago example of 150 employees and look at how this lines up:

The annual unavoidable infrastructure cost per employee (ICE)	$21,085
Additional Home ICE (for 100 employees working from home)	$3,156
Total Premises and Home ICE	
(allocated across all employees, per all employees)	$24,241

Now let's calculate productivity gain. Go back to the earlier example and assume that the average employee compensation (wages/benefits) is $40,000. In devising the plan, you found home working improves productivity by 5 hours per employee per week. That amounts to about $19.23 per hour per employee. Again there are 100 employees eligible to exclusively home work.

Let's run the numbers:

$19.23 x 5 work hours per week x 52 weeks/year = approximately $5,000/year in productivity gains.

$5,000 per employee in productivity gains

–$3,156 in additional Home ICE (home working infrastructure costs

$1,844 in net gains per home working employee

x 100 home working employees

$184,400 in annual benefits from home working

Note that by applying just the compensation as a parameter, the benefits are if anything *understated*. The amount does include turnover costs or additional employee contribution gained.

The Home Working Program Manager

As you can see, there is a fair amount of work involved in setting up and running (especially a large scale) a home working program with defined goals such as real estate savings. But if your program is small, such as supporting occasional home workers who must be at home for the phone installer, or to support valued employees then your line managers, your existing HR and IT staff may be able to can handle it.

Chances are, though, your home working program will need someone to take responsibility for it. That "someone" (and possible support staff for that someone) will need to: put together the program, including working with other departments, *especially* HR and IT, monitor its performance, and suggest any necessary changes to the program.

The home working manager position must have the same stature as the facilities manager: the jobs are parallel. Just as facilities managers put together building policies, your home working manager should write up your home working policies.

The home working manager also must work with your facilities managers, especially when there are employees who need to hot-desk it or if your firm's policy calls for you to supply the home workers with furniture from the premises. Facilities people need to know how many people, and be able to track furniture allocation.

Other departments the home working manager needs to work closely with are HR and IT. The issues faced include logging in/out procedures, ID tags and whether to issue them for exclusive home workers, laptop allocation, and termination.

The home working manager must troubleshoot the program, run interference with IT and HR in case there are problems, like a home worker taking too long of an absence during the day. They also should insist on being in the loop if there is a phone system upgrade to ensure home workers are not left out.

Most importantly, home working managers need to train supervisors and front-line employees on home working methods including home office setup, responsibilities, and management. They need to check in on supervisors and employees to see if the program is working well, and if there are problems, bring forward to line managers suggestions on how to resolve them. Similar to the way Facilities, HR and IT works.

Home Working Advocacy

Your home working manager must be the home working program's *advocate*, just as department heads are those departments' advocates. Your chosen home working manager must be convinced that home working benefits your organization, your clients/customers, your employees, and society.

Having a home working advocate is *vital* for a home working program to succeed. In every organization the internal dynamic is kill or neuter new programs that are fundamentally different from the "traditional" ways of working, because people, at all levels, rarely want to change. That dynamic exists to limit or prevent damage to that organization caused by changes that could harm it. But by having someone there who pushes for home working, who tracks results, troubleshoots problems, and proclaims the program's successes, the better the chances of home working succeeding.

Home Working Manager Qualifications

Ideally, the home working program manager is an existing and very capable manager who is respected by and has great working relationships with other departments and senior executives. The job is half-manager, half-salesperson/politician.

If your organization's media/P.R. policies permit managers to ask act as spokespeople, the home working manager needs to be media-savvy. Home working can gain a lot of positive press because it is family-friendly (see Chapter 14) by enabling workers to spend more time with their loved ones, it reduces traffic congestion, and it lowers pollution. But the person should know, or be willing to be trained by media/P.R. staff, to respond to negative questions, such as impact on local businesses if more people worked at home. (The answer: these employees will be more likely to come into the local communities at their leisure if they don't have the hassles of commuting.)

Alternatively, you can have your home working manager as a secondary source to your spokespeople—the person they go to, or to direct media to, for answers. That may be a more appropriate strategy for your organization.

If the plan is small, and there is some slack in management, the individual could undertake the home working program in addition to their other responsibilities. But for other than the small program , there should be a separate paid position created for the home working manager.

Chapter 10:
Special Circumstances

There are unique home working circumstances that present their own issues, and they require different solutions to handle, which you will need to work out with your employees. The object here is to strike the best balance between your employees' needs, your needs including the need for you to keep a valuable employee, your budget, and the bottom line (e.g. keeping or enhancing performance). Remember the mantra...

○ INTERNATIONAL HOME WORKING

With today's globalizing economy, international home working is increasingly becoming a reality. There is very limited free trade in labor in North America; in Europe, however, nationals of one European Union nation are able to work in another.

Sometimes your employees may need to move to another country due to their spouse's needs, family, homesickness, lower cost of living, better quality of life. Sometimes you may need to hire someone who lives in another country, but cannot bring them to yours because of immigration hassles. Or perhaps you need a person to cover a specific foreign market, such as an editor/journalist, trainer, salesperson, or technical support. If that country's languages are not English or Spanish, you may find it challenging to find an American employee(s) who can speak those tongues. For example, if you need employees to serve all of Canada then some of your employees must speak French.

In all of those cases, you may want to support those employees from their home offices, located in another country, especially if it is only one or two people, and they are answerable to your outfit. That avoids the costs of setting up satellite offices in an offshore location.

Home working has taken off in several countries. A 2001 EKOS study reported that 11% of Canadians primarily work from home.

Home working also is becoming popular in the UK. The government reported that home workers comprised 7.4% of the work force or 2.2 million workers in 2001. Around 1.8 million of them needed computers and phones to work. *IT Week* reported in April 2003 that over 1/3rd of Canadian voice/data products giant Nortel's 35,000 UK employees work at home; 71% of those home workers felt more empowered while 41% felt more motivated by working from home.

Home working is even more popular in the tech-edgy Scandinavian countries (e.g. Finland, Sweden and Denmark), and in Holland and Switzerland, reports a 1999 European survey. It is less popular in Germany, Ireland, Italy, France, and Spain.

There will be more home working across Europe in the future. That is because the European Union's Flexible Working Directive allows employees with children ages 6 and under to apply for flexible working like home working.

But international home working has its own locations and facilities considerations over and above those for domestic home working. They include:

Working and residence papers

Chances are if an employee approaches you about working from home in another country, or you are considering a home worker that lives in another country, that person will have the necessary papers to live and work there. But you **must ask and make sure** because they are your employees.

There could be legal and other hassles in approving home work from another country—if those employees do not have the papers to work from that offshore location. If your firm has a business presence in that nation, the authorities could go after your organization. Check with your legal department on these issues. .

Visitors' visas

The exception here is if the employee is in a foreign country on a visitor's visa with a spouse or other family member who has legal status in that country and who is permitted to apply for sponsorship of that employee. Depending on the country, they could work on their (or your) laptop and make and take calls just as if they are on vacation; they file taxes to their home jurisdictions.

The catches are that visitors have to go back to their home country, and they may not be allowed to return to the country they are visiting if authorities suspect they are attempting to live there illegally; they also can't get local benefits like health coverage. Visitors also cannot get employment even if it is with the local branch of your company.

Employees who go the visitor route must check the laws *very carefully* before doing so. Otherwise they and their employers could end up in a real mess.

Permanent residency in new country

Depending on the country, once and if the employee files papers to become a permanent resident in that new nation, if they leave they may not be allowed to return to their new country until if and when immigration authorities approve their permanent residence application. That means you can't ask such employees to come back to your home office or otherwise travel back to your home country until they have legal status in that new country.

Employees who are citizens of that country and who are returning home do not have that problem. But they then become visitors to your country, unless they have dual citizenship.

International travel home

International travel, except perhaps between Canada and the US or Mexico and the US, can be very costly. Unless the employee is extremely valuable, and there are very good reasons for them to live internationally and there are situations where you really do need to see them or they need to see others in your country in-person, avoid travel to your premises at all costs. Audio-, data-, video- and Web conferencing (see Chapter 11) is much cheaper.

If those employees do not have to travel to your premises offices often, if at all, then international home working can work out. But if you require such employees to travel to your home country frequently, think very carefully about approving them to work from home while abroad.

If your international employees had previously resided in the US or Canada, and they are going to the US or Canada on business I *strongly recommend* that they carry their US and Canadian passports and permanent residency cards along with the passports and work/residence authorizations of the countries they are living in. Otherwise they may not be allowed to enter—especially the US—creating havoc for them and for you.

I speak from experience here. I am an American, Canadian, and Briton. I was born in Canada, to English parents who later became American citizens. (I have American, Canadian, and UK/EU passports.)

I got chewed out by American immigration officials in Vancouver, Canada and almost missed a flight to the US because I showed my Canadian passport. The official checked and questioned me on all my business trips for the past three years. "You are an American! You have an American passport. Use it!" Or words to that effect. So when I cross the border I show my US passport, the officer runs it through the scanner, asks the purpose of my trip: they sometimes ask me who I work for and what I do, and I am on my way.

There is also another benefit of having and using multiple papers: shorter lineups and quicker processing. When I go to Europe I pick the "EU Nationals" line, and when I go to the US "American citizens and resident aliens" queue. Traveling to Canada is far quicker that way too.

Communications hassles

Global communications: voice and data, especially wireless, is improving rapidly and is becoming less expensive. This is another reason that international home working is potentially doable. Some parts of the world: Asia and Europe are ahead of the US on wireless technology. Canada has a greater Internet broadband penetration rate than the US.

Note the words: "potentially doable." Many countries still have very high domestic and international long distance rates; in many countries residents pay for individual domestic wireline calls as they do for wireless calls. When I worked from home in the UK, I itemized and sent back to individual clients all of my local *and* long distance calls I made on their behalf.

Also, many country are still behind the US and Canada on domestic wireline penetration rates. Getting new lines put in can be very costly and there are long waiting lists. The same goes for Internet access. Many countries and locations in them lack residential broadband.

Not surprisingly many countries, especially developing countries have gone wireless, for calls and Internet access. In those countries, you may find cellular data and cellular broadband more feasible for home workers.

Global communications recommendations

If your home workers live, or are going to live, outside of the US and Canada ask them to get you information on wireline, wireless, voice, and data costs, including installation and lead times. That way you will know what you could be getting into.

Look into making those employees' wireless numbers their contact numbers. Have both you and them examine voice over IP, even over dial-up. That could cut the costs of long distance calls dramatically, especially to and from the US and Canada. The lower expenses could make the investments into gateways at both ends worthwhile.

Alternatively, if there is more than one employee located in the same area, look at satellite premises offices (more about that in Chapter 11). It may be easier, quicker, and cheaper to get phone service and broadband access into a premises office than into homes.

Accounting and organization

Having employees work from another country may require your outfit to make changes in your accounting and organization. You may be required to have a legal presence, such as a formal subsidiary incorporated in that nation. Check with HR and Legal on that.

Even though your International home workers report to you and your supervisors say back in the US, they are not *legally* your employees. Those employees come, instead, under the labor regulations in their countries. That means you have to make arrangements to deduct income taxes and social security, etc, in accordance with those countries' laws.

Tax-deductible benefit plans in your country, like 401(k), medical, and dental in the US, are likely not transferable to employees in other countries. Pension plans like the 401(k) stay on ice until the employees retire or leave. If employees had borrowed against their 401(k) you cannot make deductions for repayment purposes, if they become foreign employees. It is noted, however, that on Social Security there is reciprocity between the US and Canada.

Your firm will also have to cope with currency differences for payroll and expenses. Fortunately, that process can be automated.

Holiday differences

You also will need to abide by local holiday and vacation requirements. For example, many European countries require that you give four to five weeks' vacations: twice as long as the average in the US.

Americans often think that because Canadians look kind of like Americans that they share similar customs and holidays. Not so. Just one example will suffice: Canada's Thanksgiving takes place the second weekend in October; it has a different meaning than the US Thanksgiving in that the day "gives thanks" to a bountiful harvest.

You will need to remind your international employees: if they don't do it already, to let their colleagues know of upcoming legal holidays in their countries. That includes putting those dates on the timesheets they put in. If you are doing forecasting and scheduling involving those employees you will need to enter those dates.

Healthcare differences

Healthcare, social security, and employer obligations vary from country to country. Many countries have basic government-sponsored healthcare. In Canada, many provinces (like British Columbia) require residents to pay healthcare premiums, but many others do not.

You may to cover such premiums in your benefits package. You may be able to offer your employees supplementary medical and dental on top that covers what the government plans do not. Canada permits this.

Taxation

Some countries may require you to fill out forms stating what home office expenses you pay for and which ones your employees pay for. Canada has such a requirement.

If your employees have been American residents: either US citizens or resident aliens, they may have to file with the Internal Revenue Service as well as with their country's taxation authority. For example, the IRS has a Form 2555 Foreign Tax Exclusion that allows them to claim their foreign taxes. If they received no income other than from foreign sources, have no other business the result should be a wash.

○ SMALL CITY/TOWN LIVING

For some years there has been a movement of people from the congested cities and suburbs for the seemingly idyllic and less-expensive-to-live-in country towns. That has saved them money and has afforded them an overall better quality of life.

The influx also has helped many small towns prosper because the newcomers create demand for construction and buy goods and services. These new residents often put money into declining downtowns turned ghostly when the local XXXYYYZZZmart opened up nearby.

Technology has enabled and accelerated this trend by permitting people to work from home. There are communities with as few as 900 people with cable broadband.

Some of these small communities have great transportation access. I live in a small city on Vancouver Island, Canada that has commercial air service to Seattle as well as to Vancouver, BC: the principal transportation hub on Canada's West Coast. A limited-access highway links us with the ferries for Vancouver in less than 2 hours.

Here is an example of what you or your employees can get with living in a small city or town. My trade is not exactly known to be high paying. But by moving to a small city

I was able to buy a house with four bedrooms, two floors, two kitchens and three bathrooms spread over 2,300 square feet, plus a deck with a view of the mountains for $140,000 US.

There was no way I could have done this anywhere else. Not with the aforementioned technology and transportation amenities, plus a wide range of shopping including a Future Shop (owned by BestBuy), Home Depot, Staples, Radio Shack, and of course, a Wal-Mart. But then again I also write about site selection, and more importantly I discussed the proposed move with my superiors and got their permission and support. I also told them they were welcome to visit us at any time. After all, it is *their* office.

I'm also heavily involved in my community politically. It is a great place to live. It is my home and I do my part to look after it, including correcting problems and making improvements within our means.

But there are caveats to small town and small city living that you and your employees should be aware of *before* they make that move or want or be approved to work from home:

Poor communications and access

The farther out someone goes, the less likely they can get from there: in-person and electronically. With airlines cutting back service to smaller communities that type of access is not getting any easier. If your employees have to travel frequently, don't like long drives, and/or are the least bit squeamish about small planes (I prefer them and small airports), and you aren't happy about paying high mileage costs and airfares then maybe living in small communities isn't a good idea.

Some small communities still do not have high-speed Internet access. If your employees need that to work then they should cross out those towns that do not have those connections. But as airlines cut back on their services, cable companies and telcos are expanding their broadband networks—especially into small cities and towns.

Most communities have post offices; post offices have courier services. But other courier companies may not have a presence in a small town. If employees need those services to work, they should check out whether courier service is available before relocating there.

Occasional power outages

When power outages from storms, or from utility maintenance hit small rural communities, getting the power back on may take longer than in an urban area. Many power lines are located in hard-to-get areas.

BC Hydro, the local utility pulled the plug on northern tip of Vancouver Island late one summer evening—with plenty of advance notice—to upgrade the generating station. The local electric system is not hooked into the rest of the provincial and international grids.

The locals live with it by having water and dry food, batteries, candles and small generators on hand. Many homes where we live have woodstoves.

Lack of amenities

You/your employees will get the basics in smaller communities, but the prices may be higher. They may have to drive to the next town, which could be an hour away, to buy items such as computers, phones, computer parts, peripherals, and printer supplies (ink cartridges and paper). Or depend on catalogs and websites, and perhaps be required to pay outrageous shipping and handling charges for stuff that takes may take weeks to get there, especially if the community is outside of the US.

One of the dirty secrets of direct marketing is that much, if not all of the profit, is in the "shipping and handling." Sometimes you will see products flogged for zip "plus $19.95 shipping and handling." Gee, how nice of them to "give" away that stuff!

Repair hassles

Employees who live in small towns must be self-sufficient when it comes to technology, bug fixes and computer repairs. If their machine dies and IT support can't help them by phone or online there may not be a local firm nearby that can fix the problem, or a computer superstore in a quick and easy drive to get replacement components.

You should ask about computer repairs when discussing a move to the country with your employees; both of you will need a chat with the IT department. It may be a good idea to put together a troubleshooting FAQ on the computers' desktops, supplemented by printed handouts to enable employees to fix problems without FUBARing the computers, peripherals, and phones.

Lack of services

Your home workers can't get the array of personal and medical services that they get in larger centers. They will have a couple of doctors, a dentist, and perhaps a medical clinic (if near a decent sized community). But don't expect Massachusetts General or Sloan-Kettering. Major surgery will require long trips out of town.

Potentially slower emergency response

One of great aspects of small city and community life is that nearly everything is close by—if you live close to town. That includes fire halls and police stations.

But the more remote and smaller the town, i.e. the more "in the boonies" the employees live, the slower the response times when there is an emergency, such as a fire. One reason is that many communities have and depend on volunteer fire departments. Volunteers have to be called up—and then they have to respond.

Social problems

Just because a community is small and remote that doesn't means it lacks its fair share of crime, vandalism, drug, and personal abuse. People are people.

Isolation

Small city and rural life can be great, but it also can get lonely. There is often less to do.

People are friendlier in small communities than in larger cities, but they are not necessarily *more friendly*.

There is a difference between being friends and being friendly. They are nice, initially, in order to check you out. But they won't necessarily want to become best buds.

Sometimes small town people will not be hospitable at all. There is one town near mine that is so insular that one person who lived there said their neighbors didn't say anything to them for years.

Small city/town people can be less tolerant to other ethnicities, marriages, and partnerships than in larger centers, but those attitudes slowly breaking down. Some communities are more open-minded than others.

When one gets past child-rearing age it is increasingly difficult to make friends in new places; kids make great icebreakers. But if your employees and their families are open-minded and adaptable, and are self-sufficient in their own company then small cities and towns can work.

A word about kids: they may or may not take to such moves especially if they are teenagers. Urban/suburban kids often have a rough time of fitting into small cities and towns.

Small City Recommendations

If your employees are contemplating such moves, raise with them all of the issues set out in this section with them. Be sure you are comfortable with the answers they give you.

Supporting them from home is a big enough step. Supporting them from a small city and town is an even bigger move. The last thing you or your employees want is for them to make such a move, enable them to get set up in their new home offices only for them to come back a year or so later.

✪ FACE-TO-FACE AT HOME

Your organization may want your home working employees do work other than the knowledge variety type of tasks, such as small-scale repairs, or mailing and distribution. Or your outfit may require that some of your home workers meet face-to-face with clients, customers, sources, colleagues, and supervisors. That can be done at your premises, the clients' or customers' premises—or your employees' home offices.

Visiting at a home office may be much easier traffic-wise for customers/clients than crawling on overpacked Interstates to so-called "office parks." There can be nothing like one's residences to put someone at ease and make them feel, well, *at home*.

On the other hand, you employees may not want some customers, clients, and sources in their homes. Your employees' rights must be respected. Their home *is* their castle. You, therefore, should not insist on having them meet people in their home. Instead, offer it as an option.

If your employees are agreeable to meet customers/clients at their homes, you and your employees must look at and work out the following considerations:

Getting there

Are your employees' homes conveniently located to where the visitor(s) are coming from?

If they fly in, is there a major airport an hour or so away? If they come on mass transit, is there a bus or train stop nearby?

Security

Are your employees' homes located in safe neighborhoods? You don't want to risk having visitors being mugged or their cars broken in. But be *extremely careful* when applying this consideration; it can be *easily* seen by your employees as discrimination. Consult your attorney.

Insurance

Chances are there will be a need for additional insurance, such as home business or more expensive commercial insurance, to cover any possible accidents that might occur while a visitor is at your employees' home offices. Remember, the less costly the insurance, the tighter the language on what the carriers cover.

Parking and zoning

Many visitors will arrive by car. Is there adequate parking for the number of people (and vehicles) your employees expect to receive at any one time? Do your employees have off-street parking?

Many communities have Residential Parking Only restrictions on local streets. Your employees' neighbors could get mighty upset if "their" parking spot has been "taken'" by a client or customer.

Related to this issue is zoning. Some communities have cracked down on home-based businesses because of parking, traffic, signage and noise problems, and items left stored in yards, but they have left home employees and contractors (like freelancers) alone. But if your employees are expecting to get a fair amount of traffic and if they handle goods at their home offices they should check to see if the zoning allows home-based businesses, scope out the regulations and report back to you and Legal.

Facilities

If your workers are to see people on business at home, then they should have enough space, chairs, and possibly tables for those guests sit and use. In such situations, the home office should be in a separate room, with ideally lockable doors.

There must be adequate heating, air conditioning, and ventilation in the home office. Your employees may like working naked in 100-degree heat or wearing four pairs of longjohns and three pairs of thick socks and seeing their breath, but chances are their visitors will not.

Heating can be an issue if your employee rents an apartment in someone's house. I've had landlords who turn down the heat during the day and turn it up at night; I wouldn't trust the wiring to take an electric heater. So I coped. But then again I didn't receive work-related visitors. Fortunately, I have thick skin and a wide temperature tolerance.

There should be little or no noise: from other occupants: children, spouses, roommates, visitors. No knocking on the door for glasses of milk or wanting to borrow the cars. Home working is no substitute for day care.

No critters. Your employees may have to tolerate mice, roaches, silverfish, spiders, and in many parts of the world, scorpions and centipedes in the homes. But their guests will not want to; they don't expect such creepy-crawlies in premises offices and they shouldn't expect them to see in home offices.

Ask your employees to critter-proof their space if they expect guests. Hire an exterminator, if necessary.

There also should be a washroom that visitors can utilize. That means no pantyhose hanging up to dry or dirty socks on the floor. If your employees offer visitors hot and cold beverages at premises they should be prepared to offer the same at home. *Remember the mantra...*

Communications

Chances are visitors will ask to use the phone, though increasingly less so, as more individuals go wireless. Home employees should let them as a courtesy, but you should pay for the calls. For that reason, for home workers that are expected to receive visitors on your behalf, it is probably a good idea to supply the phone lines and be billed directly.

Pets

Unless your employees and customers/clients are in the pet business, it is probably not a good idea for your employees that have pets to receive customers/clients at home. Many people have pet allergies; also many people find a little difficult to believe "Rodney the Rottweiler" is as friendly as your employees say he/she/it is. Especially if he/she/it is targeting their throats.

The only exception should be if the pets are banned from the home office, like if the work facilities are in a shed or above the garage; or if the visitors are well known to the employees and are used to the animals. But the minute there are complaints either Rodney, or your employee, goes to the doghouse.

There are also some pets that freak people out. Your employees might think lizards, mice, rats, scorpions, snakes, and tarantulas are cute. But chances are not many clients or customers will agree. Less objectionable creatures, like birds, are noisy. If your employees have such creatures as pets, insist they are caged and nowhere in their working space.

On the other hand, goldfish and tropical fish can add to a home office's ambiance. They are calming to watch and can help break the ice. And they are quiet, don't need to be petted, and do their "business" in the aquarium.

Smoking

If your premises are smoke-free, then your employees' home offices *must be* smoke-free when receiving clients and customers. The fumes from cigarettes are deadly; lit substances can cause fires. You risk lawsuits that way from dangerous secondhand smoke.

Your employees don't (or shouldn't) go around spraying dangerous chemicals like benzene or hydrogen cyanide into the air, and no person in their right mind would deliberately allow carbon monoxide from a running car into their home, therefore they should-n't allow the toxic chemicals such as benzene and hydrogen cyanide and gases like car-bon monoxide from burning cigarettes poison the air that others breathe.

✪ EMPLOYEES WORKING FROM OTHER EMPLOYEES' HOMES

There may be situations where it is desirable for other employees to visit and work from another employee's home office. Examples include editorial: a writer or managing edi-tor visiting and working from an editor's house; secretarial: where the secretary takes notes and does filing for a manager or executive.

Be careful before allowing this. Check with your HR and legal departments. There may be liability and workers' compensation issues. For example, what would happen if your employees were injured slipping on some ice on their way in, or got carpal tunnel syndrome while working on that employee's laptop?

CUSTOMERS/CLIENTS VERSUS SUPERVISORS CAVEAT

If the people visiting your employees are your supervisors, it is less important that your employees meet all of the considerations set out in the Face-to-Face at Home section.

You, through your supervisors, are entering the employees' homes. Thus, you must accept the conditions of their homes. Your only considerations should be factors that affect how your employees perform their work from home—nothing else should matter.

Chapter 11:
Adjuncts and Options

There are several adjunct strategies and options to home working that you should consider before deciding on a home working program. The key adjuncts and options are chiefly *self-employed home workers, outsourced home workers, satellite working, mobile working, and conferencing.* These strategies and options can make your home working and your total working environment more efficient. In turn, home workplaces can enable you to get more out of these methods.

✪ SELF-EMPLOYED HOME WORKERS

Self-employed home workers are exactly what it says, workers who are self-employed and who work from home. They can be unincorporated or incorporated. Examples include computer programmers/analysts, consultants, customer service/support/sales, freelance editors/writers, graphic artists, physicians, proofreaders, and transcriptionists.

You ask them to do a job: handle a customer, write, edit, or proof a program or a story, or undertake some consulting. They do it, they bill you, you pay them, and they pay for expenses, like voice/data to their homes, equipment, and taxes. You may or may not reimburse them for expenses they incur on your behalf, like long distance phone calls, faxes, mailing, and travel. If you don't reimburse such expenses, they write them off.

But, unlike employees, you don't control these self-employed home workers. You don't tell them when they can work or ask them to undertake tasks outside of the work for which you've agreed to pay them.

The Internal Revenue Service's website, www.irs.gov, makes it very clear who is self-employed, or in its language an "independent contractor." The IRS makes no bones about how you should treat and regard independent contractors.

Who is an Independent Contractor? A general rule an Independent Contractor defined as a person that you, as the payer, have *the right to control or direct only the result of the work* done by the Independent Contractor, and *not the means and methods of accomplishing the result.*

Let's examine the two most common categories of self-employed home workers:

Freelance

Freelancers (like I was, for several years) are independent business people. You can hire them as you need them, singly or any number of them—whatever it takes to do the job. I freelanced for some clients for several years.

If you prefer, you can ask your freelancers to hot-desk their work on your premises. You may give them permission to use your phones—but only on your business, unless they use calling cards. You also can grant them permission to hook in their laptops to your network to access e-mail through their own accounts, or to surf the Web.

If a freelancer has been with you for some years and they meet clients and customers on your behalf, you may want to issue them business cards. I have one from *Iron Age/New Steel* magazine, then owned by Capital Cities/ABC that listed my home office in Jamaica, Queens, New York NY. Anyone who didn't know better would think the magazine was based there: in some gleaming glass-and-metal high-rise. Instead the "office" was a one-room upstairs apartment in a wooden house.

Contracted

There are contractors—not to be confused with the IRS term "independent contractors"—who arrange, train, and monitor self-employed home workers for you. If you need their workers to be in touch with your customers and clients in real-time, these contractors arrange for voice and data links from your phone carriers and servers to the workers' homes.

The contractor model has long been deployed in the taxi and limousine industry, which I used to cover. Drivers own their vehicles and pay dispatch fees to the taxi/limousine firm, which specifies the make/model/age of cars and driver attire. That firm then puts the work out to the drivers based on their availability and proximity to the pickup locations when they get the calls from the clients.

My landlord in Queens owned and operated a Lincoln Town Car, contracted with a "black car" firm that served corporate clients: picking up employees from offices, the airports and train stations. I occasionally did business with him. At least I never had to give him directions!

One of the biggest users of these contractors are organizations that require call center-styled customer service, support, and sales. They hire specialty contractors that promise to bring on and pay for self-employed home workers to be on the phones and terminals for time increments as little as 15-30 minutes.

Such small time increments enable contracting organizations to precisely match the workers to the call and contact volume. In contrast, their competitors—non home-working service bureaus—typically will need to bring in and pay employees for at least an hour or so to make it worth their while (and the employees') to commute in.

Some of the leading specialty contractors e.g. WillowCSN, West, Working Solutions have pools of self-employed home workers. These home working contractors have tools that enable self-employed home workers to obtain the work.

One tool is bidding software, which in Willow's case, was developed internally. (The company applied for and received a patent for the scheduling system embodied in the software.) With bidding software, contractor's clients generally post schedules a week to a month in advance. And, as long as the self-employed home workers have completed training to answer calls for a client, they can indicate if they're available to take these calls.

To ensure all workers each have a fair shot, there are limits on the number of hours they can devote to a client per week. Conversely, contractors' clients refer to the software to look up the number of self-employed home workers they've requested for a given hour and the number on call during that hour.

Some contractors may, in turn, outsource the data connection hosting to application service providers. Others, such as WillowCSN, may have their own data servers that they may choose to co-locate in a carrier class telecommunications facility. Others use their own technology. West's contractors use the same platform that it had developed and refined over many years for its premise-based call center agents; clients can use it for their own call center agents.

Some contractors (e.g. WillowCSN) charge its self employed home workers a service fee for their participation in the contractors' service pool and access to the bidding system. The contractors then bill the clients and pay the self-employed workers. At tax time, the contractors issue miscellaneous income statements (IRS form 1099s) for those workers not incorporated.

Self-employed home workers are not limited to providing their services as a sole proprietor or to use the IRS term "independent contractor." Willow says that should the self-employed worker desire, the individual can establish a corporation to contract to provide the services and be an employee of that corporation. The corporation, as the vendor, invoices the organizations that it has been contracted to provide services.

Self-Employed Home Worker Advantages

The big advantage of using self-employed home workers is flexibility. You can bring them on for one campaign, program/project and story. Then when the work ends, so do your involvement and your financial commitment.

You can easily hire these independents to do the work and fire them if they do not do it well. Because they work from home, they are out of sight, out of mind. There are no personal attachments unlike with consultants and temporary office personnel who work as independents or for contractors on your premises, and which your staff gets to know.

With self-employed home workers, you get the productivity, quality and disaster recovery benefits of home working without the up-front costs and time of creating a home worker program. These independent workers own/lease, pay for, and write off, their computers, phones and lines, not you.

While you may not have to train them on the basics, you may have to teach them a little about your organization, product, service, market, purpose, and/or readership. You hired them based on their skills to do the job including their ability to pick up the knowledge of your field very quickly.

If you hire contractors, they handle the screening, training and voice/data technology setup. They train the self-employed workers following your internal training program—either instructor-led and/or technology-based such as online, CD and video. The contractors stay in touch with these individuals such as through e-mail, newsletters, voice mail alerts, instant messaging, and secure chat rooms.

Also self-employed home workers possess a great potential talent pool. You may want to recruit the top performers to work at your premises or at their home offices as your employees.

Other big benefits of doing business with self-employed home workers are lower costs. Because they are self-employed you avoid paying for their voice/data, equipment, and furniture.

There are also fewer regulatory and jurisdictional hassles. You don't have to comply with OSHA and workers' comp or be involved with cross-state tax issues. It is up to the self-employeds to be responsible for occupational safety, building codes, and local regulations, like zoning. It is clear where they pay their taxes to: where they as independent business do business in.

Self-Employed Home Worker Downsides

Perhaps the biggest downside to hiring self-employed workers is lack of control and management compared with employees. As noted earlier the taxation authorities *are very clear* who's self-employed and who's not.

You can't ask self-employed people to do any more than what you and they have agreed to. They can tell you to go jump in the lake if you rang them at midnight and demand that they fix some problem with the website; if it isn't in the deal, then tough. If they're sufficiently annoyed with you they can call the IRS—and this isn't the first time someone in your outfit have attempted to push the rules—what comes next may make you and your organization truly regret pushing their buttons.

There's also the flip side of flexibility. Self-employeds are not loyal to you. Many of them have other clients, as I did when I freelanced. They do only what you agreed to pay them to do. Nothing more. Nothing less.

Let's examine a bit closer some of these disadvantages:

Lack of loyalty

Self-employeds are not loyal to you. Many of them have other clients, as I did when I freelanced. They do only what you agreed to pay them to do. Nothing more. Nothing less.

You also risk self-employed workers dumping *you* on short notice if they get a staff job or lucrative long-term contract. When I got hired with *DM News* in 1995 which began my career covering call centers, I just made a few quick calls to my freelance clients and that was it.

Since employers usually provide medical benefits (at much less cost to employees than an individual, self-employed workers can obtain by themselves), in the US there is a disincentive for self-employment, especially for people who are older and have families.

(However, in contrast, in Canada where there is government supported basic health insurance, there is no such hindrance: you pay premiums in those provinces that require them whether you're an employee or self-employed, though the larger firms often pay for added coverage, like dental. Not surprisingly, there are many self-employed people and home-based businesses in Canada.)

You also may find that you need to mesh managing self-employeds with employees.

In many instances the self-employeds may feel like second-class citizens, shut out of the loop. This can create another reason for a self-employed to not stick around.

Less knowledgable

Chances are self-employed home workers will have less knowledge about your particular field, and less incentive to obtain that knowledge than your employees. Self-employed people tend to work for more than one client at the same time and have to be up to speed on all of those clients, whereas employees work for one employer at any one particular time and can focus on that employer. When I freelanced, and the phone rang I had said: "Hello, this is Brendan Read". "Now I say Hello, this Brendan Read, Call Center Magazine. "

The exception is when those self-employed have worked in your field before; in those cases they may know more than your employees. The same goes for any contractor or outsourcer.

There also is less incentive for you to train self-employed workers.

Less data security

You risk some of that information going to competitors. Just as you do when employees leave your firm. For that reason, and to avoid the appearance that these contractors are employees, you cannot supply (though you can sell) equipment to self-employed people, which means, in most cases, they should not gain access behind your firewall to your network.

Enabling Self-employed Home Working

Self-employed home working can be a success. Freelancing and consulting have long traditions; organizations that use them effectively have policies and procedures in place to deal with issues such as complying with tax laws. There are reputable and experienced home working contractors.

You will need to know ahead of time what your staffing needs are and budget for them. That includes voice, data, travel, and any other expenses that they incur on your behalf that your policy is to reimburse them for; otherwise they write it off as business expenses.

You may need to have these self-employed workers connected directly, by phone and data, to your clients and customers. If that is the case, you must arrange for that (see Chapters 2, 4, and 7). You also may require them to come into your premises from time to time and hot-desk (see Chapter 7). In such case, you will need to set aside hot desks and arrange for temporary passes.

But leave it to the self-employed person as to how they set up and organize their home office. Avoid giving or lending them equipment. Also *do not* ask them to undertake tasks other than what you have specifically contracted and agreed to pay for, like additional assignments.

The more involved you get with the self-employed worker—in setup and in tasks—the more the taxation authorities are going to look at this relationship as employer-employee. The tax departments will go after you for avoiding social security and other taxes.

Never use self-employment as ways to cut staff to avoid paying into social security or unemployment insurance. Taxation authorities take a *very dim* view of this practice. One employer did that to me in Canada and soon found Revenue Canada breathing down their neck.

But if you are working with a self-employed home worker that is extremely good at what they are doing and you find that may have room and budget on your staff for someone like that, track their performance, and make overtures to them. You may get more out of them if they become your employee.

But make sure that you're willing to have them continue to be working from home. One of the reasons many people became self-employed home workers is that they liked working from home.

Also when deciding between independent contractors and corporations, check with your accountant to see which is to your benefit. Tax laws do change, so stay on top of that issue.

Independent Contractors versus Corporations

Self-employed home workers are not limited to providing their services as a sole proprietor or to the use of the IRS term "independent contractor." WillowCSN says that should the self-employed worker desire, an individual can establish a corporation to contract to provide services to you (and others)—the self-employed worker becomes an employee of that corporation.

The corporation, as the vendor, invoices the organizations for which it has contracts to provide services. Check with your accountant to see which is to your benefit, independent contractors or S corporations.

✪ OUTSOURCED HOME WORKERS

Outsourced home workers are similar to self-employed home workers in that you have a contract. Only this time the contract is with a firm that has home worker employees. In the call center field there are numerous examples of this type of contractor, including Alpine Access, ARO, Convergys, and O'Currance Teleservices.

The advantages of outsourcing are that you gain the benefits of home working without dealing with the minutia of 1099s and other tax matters. Your business dealings are with the outsourcer, not with the employees.

The disadvantage is that you have much less control over these workers. You set up the program, requirements, qualifications, and the training with the outsourcer; but the outsourcer does the hiring, screening, basic training, and supervision. While the home workers may be working on your contract, they still work for the outsourcer.

✪ SATELLITE WORKING

Satellite working is the providing of mini premises offices near where employees live. These offices have shared peripherals like copy machines, faxes, and printers.

Employers directly control and manage equipment and communications. Satellite working employees also may be required to conform to employers' dress codes.

Satellite working has both electronic and in-person communications, though employers will "see" their employers less often than if they were in more central location. But it is more feasible to visit three satellite centers than 30 home workers.

Typically satellite working premises are at the periphery of major metro areas, where there are long commute times to the major or HQ premises offices. Kitsap County, in Washington State, which lies across Puget Sound from Seattle, has been studying satellite working and home working.

Advantages of Satellite Working

Satellite working has several key advantages. It reduces the commuting time, costs, and resulting productivity losses from tardiness and lost hours. Employees get more of their life back by reducing commuting. It also supplies in-person camaraderie for employees, even if they in different departments.

You expand your labor pool geographically and qualitatively, by providing premises closer to homes. Prospective employees who shudder at, and existing employees who despise, the long commutes may be game to hire and stay longer if the drive and transit times are shorter.

Often satellite premises offices cost less per employee than premises offices. The smaller the space needed, the greater the choice. Satellite locations also typically cost much less per square foot—depending on the communities they are in—compared to central business districts and office "park" conglomerations, says Bob Fortier, president, Canadian Telework Association.

Satellite premises offices provide economies of scale in equipment and technology. An organization needs only wire one building instead of several, supply one printer, copier or fax instead of several. These facilities have fatter pipe than many homes, though as wired and wireless broadband networks expand into more residential areas that benefit is becoming less important.

Perhaps, most importantly, satellite premises offices still provide employer control, while reducing the commute. Such offices often have an on-site supervisor or manager who can help employees with problems and it is easier for employees to find out what is happening with the organization.

Downsides of Satellite Working

Satellite working still requires you to pay rents and lock in to leases, albeit with smaller and typically easier-to-sublease rooms. You still must outlay money for voice/data connections, workstations and equipment, plus IT, cleaning, and security.

You could end up paying more for several satellite premises offices strung out over a metro area than in supporting home offices. If you need to lay off large numbers of staff, it is harder to shut down satellites and get out of leases than it is to kill the lights on floors or terminate those who are working from home.

With satellite worker you can only tap into those workers who live in commuting distance of the satellite premises offices. It will not help you retain employees whose family had to move farther away; or draw on talent not available locally.

Satellite working can be a pain in the tail if the workers on particular teams are not from the same area. The commuting advantage shrinks real quickly if members of those teams have to make a Phineas Fogg-like trek to the satellite premises. Though in Kitsap County's case it isn't that bad. On the Washington State Ferries that ply Puget Sound employees can have a coffee on the way out, and a beer on the way back, and if the weather isn't too rainy they can hang out on the deck and enjoy the most fantastic scenery of any major American urban area.

The only way around that issue is for the team members to work virtually: in the satellites and at the main premises. In which case, why have satellite premises? Why can't they work from home instead?

There are options to satellite working for services like faxing, printing and computer use. These include "virtual office" chains like Fedex Kinko's that provide shared peripherals like copying and printing; employees can rent time on computers and do printouts. The major office supply chains: such as Office Max and Staples have fax machines and self-service copiers, as do local business supply firms like Monk where I live on Vancouver Island.

There can be issues in rigging voice call handling. If your call routing is on premises you will have to reroute to the satellites, which, if they are located outside of the local calling area, could lead to long distance charges. The only ways around that are for the calls to be routed by the carriers, or deploying VoIP.

Satellite working can pose hassles—especially if you have employees who exclusively work from home, and drop in only once a week or so. If they get calls directly, must you have your IT staff reroute them to the satellite premises instead of from their home offices? If so and if the home workers need data like who is calling them or screen-pops, they may lose it if they have the calls rerouted from the premises officers to the home office then transferred to the satellite offices.

Enabling Satellite Working

The key requirements for satellite working are small premises offices with plenty of parking and excellent voice/data links located near where clumps of your employees live. You need a large enough pool of those employees to justify each satellite premises.

To determine the pool size, you must find out how many of those employees can work at the satellite and how often. There will be employees from that clump who need to work exclusively at the central premises office, or at home. You will need to decide whether those employees are going to occupy workstations that are dedicated to them, or work in shifts, or use the hot-desk option (for those who come in one or two days a week from the road or from home).

You will need to arrange for voice and data connections with your providers. If so, calls should be routed off your switch through off-premises extensions (OPXes), voice over IP gateways, or by network routing. (See Chapter 2.)

Or you may want to consider having your calls routed to employees' cellphones especially if those employees also mobile work (see later on). That saves you installation money at the satellite and gives one-phone convenience to your employees.

You also may want to consider private leased lines for data handling. Chances are virtual private networks (VPNs) are probably a less expensive bet (see Chapter 7). You can also app host your secure gated data access. To cut cable installation costs and increase flexibility consider having secured wireless LANs instead of wired connections.

IT is also a must. You will have to figure out whether you need an on-site IT person or can live with waiting for a support tech to drive out from the HQ or from their home office.

The satellite facility must be equipped like your premises; with ergonomically-sound chairs and workstations that meet the same standards as those used at your premises offices. If you stay with PSTN or go to VoIP, the phones should be your own to enable the same functionality.

You will need to decide whether to fit the workstations with the same computers with the same features, e.g. authentication devices you use at premises, or have your satellite-assigned employees hot-desk it with their laptops. The laptop route may be the most practical if your employees also work at home and on the road.

You also need to outfit your satellite premises with the same basic features as your main premises. These can include copiers, fax machines, scanners, printers, coffee/espresso machines, and vending machines. You will need to arrange for cleaning and security.

The Hoteling Option

There is a variant of satellite working, called "hoteling" where you or third parties provide workspace and shared services such as printers, fax machines, and in some cases, meeting and conference rooms for home and mobile working employees. The "hotel" accommodations range from basic offices to executive suites. These spaces often provide conferencing services (audio, data, and video) that reduce time-wasting and costly employee travel. Employees book them by using automated (typically web-based) reservations systems.

There are firms such as HQ Global Workplaces and Regus that provide and manage hoteling space, partnering with property owners. These third party firms can also supply administrative support staff.

Pros and Cons of Hoteling

The key benefit of hoteling is that your employees work in professionally set up offices and have access to resources like conference and meeting facilities without the costly time and investment of establishing and operating your own satellite office. Hotels can be rented by day, week, month or year, up to 5 years.

One creative way of using hoteling is for training. If your firm tends to hire significant numbers of new employees, in spurts, or have new products, services, or technologies and you need to train staff—you may find it more economical to hotel it rather than provide for and pay for training rooms on your premises.

Hoteling also may save you money on conference facilities; many such spaces have advanced videoconferencing technologies.

The downsides of hoteling are that you are limited to the locations where the hotels are available. This means that your employees may still have to commute—consuming time, money, and productivity. You also are still paying for space—space that you do not control—and you use facilities, furniture, and voice/data links chosen by the hotelier, not you.

Hoteling Recommendations

If you "hotel" pick those facilities that are easy to get to by your employees. For example, SuiteWorks has its hotel, or 'telework centre' located in Barrie, Ontario, Canada: a small city lying north of the Toronto metro area. Many major roads funnel into Barrie, connecting bedroom communities located around and to the south of it. The Barrie center also enables employers to hire workers who live farther north: which could not have commuted all the way to Toronto. The center cuts down on total commuting: reducing pollution, congestion and on taxes wasted in road upkeep and in emergency services.

One creative way of using hoteling is for training. If your firm tends to hire significant numbers of new employees, in spurts, or have new products, services, or technologies you need to train staff on you may find it more economical to hotel it rather than provide for and pay for training rooms on your premises. Hoteling may also save you money on conference facilities; many such spaces have advanced videoconferencing technologies.

Virtual Office Services

There are firms that offer all the benefits of hoteling sans the rooms, i.e. phone numbers, voice mail, mail and fax forwarding, and answering services. Regus offers such virtual office services for senior managers and other professionals who are working from home, or missing calls because they are in meetings, and/or who are testing new markets.

○ MOBILE WORKING

Mobile working is working on the move. Wherever that employee may be—plane, train, bus, streetcar, monorail, gondola, ferry, kayak, car, bicycle, rickshaw, spacecraft and on that rarest of modern travel modes: on foot. This includes being in an overpriced espresso bar; an airport lounge jammed with delayed mobile-working passengers on worn seats; in a thin-walled, cookie-cutter hotel room; on a noisy trade show floor; dust gathering in a client's office reception, a client's representative's bedroom, or in your employee's family doghouse.

With mobile working the employee can be connected or offline. The work can be delivered from where they are, in time and space, or when they arrive back on earth. Improved cellular data bandwidth and networks and growth of Wi-Fi (see Chapter 4) networks make it possible to be connected on your laptop in airports in the compared-to-aircraft "comfort" of waiting area seats. That avoids lining up to use the payphone datalinks. Many aircraft and trains now have power outlets at seats, saving precious battery energy.

Home working is an ideal adjunct to mobile working. By making the road warrior's base their castle, you save money by not having to accommodate them at your premises offices. You save them time and aggravation—they can hit the road and come home again without making time-consuming intermediate stops at your premises offices.

The mobile workers' families and loved ones will appreciate having the mobile workers based at home. That makes mobile working somewhat more bearable. A happier mobile worker is a more successful mobile worker.

Advantages of Mobile Working

The key reason why employee must travel is to conduct work that has to be done in-person. There are many tasks that need that face-to-face interaction or are best done in person. These include:

* Negotiations
* Touring premises
* Demoing and checking out tangible equipment
* Field sales
* Field support
* Building and renewing business relationships through dining and recreation,
* Attending, speaking, and moderating at conferences, and seeing vendors at trade shows.

When the employee is optimally equipped, mobile working saves you and your employee time by getting as much done as possible when the employee is travelling. If employees have to go somewhere, they should be productive. Being prepared for mobile working also means that performance-killing, stress-building catch-up time when they get back is a thing of the past—be that at a home or a premises offices.

Downsides to Mobile Working

Mobile working still has its limitations. Work-related travel cuts into employees' personal time and can eat up work time; an organization's budget can balloon due to travel expenses and the need for additional technology purchases and support costs; employees can be out of touch when most needed; productivity is sapped because your mobile employees are put in circumstances where they can't avoid "foreign" germs; and many employees dislike the hassles that come with travel. Let's now examine some of these downsides in a bit more detail.

Time and Productivity Consumption

Key among the downsides is the act of traveling. Work-related travel eats up work and personal time. Time wasted in making hotel and transportation reservations, in packing and unpacking, in the hurry-up-and-wait of getting there and back through airports, train stations, car rentals, taxi and shuttle bus lineups. (Eating on the road also doesn't make the top ten favorites list of most mobile workers.)

With travel good news *is* bad news. What do I mean by that? In a down economy when budgets are tight travel is more reliable and comfortable: there is more capacity, fewer delays, greater comfort and lower prices. In a booming economy, when budgets are looser: airlines, aircraft and terminals get packed, resulting in wasteful delays, rising prices and bumped passengers, and road congestion.

No matter how good the technology is it can't substitute for being in a fixed location, like a premises or home office. No one can work on the road all the time. In many cases working while traveling (e.g. flying in economy) is not practical, and may be unsafe (e.g. driving and talking, let alone computing).

An Inefficient Use of Scarce Resources

Travel can take a big bite out of organization budgets. Is travel a good use of resources, including employees traveling on "company time?" Are there other expenditures that could get more returns—like arranging to have employees work from home?

Many organizations have cut back on travel. Not only because it is an expense and there are questions about the value created by having employees travel, but also because mobile working often is only for the pain-and-expense tolerant. Hotels are still notorious for their outrageous phone charges; and carpal tunnel and tendonitis from working on too-high or too-low desks and rigid chairs plague mobile workers. (Hello, hotel management, this is your wake-up call from the 21st Century! Your top-paying guests most often work in their rooms on computers, not paper and pen, so get with it!).

You can't use your cellphone in flight (at least at this writing), which puts you at the mercy of costly aircraft phones. There is woefully limited room in aircraft economy passenger "compartments" to work. It takes the agility and balance of a Flying Wallenda to use a laptop without getting your elbow in someone's face. The same goes for shuttle buses and taxis.

Mobile working in hotel rooms can be chancy. Traditionally most rooms, even in five-star hotels had old-fashioned high writing desks and nonadjustable chairs. Hotel management are finally getting smart and are installing adjustable chairs, like this one in the Hyatt Regency in San Francisco.Credit: Brendan Read

Air terminal seating for the hoi polloi that don't have airline club membership are more uncomfortable than flying economy if you're trying to work. Mousing and keying on a laptop while in a waiting area seat can require you to make so many awkward (and painful) bends you swear you're training to become a contortionist. While many airports have WLANs many don't. I've given up juggling payphone dial-ups. Also there are far too few power outlets; your computer needs that AC fix.

Organizations are realizing, for example, that they can get more results for a lot less money by having their employees work from a fixed location than going on the road. It is better for the bottom line, for example, to have sales reps call and e-mail clients and prospects from premises, satellite, or home offices than gathering cobwebs in reception areas or pouring away the T&E budget on the 19th Hole.

As the next section on conferencing explains, voice/data and communications technology improvements, and public acceptance and adoption of them, are shrinking the functions that need in-person interaction. For example, people don't think twice about making and engaging in conference calls, registering and text chatting on Web conferences (hosted and moderated meeting conducted over the Web), and seeing demos and presentations conducted by presenters taking over their computers. All these operations have displaced face-to-face meetings.

Communication Problems

Mobile working has a long ways to go before it becomes "seamless," if ever. There are voice gaps on the road, on mass transit, and in buildings and transportation terminals. Wireless data transmission is even more problematic; if you lose a connection you may lose the data that you are transmitting and receiving; you may have to reboot your computer and applications: a pain in the butt.

Airlines still, at this writing, restrict cellphone use on board but you can expect that to change over the lifespan of this book. The reason: customer convenience and demand. CNN reports there are new technologies, such as WirelessCabin, which promises to mitigate safety risks. WirelessCabin features on-board base stations, enabling phones to use minimal power.

Conventional aircraft 'seat-back' phones use a different frequency than cellular and are shielded, unlike cellular, but they have high connection charges. So do wireless data via aircraft phones. That is if you can find them. Some airlines have removed them altogether; their use has dropped dramatically over the past few years.

But many of those seat-back phones that are available are amazingly full-featured. Verizon's Airfone provides e-mail and IM as well as voice and allows other people to call Airfone customers in flight. You can find them in North America on, at presstime at Continental, Delta, Midwest Express, United, United Express, US Airways, Air Canada, Aeromexico and Mexicana.

All of this means that there are plenty of times when a valuable employee is out of reach. This can cost you a client, customer, or contract.

Employee Resistance

You may find that employees are more reluctant to travel than in the past. Increasingly they want to stay close to their families and loved ones especially with the persistent threat of terrorism. The ongoing "war on terror" and communicable disease outbreaks like SARS in 2003 has put people on notice that travel, especially air travel, is no longer perceptibly safe and risk-free.

Air travel stinks!

Gone are the days of the wide-open "friendly skies." In their place is a squalid airborne misery known as "flying." Hey, they don't call them "Airbuses" for nothing.

Nowadays, air travel consists of:

Being corseted into seats with pitches that only a weasel could love, cabin air quality that makes a walk around downtown L.A like a country stroll; being de facto black-mailed to buy overpriced food landside or on the plane, or suffer with salted snacks that tightens the airborne stays as your knees swell up.

Then there's your fellow travelers. Like the ones who yap loudly, insist on making unwanted conversation, (or worse) spill food, get drunk, think the plane is an extension of their romper room by feigning deafness as their kids assault others' eardrums and crush your knees by reclining their seats. Plus the ones who carry and spread their next debilitating or deadly "bug" and inadvertently or carelessly spread it. Or who may want to use your flight and your mortal remains (and everyone else's) to make a political "statement"...

Then there's the baggage. You honestly think it would get there at the same time as you, without being pilfered enroute? Serves you right for not going carry-on. Of course, you're lucky if the bag you're allowed to bring has enough room for a supermodel's unmentionables. Do you think it was coincidental that you can't pack as much on board as you once did? The more paranoid of us suspects the airlines put those restrictions on so they make a mint on freight charges.

And did I mention delays, the having to change at hubs and the 'heart-attack-dash' to the next connecting flight? My wife will *never* transfer in St. Louis again for that reason.

Or how about the putting up of passengers in hellhole hotels when airlines fail to make connections on the last flights of the day? Several years ago I once had to stay in the most rundown supposedly quality chain hotel (I'm not disclosing the name of the airline or the hotel to protect the guilty) I had ever come across when my airline didn't hold the last connecting flight from Denver to San Jose; my flight had delays leaving New York. I've stayed in bar-top hotels and youth hostels that were more comfortable—and inviting. That caused me no end of hassles in rescheduling appointments the next day. Some years later I ran into another traveler who endured a similar experience: with the same guilty and with that same hotel...

God Forbid should have you a sudden change of plan, or have to rebook, if you missed your connection. Unless you have a corporate or personal travel agent lobbying for you, your chances of getting through to a human being that has authority to act is like winning the lottery...without a ticket.

Then there are the security humiliations. Airports have taken on the appearance of being secure, but at the price of potential humiliation through pat-downs, revealing X-rays, belt-unfastenings, and shoe removals with weapons-carrying cops and troops standing watch.

Then there's the laptop drill: unzip, unstrap, lay out, zap, boot up, boot down, strap in, zip up—with a hundred other people behind you having to do likewise; and of whom have just five minutes to get to the gate.

And how about getting to the airport? The same traffic that makes commuting a nightmare does double-duty on making the plane. In fairness a growing number of North American cities are belatedly putting in rail transit systems to airports from their downtowns: which are handy—if you're one of the comparatively few people going downtown.

Some of these lines, in Atlanta, Chicago (my favorite), Cleveland, Philadelphia, Newark, Portland, Oregon, St. Louis, San Francisco, and Washington, DC (to Reagan) are direct, frequent and comfortable. Others, like New York City's AirTrain, from Jamaica or Howard Beach, Queens to JFK, are a joke.

AirTrain forces you to schlep, bags and all, onto either a notoriously overcrowded New York City Transit Authority E subway train to Jamaica, or an equally packed not-so-speedy A train via Lower Manhattan and Brooklyn to Howard Beach, one hand holding the metal pole or strap and the other clutching your luggage, staring at the ads for zit treatment. Or you can have your luggage wheels accidentally crush the $200 wingtips or pumps belonging to the snarling, lousy-day-at-the !#$%^*&() office "dashing Dan" or "darting Danielle" commuters on the Long Island Rail Road from Penn Station: a shorter but no less aggravating ride to Jamaica. And then you have to change trains...

In fairness AirTrain avoids the joys of crawling through Manhattan traffic, in steamy tunnels, over rickety bridges, and along garbage-strewn "expressways" and "parkways", potentially missing your flight. But at least you're in the safety and security of a cab: and you can get some work done with your laptop and cellphone. In which case why travel at all?

Frequent flyer? Not if I can help it!

To add insult to the injury of miserable flying experiences, and getting to and from the airports the airlines try to induce you to stay loyal to them with frequent-flier programs. On the surface these programs are good marketing tools: by promising upgrades and free flights if you choose them and travel often instead of the competition, like trains.

But try and cash them on flights that are convenient to you. Try dealing with them on disputed miles. You can't blame the airlines, though. They're there to make a profit in a tough business and they really would prefer not to give away seats that they can sell.

Airlines can always change the rules on short notice. Like raising the number of miles you need to get that free trip to Hawaii. If the carrier goes out of business, tough. You may be better off just buying the ticket instead of cashing in the miles.

There are also other problems with frequent flyer programs. The carriers may not go where you want to go and at the times convenient to you. Also your organization may have a preferred-carrier program, but that carrier may not be the one that has the best frequent flyer program or go to your preferred leisure destinations. And if you live in, or need to travel to, a small city you're restricted by the local carrier (which often has a monopoly on that market), which may not take you all the way to where you need to go.

Christopher Elliott, editor of *Elliott's E-mail*, a free weekly newsletter aimed at travelers urged, on Microsoft's *bCentral* website that people cut up their frequent flier membership card.

Frequent flier programs are, according to him:

* A waste of time, prompting to add useless time-eating legs to trips to add points
* Hard to turn into a ticket
* Practically worthless
* Habit-forming, making you spend money that you shouldn't
* Can "seduce" you into making stupid decisions based on collecting miles
* Providers of nonexistent perks, like first class upgrades, which he found out through his experience with United

The only program he thinks is worthwhile is Southwest's Rapid Rewards, which doesn't limit the number of seats available for awards and which has few blackout dates. Getting a free ticket just takes eight trips.

So with all these hassles who needs the so-called "benefits" of frequent traveler miles? I prefer to take a carrier that has survivable seating and which takes me where and when I want to go. I don't care who it is. And if driving, the train or the ferry is more convenient and cost-effective, I'll take them instead.

Better yet, if I can conference instead of traveling I save my company a lot of money and time and me a lot of aggravation. Virtually there, in many cases, is better than getting there.

I compare frequent flyer programs to the grocery loyalty cards we have in Canada. We have three in my wallet. Fortunately all three, plus a big-box superstore and a WalMart are within ten minutes of each other (we live in a small city). Each store has different mixes of products and prices. The shopping experience is equal; for most men food shopping ranks somewhere near taking out the garbage.

Do any of these stores get our "loyalty?" Only for that transaction, if another store has what we want at a better price we buy it there. Travel is no different. It is a commodity, like grocery shopping.

Road travel stinks!

The good old days of freeways that are truly free of congestion, comfortable, and spacious are gone forever. Today's roads are slow and crowded with homicidal maniacs in their flipover-prone air-murdering tanks.

Some highways are so full of potholes that you begin to wonder if there isn't a conspiracy between the local governments responsible and the body shops, car repair joints, tire dealers, and tow truck operators. My mechanic in New York City loved rainstorms; the water disguises the potholes. He bragged that one nearby crater alone during one storm led to 15 tire sales and installs.

Then there's the "joys" of eating on the road. The signage "Eat Here and Get Gas" is one of the few examples I've come across that is truthful advertising. Filling the gut with so-so-quality of food from Interstate/exit or "service area" chains and getting back in the vehicle for the next few hours of sedentary activity, without exercise is a sure fire way to get that spare tire back, and I'm not talking about the one for your car.

Other transportation options for the mobile worker

Thanks to lobbying by the bus, road, and air lobbies (that not coincidentally, get enormous direct and indirect infrastructure, emergency services, and bailout subsidies), few mobile workers have any other option. But for the lucky few, there is a decent, comfortable, if slightly slower option—passenger trains. But trains, too, can be a schlep. Gee, I always wanted to count the bricks left in broken-down (and broken-into) North Philly tenements...while waiting for the signals to change or the power restored.

Lack of sleep

Travel is stressful and sleep is dispensable. How many people really get a good night's sleep before a trip? Or on a trip? Many people find it difficult to get used to hotel beds, especially during the first night.

Then there's jet lag; it takes one day per each hour changed to reset your clock. And then you have to re-set it when you return.

The net effect is a loss of employee productivity. Groggy, jet-lagged and road-surviving travelers are not going to perform as well and will make more, potentially embarrassing mistakes than those who are bright-tailed and bushy-eyed.

Obnoxious fellow travelers and guests

After a few mobile trips, the charm wears off staying in hotels and restaurants. Mobile workers learn real quickly about the creeps that are looking to get their jollies and the perps that are looking for a mark. Someone tried to break into my wife's room when she was at a training session in Phoenix.

There are the all-too-familiar noises in hotel rooms. Who *hasn't* heard the "happy honeymooners" in the room next door who decide to have a replay of their primal acts at 2am? Followed by the quarrel at 3am? Not to mention the makeup at 4:30am? So much for being bright eyed and bushy tailed for your 7:30am breakfast meeting.

Then there's the screaming kids tolerated by indulgent selfish, tone-deaf, brain-dead, inconsiderate parents on planes, in hotels, and restaurants. These parents are the same ones that let their brats run around hallways, lobbies, and dining areas. Unfortunately they breed...

My wife raised our son. If he started acted up in a restaurant she used to give him the look of death and he stopped.

Unhealthy Environments

Being packed into transportation, especially planes, increases the risks of catching diseases from other passengers. There are reports of inadequate ventilation on planes; when they are on the ground they suck in exhaust fumes, deicing fluid, and other nasty substances.

Air travelers especially are at risk at developing blood clots that, allegedly, are caused by sitting for long periods of times—this health hazard even has a name: "economy-class syndrome." Some, but not all airlines, recommend that passengers to stretch their limbs; but in the tight, wretched confines of economy, there is little room to do so on most aircraft.

If you're staying in a hotel are you *really* sure the room is clean? Experts advise that you don't lie on the bedspread—they get washed infrequently.

Additional Support, Losses, Risks

You also have additional technology purchasing and supporting costs. Laptops, phones, PDAs, and wireless devises get lost, broken, stolen, and break down. Such events are more likely to happen when employees travel than when they work in premises or home offices.

There are safety issues that limit mobile working. Safety experts do not recommend working with voice and/or data while driving, even if it hands-free.

Many jurisdictions have banned non-hands-free cellphone use while driving. All it takes is a few careless idiots to compute or text-message while driving and then injuring or killing themselves and/or others to get *all* mobile working while driving—stopped or not—prohibited.

Sorry, but people who focus on anything but the road when driving are idiots. They deserve to have their drivers' licenses ripped up for life.

Take your mind off the road and you'll risk getting a ride from—if you're lucky—with the likes of my son. He is a paramedic...

Enabling Mobile Working

Despite its downsides, mobile working can work if you have employees go on the road for the right reasons, and equip them appropriately. When enabling mobile workers, ask yourself, and encourage your employees, to ask *"is this trip necessary?"*. See if the work can be done by phone, e-mail, or audio or data conference (see below) instead. Only if all of the alternatives are exhausted, should you approve the trip.

Take a close look at your employees' mobile technology needs. Many people take computing devices like laptops on trips out of habit. But do they really need to use them, especially for day trips? Considering that laptops can easily get lost, stolen, and broken on a trip you can probably save a fair amount of money asking your employees: "Do you really need to work on your laptop?"

Enabling mobile working is a book by itself. CMP publishes just the tome *Going Mobile* by Keri Hayes and Susan Kuchinskas that explores and examines the hardware, software and applications needed for mobile work.

But for this book's purposes, the essentials are: a wireless communications device with voice and/or data (e.g. a cellphone, PDA, etc.) and a computing device (e.g. laptop, Tablet PC). Those devices may be one and the same.

There is now seemingly a search for the "ultimate wireless device" (UWD) that combines voice, text and computer. Today's wireless communications and computing combo devices suffer from compromises—they either make great phones or great computing devices. No device, yet, provides a good combination of both because the keyboard, pointing device, and screen require miniaturization of *your* eyes and fingers (while such miniaturization may be in the works, so far is not in beta testing). Or the keyboards, pointing devices and screens are great, but the communications end leaves much to be desired and you're better off juggling a separate phone. Until the UWD comes along, you're looking at buying and supporting additional technology.

For instance, you may need to connect those road warriors to your voice and data networks to enable real-time access by those outside of your organization. Chapter 2 outlines the use of phone switches to route calls to cellphones. Still, there are issues with transmitting large amounts of data to mobile computing devices.

However, many of the same devices can be used at home and premises offices. Your employees can work on laptop computers anywhere; they can have calls go to their cellphones practically anywhere. There are ergonomics appliances like Laptop Desk that enables users to adjust the height of their machines for comfortable, injury-free working: at home, on the plane or train, in hotel rooms or at premises offices.

Let's now look at how you can make mobile working easier for you and your employees.

Pack surge protectors

No matter what equipment your road warriors take with them, make sure they protect it electronically by packing and connecting portable surge protectors. Power surges can happen anywhere, including on transportation—the onboard power supplies exist primarily to run air conditioning, lighting, heating and cleaning equipment, not hypersensitive computers.

In the "early days" of mobile working with computing I used to get weird looks hooking surge protectors into the cleaners' outlets on trains, so I could plug my laptop and printers into them. Nowadays no one thinks twice.

Prohibit working while driving

Write into your organization's policies that employees are prohibited from using communications or computing devices while driving. They must pull over to answer contacts or to work. Make disobeying that policy grounds for instant dismissal.

It takes a split second of carelessness behind the wheel of a vehicle to kill or injure someone. ***Do not*** let the excuse be that your employee was working.

Don't work, relax

Consider restricting wireless working en route. As discussed previously, chatting and computing while driving should be prohibited because it is unsafe and may be illegal. But that is the *least* of possible consequences.

Savvy road warriors also are giving up the futile exercise of trying to work on planes. Instead they are using this time to relax—if possible—by bringing along CD and DVD players with quality headphones that drown out the noise from the engines and from other passengers. Such devices beat the in-flight movie and music (if available) schlock and avoids the need to use those annoying headphones that you *hope* are clean.

Fly into small airports

If your employees have a choice, there is little difference in airfares, and they don't need to change planes suggest flying into small airports. They can get closer to their destination and the time it takes to get on and off planes is much less than in larger terminals. The airport staff is often more friendly and helpful. Customs is often quicker too.

Fly with discount and local carriers

With the commoditization of air travel, there is very little difference in comfort between the big carriers and the discount airlines. Expect yourself to be cramped, with no food service, whoever you fly with.

But the discount airlines, e.g. JetBlue, Southwest and Westjet, offer one quality that the bigger outfits don't. The employees genuinely care. They have excellent customer service (JetBlue's and a fair percentage of Westjet's reservations staff work from home). Their frequent flier programs also are effective.

The same goes with many locally-owned airlines. Two of my favorites are Central Mountain Air, and Helijet Airlines, which in addition to helicopters, operates 19-seat Beechcraft turboprops.

One evening I was rushing to get to the CMA counter at Vancouver International; there was a long Customs delay that cut short my connection time from my flight from Seattle. But when I saw the calm, reassuring face of the CMA agent—who would greet me at the gate—my stress disappeared.

When my son flew to see us from New York he changed to Helijet at Seattle. He tried to make an earlier connection but he could not get on; the flight was oversold.

So instead of twiddling his thumbs for a few hours the Helijet shuttle staff (the airline runs a shuttle from SeaTac to King County International Airport, where it flies out of) took him to downtown Seattle, where he toured around, the driver then picked him up and brought him back to the terminal. The driver said he was going home for dinner, so why not?

Try trains and ferries

Look at using trains, if possible, for short-distance trips. You or your employees don't get the same strip searches at train stations as are performed at airports, which means they can board with minutes to spare. Trains also have much more room than planes. Passengers can even get out of their seats, walk and use the washroom—an exercise confined only to star anorexics on planes.

Better yet, your employees can work more readily on trains. They can use their cellphones and they have room to key around their laptops. The most modern trains, like Amtrak's Acela, have at-seat electrical outlets. In fact, road warriors can get a lot done working on a train because there is the room to sprawl out, and there are fewer interruptions. I've spent an entire day working productively on older slower trains across New York State and Michigan.

Trains have the convenience of stopping closer to destinations that commercial or affordable commercial aviation. How many flights are there to Stamford, Connecticut, Dedham, Massachusetts, on Route 128 or the Guildwood section of Toronto, Ontario, Canada—which are stops on major train routes?

There are also overnight trains that save on hotel costs and allow people to arrive fresh the next morning, like between Washington, DC and Chicago and between Toronto and Montreal, Canada. The sleeping car rooms are remarkably comfortable. I've used them also for offices with my laptop and my surge protector plugged into outlets.

Many cities have commuter trains that run in the off-peaks and in the reverse direction from the downtowns. This saves rental car and congestion hassles, though you need to make sure there is a taxi service and arrange for a cab at a station closest to your destination. New York, Chicago, Philadelphia, Boston, Baltimore-Washington, San Francisco, Toronto, Montreal, and Los Angeles have commuter train systems with some off peak and reverse commute schedules depending on the route.

If travel takes you or your employees to coastal communities, consider taking ferries instead of flying or driving. Waterways, unlike airways and highways, are not congested; there are now commuter and intercity passenger-only fast ferries that clip along at 35 mph or so.

I live on Vancouver Island, Canada where winter weather makes flying problematic. When we have heavy fog and rain the only way off the island is on conventional vehicle ferries and passenger-only fast ferries.

Leave plenty of time, on all legs of the trip

Travel is too insane, with too many opportunities for delays, such as traffic jams, air terminal hassles, and security at the gates for the "last-minute Larrys." Therefore, when planning your trip, leave yourself and recommend to your employees to leave, plenty of margin time to get to the airports, to get checked in, and to make connecting flights. Use the long wait times to work, or relax, or eat. Also, when planning your trip ideally have a backup schedule, i.e. don't take the last flight of the day, take the one before in case it gets cancelled.

Check waiting areas for connections

Mobile working is becoming easier, with growing numbers of Wi-Fi hotspots. But make sure the connections are there before getting your hopes high.

Expect over the lifespan of this book agreements between the carriers and the airport and other major transportation facility owners/operators to permit universal access, like cellphones are today. In the meantime, book editor Janice Reynolds recommends plopping down near the airport executive lounges. You can be sure they will have a Wi-Fi hotspot that you can piggyback into.

However, check for power outlets before sitting down, and don't forget to plug in your surge protector. If you've done any amount of working (or games playing) enroute chances are your machine will be thirsty for juice.

Check out hotel ergonomic/noise before booking

When making reservations suggest to your employees that they ask if hotel rooms have adjustable chairs and low tables. Their backs and arms will thank you for it.

Hotels are beginning to get it on ergonomics. I stayed in the Hyatt Regency Embarcadero in San Francisco recently and it has an adjustable high-backed chair and a low height-desk. Some airlines actually deliver good service.

Here's another tip: Ask for rooms that are located away from elevators and ice machines. The noise can make for a nightmarish night's sleep. When your employees check in, check the door locks and outlets. Bad locks and electrical outlets that lack covers are invitations to trouble.

TIPS FOR MOBILE WORKERS

Eat light and healthy
Food sinks to your gut when you're traveling. Avoid heavy meals, especially if you're driving; they can kill you by making you fall asleep, along with increasing your risk of a heart attack. Stick to light breakfasts, salads (romaine, not iceberg), chicken, fish. Chuck the fries and the hamburgers.

Pack pillows and meals
Most road warriors appreciate not only the neck support that a pillow offers, but also the security of bringing their own space with them. Their bugs will be on it, not anyone else's.

BYOM (bring your own meal)
Make a meal at home, or stop by the eat-here-and-get-gas convenience outlets and buy some sandwiches or salads there—the quality is just as good and the prices and variety are far better than anything you'll find inside most (but not all) terminals.

✪ CONFERENCING

Conferencing is communicating with others as an alternative to travel. Conferencing can be one-on-one or in a group; it can be voice; data, such as chat rooms, filing sharing; whiteboarding; and/or video.

Conferencing comes in several forms, all of which are fairly self explanatory: *audioconferencing, dataconferencing, videoconferencing* and *Web conferencing*.

Audioconferencing is just that, the traditional conference call set up either on a phone or through a hosted service.

Dataconferencing is sharing data and files over an online usually Internet connection between participants; it can be and usually is combined with audioconferencing.

CONFERENCING BEATS AIRPORT DELAYS, SAVES MONEY, AND MINIMIZES TAX DRAIN

The better the economy gets the more it is going to cost employees and employers, in time, productivity, in fares and in ticket surcharges and tax dollars for infrastructure. Not to mention producing and taxpaying land lost to pavement. Not including the dollars for emergency services and for security.

The US Department of Transportation says air travel will return to loads exceeding 1 billion enplanements by 2015. The department believes that it can avoid the chronic delays in the past.

But in a study released in 2004 the Federal Aviation Administration says five airports—Atlanta, Newark, LaGuardia, Chicago O'Hare and Philadelphia—need additional immediate improvements, and a total of 16 airports and seven major metropolitan areas will need additional capacity in 2013.

Moreover the FAA says that the popular, comfortable, small regional jets may become part of the air traffic problem in the future. The FAA says they require greater separation between them than the larger aircraft they replaced. Like more people driving requires more road capacity and leads to more congestion than if they rode buses instead.

Videoconferencing is transmitting moving images of the participants either one-way, where everyone sees the host or subject; or two way, where everyone can see each other.

Webconferencing is a led or joint online conference like dataconferencing but where participants often interact through chat.

Conferencing, like home working, *takes the work to the worker*. The meeting comes to them, rather than them going to the meeting. To conference your employees and/or your customers and clients can be home working, mobile, premises, or satellite working; whatever works best.

Conferencing is also an excellent and vital way to stay in touch with your home workers. If you have home workers on your team, and you have a meeting affecting them, conference them in. That's what my team does. If you make video presentations to your premises employees, make them available to home and mobile workers. CMP places those presentations on its Intranet site, making it accessible for home workers, and for premises employees who missed them.

This section briefly looks conferencing for those reasons. But it is by no means a complete discussion. There are books such *Videoconferencing, The Whole Picture* written by James Wilcox and also published by CMP that go into this topic in depth.

Conferencing Advantages

Conferencing avoids high travel costs. It enables interaction with others without the stress of getting there and back, and the stress of employees worrying about family and other personal matters unattended to while away. For example, Microsoft reports that its Live Meeting Web conferencing saves Chicago-based Fieldglass $1.91 million annually in client support services, sales, operations and marketing. Fieldglass's customers save more than $8 million in reduced travel costs and productivity gains.

Conferencing saves employees and employers that priceless commodity: *time*. No time wasted in getting to and from fixed locations, or in face-to-face meetings. No time lost in playing phone and message tag enroute or when back from a trip.

In the words of a former colleague, *Electrical Construction and Maintenance* magazine editor Joe Knisely, "there's never enough time".

Conferencing also enables greater productivity. You and your employees get more done by conferencing. By not having to look at someone you can multitask.

Admit it. Who *hasn't* answered e-mails and messages, surfed the Web for information, wrote, edited or checked a report while talking to someone on the phone? Who *hasn't* made or taken phone calls in the middle of a chat or e-mail exchange?

Conferencing Downsides

Conferencing has its disadvantages. There are still interactions that require being there in-person. There is no substitute for a handshake on a deal, a walk around a plant or trade show floor, or the feel of a piece of hardware. There could be breakdowns with conferencing software just as there are with transportation hardware (and software).

The Video Hassle

Videoconferencing has its own peccadilloes. At best, video is a poor substitute for being there in-person. You see, but you can't touch, smell, taste, walk around, or oft-times see what you want to see. The images of people are too small and grainy to capture their body language; the resolution is inadequate to provide accurate color.

There is little value, at least for work-related purposes, of seeing just a talking head. Who cares what the person looks like? So they have pink eyes, blue hair, ultraviolet skin, five eyebrows, have their hair shaped like cat's ears, and are wearing chain mail armor? Do they care what *you* look like?

The only time video works is if the person(s) are moving around demonstrating something or if their background is conveying meaning, like an executive desk. But they have to be moving, gesturing, to make the interaction meaningful.

That's also where video runs into trouble—all too often the images are blurry and jerky. That's because to get a decent quality video connection the system has to push out between 30 and 12 frames per second (fps). The standard that people are used to is 30 fps, which is TV quality. At that high speed your brain "sees" continuous motion, reports *Videoconferencing* author James Wilcox.

You need a *consistent* bandwidth end-to-end of at least 384 kilobytes per second (Kbps); home broadband cable and DSL connections touch on 384 Kbps. Consistent is the key; if the video signal is traveling over the Internet, either naked or in a VPN connection, consistency is *not* an attribute to rely on.

If the evils of Internet connections—latency (packet delays), jitter (different packet arrival times), and dropouts—are like a mild cold for voice over IP, they are like pneumonia to video. Everyone sees all too well the results the awkward movements, the lousy synchronization with audio; you can ask someone to repeat their words; you can't expect them to repeat their gestures.

Video connections to the home are a pain in the tail. My wife and I bought the hardware and software when we moved to Canada in 2001 to try and stay in touch with our family and friends in New York; the cable companies and telcos were giving them away to entice consumers to install broadband. The stores were pushing them. We put ours on my wife's cable broadband.

We followed the instructions at our end as our son did on his. All three of us are fairly tech-savvy; my wife used to be a mainframe and later back-end Web programmer and analyst. We tried many different audio and video configurations, including rigging up speakerphones over PSTN, which added long distance charges over several days.

But the net results were terrible. The images were blurred and jerky; the resolutions awful. Now the stuff gathers dust in one of my bins.

Maybe I'm just being a Luddite here. I'm open to being proven wrong. But I haven't seen a big demand workwise for videoconferencing between home working employees and others.

On the other hand, if the only way your top management will accept home working, or conferencing, is for there to be a video element, then go for it. The technology is improving and costs are dropping. For example, Cisco's Video Telephony (VT) Advantage allows videoconferencing to be added to VoIP calls, at under $190 per user, including the USB camera. VT is part of its CallManager IP-PBX that resides on the Cisco Media Convergence Server.

Still, Wilcox reports that desktop videoconferencing users refer to the 80/20 rule to describe the benefit of adding a talking head' to a screen. Data amounts to 80% of the content and 20% of the bandwidth; the proportions are the reverse for video.

And then there are the *other hassles*. Lighting is a big one; not every home office is or can be set up like a TV studio. You see shadows on the home workers so bad you thought you might have dialed a remake of a Grade F horror flick. Home employees also have to be extra careful with makeup and hair; because their faces are inches from cameras you get horribly distracting details like zits and moles. And let's face it some people aren't photogenic even though they are great workers.

Enabling Conferencing

Audio-,data-, and Webconferencing should be in every organization's toolkit, especially if you have home workers. Audioconferencing is the next best option to "being there." It is a vital tool for staying in touch with home workers.

If you require your at premises employees to have multiple-party audio conference or data conference with colleagues, clients, and customers, then you should give your home workers and mobile workers the same tools. At the very least, they need three-way calling at home, or on the road.

Even if you require all employees to work at premises offices you can benefit from the travel avoidance: in costs, time, and aggravation that audio and data conferencing provides. In turn your colleagues, clients and customers also benefit by avoiding the same. That's win-win-win all round.

Videoconferencing is another matter. Carefully analyze whether you *really* need to see the other parties. With high-resolution quality equipment still costing in the hundreds if not thousands of dollars, plus bandwidth charges there must be a compelling business case for video.

If there is such a case, there is a growing array of tools to support it; but they rely on broadband connections. There also are videoconferencing firms such that host the interactions or sell the software and hardware. There are laptops that have Webcams built in. Some iris scanners, like Panasonic's Authenticam, enable video.

Think twice about video to home workers. You have better things to do than to see some employees' mugs on your computer screen, don't you? Why waste the bandwidth on that?

One of the great aspects of home working, for employees and employers alike, is *not* having seeing each other. The fewer distractions, like 'hmm, interesting pajamas' and 'oh, that's what their spouse looks like' the more the attention is focused on where it should be: on the employees doing the work.

If you or your supervisors can't manage without making sure home workers are not

HOME WORKING AND CONFERENCING
CAN SAVE TRAVEL

Yes, the key to gridlock and airlock lies in less commuting and flying for business. How? By reducing the congestion on the roads, in the terminals, and in the air. Both the airlines and highways are geared up to peak period travel, while demand is slack for much of the day. Get rid of the peaks and you do away with expensive limited-use capacity.

For every car that is not driven "to work," for every flight not taken "on business" means room and time savings for those who must commute, drive, or fly. This also creates space for those who want to travel for leisure, for vacations, to see family and friends.

Leisure travel, unlike business travel, is much more conducive to pricing strategies to manage costs because they have more choice when to go. Have a big inflow on a Sunday evening? Lower the fares and the tolls for Saturday evening or Sunday morning.

The airline industry would benefit from this strategy. The very efficient discount carriers have shown the way, by serving the leisure traveler. I live in a small city on Vancouver Island, Canada that is not exactly a big business hub. Yet we have Westjet, Canada's most famous discount carrier, flying their 737s into our local airport on daily flights from Alberta.

Robert Moses, New York City's famed (and notorious) "Master Builder," through his program of building parks, bridges and neighborhood-destroying parkways originally constructed his highways to take city dwellers to the fresh air of recreation spots like Jones Beach. He reportedly never envisioned those roadways to become the traffic-snarled messes of today. Perhaps by home working we can return to that ideal, when driving comes once again a joy, to get away to leisure locales and to relax, and to see family and friends.

in front of the terminals every second of the day then you should reconsider home working, or better yet, how you and/or your supervisors manage people. Few employees, unless they are really desperate for work, and then you have to wonder about their qualifications, will put up that level of micromanaging.

My supervisor could care less what I look like. I haven't seen him in years. He also trusts that I do my work. He only contacts me when he needs to. Which is the way it should be.

Conferencing Tools

The tools you need for conferencing depend on what type of conferencing—audio, data and video—you want to conduct. It also depends on the number of and location of the participants.

In General

For example, to audioconference several parties: whether they are in meeting rooms, at individual premises desks, using mobile workers' cellphones or home workers' desktop phones, you may need to set up conference bridges. If there are multiple people in the same rooms, you will need special microphones that pick up all speakers. There are vendors who sell or host these bridges and sell the microphones.

If there are only three parties on a conference call, then the three-way calling features supplied with business and residential wireline or on wireless will suffice. Three way calling is easy to use; telco websites explain how to make the links.

Data- and Webconferencing can be group or one-to-one. It takes on many forms. They include: applications and filing where users (employees, clients, customers) collaborate on creating and modifying programs and transferring them; whiteboarding, where users cut, copy, and paste text, graphics; audio and video messages. You also may need audio conferencing and screen sharing for presentations and demos.

Conferencing products can be purchased and/or hosted, depending on the firm and the conference tools. Vendors include Cisco (MeetingPlace), Gatelinx, Microsoft (Live Meeting) and WebEx. Many of these tools, including data- and video-conferencing: one-on-one and group are now browser-based through the Internet, and ideally through broadband or Wi-Fi connections. Some, like Gatelinx, allow users to post pictures or avatars if they don't want others to see them.

Many of the hosted services are very extensive and provide a high degree of reliability. For example, Gatelinx provides voice, data and video-over-IP collaborations between employees and between employees and supervisors as well as employees and customers and clients.

Your home worker could have a question from the customer that they don't know the answer to and needs assistance from colleagues or from a senior employee. Gatelinx connects, through the same broadband connection those employees who can deliver the answers by voice or online that the other employee then relays to the customer. All parties can be on the line at the same time as needed—up to 8 at a time.

WebEx provides three types of services. These are:

* WebEx Meeting Center which provides audio, data or video capabilities that can be set up instantly, regularly scheduled, or ongoing, in collaboration.

* WebEx Support Center, which hosts home' computer diagnosis, bug fixing and applications monitoring and ensure no unauthorized software had been loaded onto their computers. It can also link home workers doing tech support with customers or staff.

* WebEx Training Center, which enables instructors through any of the three media lecture training sessions and then at the end of each session provide online testing. The service can be recorded and then automatic tests and polls given after the agents complete the lessons.

WebEx does not use the public Internet. Instead it relies on the MediaTone Network, a separate network that was designed to give its services greater performance, reliability, and security.

Videoconferencing

Videoconferencing, like audio, also can be multiple-party or one-to-one. Video can be one-way; with the receiving parties seeing and hearing the transmitting party on their computer screen, or two-way. But the equipment is more elaborate: cameras, microphones, software, and lighting at the transmission ends; there also must be high-speed connections between all parties. If you equip employees with iris scanners they can be used for conferencing depending on the model, but they need to see everyone else in two-way sessions.

VIDEO TIPS

The February 2003 issue of LAPTOP magazine offers some excellent tips if you are going to videoconference. They include:

- Use the fastest and latest technology available. Infrastructure reliability is key.

- Pay attention to visual and verbal communications. Participants should speak clearly, deliberately, directly into microphones and avoid making sudden movements that will blur. They should look at the camera; ideally about two feet away.

- Ensure good lighting. Highlight the face if possible; avoid shadowy faces.

- Clean up the surroundings beforehand to avoid distractions.

- Minimize background noise.

- Designate a meeting organizer to control the meeting flow.

- Relax; the meeting needs to be organized but not run at fast pace.

Chapter 12: Deciding On Home Working

As the chapters of this book points out, there are lot of "components" that go into initiating, supporting, and managing home office working. That goes for you and for your employees, especially if you are supporting workers whose homes are their principal or *exclusive* offices.

Your homebound staff must have the right space, and sometimes the right location and work-friendly environment, to enable them to carry out their job tasks from home. They need to equip spaces in their homes with the same quality of voice, data, hardware, software, furniture, and security so they can deliver an ***equivalent if not superior performance as at premises offices***: this book's mantra.

At the same time, in order to support your home workers, you may need to equip your premises with additions to and/or perhaps modifications of your voice and data networks. You must devise means to communicate with your workers when they are working from home. You will have to consider how to best select, train, and manage this labor force, including dealing with tricky issues like career development and termination. There are day-to-day matters that must be worked out, too, like signing in and out and "hot desks" (if home workers come into your premises offices they may need work areas or "hot desks").

If you have employees that only work from home occasionally, you and those employees still need most of the same components and practices set out for the full-time home worker, i.e. a dedicated, if temporary, work area; voice/data and power connections; and sign in-out procedures. But there are other issues that also must be considered. For example, who pays for what, do employees or you supply the computers, and if the former, are you comfortable at having those machines access your organization? These are just some of the issues that must be sorted out before implementing a formal home work program.

Finally, there are alternative strategies to be examined and weighed, like self-employed home worker contracting, and satellite premises working. You probably ought to consider adjunct strategies also, namely mobile working and audio-, data-, video- and Web conferencing. Your home workers may work on the road, or use conferencing as an alternative for going on the road, or a bit of both. Conferencing may offer sufficient benefits to be deployed as an alternative strategy to home working for your premises office employees.

There are special instances like international home working and handling of customers and clients at home offices that have their own issues; these also must be worked out. You and your employees have to be flexible and imaginative to work out answers and solutions to unforeseen problems.

✪ HOME WORKING ADVANTAGES

You are probably wondering at this point if the advantages of home working are worth all the extra effort and investment. You will want to know if the downsides outweigh the benefits, and how to square the two.

You may be pleasantly surprised that having your employees work from home, even occasionally, may pay off. This is not theoretical. There are examples and experiences, sprinkled throughout this book, that demonstrate that the gains of home working are being realized by a range of employers and employees.

We will now look, in depth, at the many advantages of home working.

Reduced Real Estate Costs

This is perhaps the single biggest incentive for home working; but it pays off mainly for exclusive home employees. Fewer rooms, fewer floors, and fewer buildings needed, mean fewer expenses and capital investments.

Even if you are in a long-term lease, or got the building for "free" from a local economic development agency, you can still incur savings, for example, home workers mean not buying and installing workstations, not laying out cabling, not paying (and waiting for) contractors and their high-paid help to move or modify work floors.

Virtual working consultant Jack Heacock, who as project manager, built and outfitted Amtrak's award-winning and ergonomically sound Riverside, CA call center says it costs **over $33,000** per workstation to accommodate employees in premises offices.

In contrast, based on the model developed by Heacock, for a client with 100 employees (outlined in Chapter 1), it costs just **$7,100** in one-time setup charges for home workers. The home working expense includes $5,900 for furniture, equipment, and telecommunications (those costs can be *less* for employers if employees pay for some of it).

The real estate savings are enormous, and ongoing. The Heacock model shows an annual real estate rental cost avoidance of $500,000 on top of $1.5 million from amortized unneeded build-out over 5 years. That translates to **$16,520** in raw costs per premises office employee per year averaged over 5 years, *before* calculating in productivity and retention improvements with home working, which amounts to **$20,020** per employee. In contrast, it costs just **$2,640** in raw costs per home office employee per year, averaged over 5 years.

The bottom line—over **$10,000,000** in savings over a five year period: $1,276,000 in the first year and $1,986,000 in each of the second through fifth years.

That's over $10 million in profits and additional resources that would have been left on the table *without* home working. Who can afford to see that sum go bye-bye?

Want to find $10 million? Who doesn't? Begin by seeing about getting rid or down-sizing your building. Why miss out on the money?

The five-year period is important. In order to avoid getting stuck with property that is no longer suitable for their needs and/or being trapped in labor markets that have become too expensive to sustain operations, many real estate consultants are recommending lease lengths of three to no more than five years to their clients.

"The more the tenant improvements, the longer the lease," says King White, senior vice president of global site selection and real estate firm Trammell Crow. "Some companies are doing 15 year lease for build to suits." But a three to five year lease is doable on space that requires minimal tenant improvements.

Employers, unless they own their own buildings, are at the mercy of their landlords and contractors. In bad times you can get good deals; in good times they want better deals from you. You pay through the nose for property and maintenance on the upswing.

The Real Estate Savings are Real

Many organizations have realized significant property cost savings from home working. A 2003 study for the Kitsap County (Washington State) Public Utilities Department "Kitsap Telework Analysis, Assessing the costs and Benefits of Telework for Kitsap Commuters, Puget Sound Employers, And the Public," performed by the University of Washington's Center for Internet Studies reveals that employers such as AT&T, Andersen, and HP report 35% to 55% premises office space reductions with home working.

The report cited another study "Telecommuting in the 21st Century: Benefits, Issues, and a Leadership Model Which Will Work," published in *The Journal of Leadership Studies*. That report revealed that Merrill Lynch would achieve savings of $5,000 to $6,000 per office per employee per year if each of those employees worked from home.

Using Seattle-area real estate prices: $30/per sf and local (Kitsap County) $20/ per sf the study estimates that organizations with 1,000 employees can save as much as $300,000 annually if just 10% home worked

The savings can be had in lower-priced Midwestern markets. ARO, a Kansas City, MO-based outsourcer, switched from premises office to home working; on overhead alone it is saving $500,000 a year.

Granularizing the savings

Let's granularize the savings. Line employees: administrative/clerical, call center, IT/programming, journalists, require about 100 to 200 square feet (sf) per workstation. That includes aisle and common space like break rooms and reception areas.

Trammell Crow reports US and Canadian premises lease rents are about the same: $18 to $24/sf/year. Rents include operating expenses such as utilities, security, common area maintenance, parking, taxes, and insurance. The per-workstation rent works out to $1,800 to $3,600, which includes operating expenses such as HVAC.

The fewer people working in premises offices the less space you have to buy, rent, lease, and have decorated and the fewer cubicles you have to wire to install workstations and accompanying chairs. You don't need as many parking spaces. Fewer employees also mean lower power and HVAC costs.

If your work volume grows, you can quickly and inexpensively add employees when you have them work from home, because you don't have to add cubes or scrounge for rooms or floors. Or have to rent and network in additional buildings or move into a new one.

But if your growth spurt takes place on an economic upswing, watch out. The so-glad-to-do-business-with-you-landlords during the downturn will jack up your rent and will want get better-payers into your space: just as they did during the dot-com era.

Also expect your construction and renovation costs to soar, trades people, e.g. carpenters, drywallers, electricians, and plumbers become hard to find except at top dollar and good property likewise. There is *already* a shortage of trade workers and it will get worse as these highly skilled and experienced individuals retire and die off. The reasons: fewer offspring; worsened by parents pushing their progeny into seemingly more prestigious "higher class" college education—that too often leads to lower paying jobs upon graduation.

Small premises is beautiful

Locating in smaller premises gives you many more real estate choices, which lowers costs. Instead of having to find 50,000 sf, of which there is a limited supply, you can look for 1,000, 5,000 or 10,000 sf in a sublease or in a smaller building at the lower end of the scale. You may get better deals because landlords can rent them out easier.

With a smaller space and lower costs you could afford that nicer, better-located building. You may find that you can buy that cool and very comfortable and practical high-tech office furniture like Haworth's *Look* and Herman Miller's *Aeron* chairs and *Resolve* workstations.

Relocation eliminations

Think setting up a home office program is expensive and disruptive? Try relocating premises offices. Who *hasn't* been through the nightmare of an office move? With home working the fiddling about occurs in the computer/telecom room. No boxes in aisles and stacked in cubes.

Just as an exercise, go back and calculate how many staff-hours you and your employees spent in total on the last move—from conception to packing, moving, setup, break-in, to when everyone is back up to speed. How much time did you consult staff on locations, listening and responding to concerns about traffic, transit access, amenities, security and, perhaps topmost, who gets what space? How much money did you spend in site selection, real estate and design?

Break these total costs down per employee and compare that with a home working program. You might be surprised how much money you might have saved with home working.

Be a landlord!

If your building lease allows you to sublease to other tenants, or if you own the building and want to play landlord, then home working can be used to generate some positive cash flow. If you have a lease or own the building for one department, and another department's lease is coming up, home working may let you accommodate, in one spot, all those employees who need to work at premises offices.

You may make enough savings or profit from the deals alone to finance any additional investments for home workers up front. That's win-win for everyone.

Lower direct upkeep costs

With a home working program, you may be able to dispense with the need to lease/rent a room, floor, or building, and you can transfer the power/sewer/water, HVAC, security, maintenance, and upkeep costs from your books to your employees'. They in turn, depending on the tax laws, can deduct them from their taxes.

Exclusive home office employees usually can set up and maintain their workspaces at less cost than you can at premises. They (and you) don't have to rely on the well-paid and sometimes unionized custodial, maintenance, and repair staff selected by the building owner or management company. Home workers are not dependent on commercial, and in big cities high-wage and often inflexible, union contractors.

Instead, home workers do the renovations themselves, or pay local and usually non-union residential contractors. Or have a pal do it by barter—you fix my roof, I fix your car. They also arrange for security.

If you have employees who work late and/or on weekends and your lease terms dictate that you pay for your power or security for those times, you can save money by having your workers perform the necessary tasks from their homes. Exclusive home working can work for employees who work odd shifts, and occasional home working can fill the bill for employees who work them one or two days a week.

It is also safer if employees work from home after regular business hours. If they work after-hours, there is more risk of employees being assaulted in a parking lot on their way to and from your premises, at the sides of roads if their cars are broken down, and on mass transit.

Ending the Commuting Tyranny

Commuting has become the *ultimate* nightmare, no matter the transportation method your employees use: driving or mass transit. Who loves putting up with road-ragers touching your bumper or getting up close and personal with someone else's armpit on the 5:52 to Amagansett? Even cycling and walking are less fun these days with the growing hordes of impatient nutjob motorists behind tank-like SUVs on the street.

Commuting times are getting longer and delays worse. The Road Information Program (TRIP) a non-profit group, reports that commute times increased nationwide by 14% from 1990 to 2000—from 22.4 minutes to 25.5 minutes. Cities with big commutes include Atlanta, GA (31.2 minutes), Miami, FL (28.9 minutes), Orlando (27 minutes) and Jacksonville (26.6 minutes).

The average urban driver now spends 62 additional hours annually—the equivalent of 1.5 working weeks stuck in traffic. That's up from 44 hours in 1990, reports the Texas Transportation Institute 2002 Urban Mobility Report.

That's just the average. If your travel time exceeds two hours a day, you have plenty of company. More than 10 million Americans now face what's known as a "killer commute" (defined as more than 2 hours a day) says Alan Pisarski, one of the nation's leading experts on commuting, and author of *Commuting in America*. He made his remarks in a story carried by the Universal Press Syndicate in February 2004, published in papers such as the *Chicago Tribune*.

You know this firsthand. What manager *hasn't* had their employees show up at the office 20 minutes late out of breath and clearly out of deodorant? Perhaps this time the delay was caused by a beer truck, which hit a pothole and overturned on the I-whatsit and volunteers rushed out to "help" remove the spilled cans and bottles. Another time it can be because a "sick passenger" held up the entire rail system for the third day in the row: long enough for your employee to finish the crossword or get the phone number from the cute passenger who had gotten on two stops before?

Then there's the "getting ready for work" hassle. Think about how many times you've been late getting out the door to your office because you found your shirt or tie you had just put on had a stain? Or because your !#$%^&*() "non-run" pantyhose lied?

In contrast, for your employees working from home there are few things that can delay them getting to work on time. Tripping over the cat doesn't count.

Commuting is **Your** Problem

While commuting technically may not be your problem—it is the employees' responsibilities to get to work on time—you must face reality: commuting *is an issue for you* because it is your organization that pays the price—through lost productivity, and/or lost employees who are tired of working for you because the commuting stinks. Commuting also may hurt your chances of recruiting the top performers that you need to make your outfit the best it can be. Many potential employees won't want the hassle—they would rather work for less nearer to their homes, or from home.

The hidden costs

Commuting also hits you—and everyone else—in the pocketbook. More traffic, more wear-and-tear on roads and more packed the buses and train means more new roads and transit systems and the more land ripped out of the tax base for asphalt, concrete, and steel.

Furthermore, the more demand for transportation, the more money spent on law enforcement and emergency services due to accidents and crimes. These include collisions, derailments, fender-benders, flipovers, hit-and-runs, impaired driving, derailments, plus assaults, breakins, carjackings, drive-bys, murders, rapes, road-rages, shootings, theft, and vandalism.

Justice, healthcare, and insurance costs climb up the charts, which means more of your taxes goes to pay for the commuting mess. And, in turn, that means, guess what, more taxes that for you and your employees must pay.

Early departure

There isn't a premises office anywhere that hasn't had the issue of employees leaving early to pickup Rugrat and/or Rugrat's co-parent and/or their sports team. That causes two issues: lost productivity from those employees that have had to leave (they will have to play catchup instead of being on top of their jobs), and resentment from single/obligationless co-workers who must carry the extra weight.

Employers have long struggled with both problems. With commuting times getting worse expect these matters to deteriorate.

Home working remedies the problems. The chances are that where Rugrat has to be picked up is far closer to where your employee lives than your premises, reducing if not eliminating time and productivity lost. If you determine, after examining their tasks, that you have employees who can work at-home occasionally, perhaps those days can be scheduled around their parenting and caregiving duties.

Automobile costs

Home working minimizes time off wasted in car breakdowns, repairs, and maintenance. Less driving means less wear-and-tear and fewer breakdowns.

Who amongst the car-owning public *hasn't had* their clunker get a flat, snap a fan belt, lose a muffler, or overheat on their way to and from work? Who *hasn't* eaten up their cellphone minutes in the car repair joint's squalid waiting rooms with the rancid coffee?

With home working employees may be able to get rid of the air-killer altogether. That saves them even more money, and time.

Improved Employee Productivity

Home working is more productive than premises working. It liberates you and your employees from the commuting tyranny: from wasting time in transit. Time that could be spent getting and making calls and contacts instead of waiting until the next day, which swallows up the time one had planned to allocate to other tasks. Time employees could use to recharge their batteries to be more productive the next day.

There are many examples to illustrate this. The Kitsap County study cites International Telework Association and Council data showing gains averaging 22%. Other surveys show 10%-20%.

Call centers are perhaps the leaders in measuring productivity and getting the most out of workers, with employees sitting in cubicles and handling massive volumes of calls and contacts. But home working has enabled even top call center operators to obtain more output from employees.

For example, if you call in about Procter and Gamble products, such as Bounty or Ivory Soap there is a better than a 50% chance that the call center agent will be responding to the call from their home office. P&G saw productivity climb between 5% and 10% when it implemented home working for its consumer products call center in 1999. The gain also allowed it to close one of two call center floors.

Here's why:

Home employees begin their workday refreshed, without feeling worn out and drained from commuting from their home to your premises offices. They don't have to hurry back to try and catch what was left of their waking day to relax, to be with family, friends, and loved ones.

Home working enables employees to limit message tag—they can be there first thing in the morning and later at night to catch important calls and messages. How many times have you had to play time-zone roulette with people because you were in transit to your premises?

For example, when I worked in Manhattan, I missed calls and contacts from Europe on my commute into work, and from the West Coast on my commute home. To ensure I was available for a specific call, I sometimes would be forced to arrive at the premises office far earlier than my normal starting time, or stay well beyond my typical quitting time. But when necessary I did just that, which gobbled up my sleep and family time and increased my stress. I wasted two to three hours a day in commuting round trip when I could have taken those calls and contacts instead.

While some employees can mobile work—poor signal coverage, cramped buses and trains, and if driving, the need to concentrate on that task—places limits on what they can do during their commute. As Chapter 11 on adjuncts and options points out, many jurisdictions have laws that clamp down to phone use and inattention while driving. And for good reason, for even with voice relays and speech-to-text to computers, if your eyes are off the road chances are you will be too.

Greater Staffing Flexibility

Home working permits employers to meet demand spikes; employees can be on their computers and phones in 60 seconds, rather than the 60 minutes or more that it typically takes an employee to commute to a premises office. For those reasons, customer-savvy discount airlines like JetBlue and WestJet have most or a portion of their employees work from home. That allows them to closely match labor supply to work demand and enables them to avoid paying for employees to do nothing.

Home working makes it easier for employers to hire and deploy part-time staff to meet work demand easier than at-premises working. It is less onerous on employees to be available to work for three or four hours if they don't have to consume one hour, and carfare or gas to get to and from.

There are specialist outsourcers that have sophisticating bidding systems by self-employed home workers that allow them to schedule home workers for time increments as low as 30 minutes (see Chapter 11). You pay for only those increments you buy. In contrast, conventional outsourcers need to pay even part-time people to come in for three to four hours, to make their travel worthwhile.

Free (to you) Employee Benefits

Home working gives your employees indirect pay raises and bonuses that cost you little on the salary, benefits and tax side. Your home workers save money on cars, carfares, ties, pantyhose, cosmetics, toupees, and dry cleaning bills. Jack Heacock estimates home working "gives" each employee a $4,000 to $5,000 annually.

These savings add up even for occasional home workers. One less day having to "go to work" means one less vehicle to fuel and wear down, one less bus or train fare, and one less set of expensive work clothes to rub, rip or spill.

Employee Attraction and Retention

Employee attraction and retention is becoming one of the hottest issues amongst employers. You need to attract top people; you cannot afford to lose good people.

Employee turnover, especially of highly skilled employees, costs employers more than they realize. To hire, train, and bring a replacement person up to speed costs as much as 150% of the replaced worker's annual salary. The more uniquely skilled and talented and the more loyal contacts the departing employee has, the higher the costs.

The labor supply of experienced workers is going to dry up rather than gush forth. The US and Canada is facing a diminishing labor pool as the baby boomers retire and die off. That generation had benefited from and used modern birth control, which left fewer progeny in their place.

The Employment Policy Foundation (EPF) (Washington, DC) predicted in its August 2003 report that demand for labor will outstrip supply by 22 percent in the US over the next 30 years with most of the unfilled jobs likely to be in highly paid managerial and professional occupations. Every job unfilled will cost the economy $100,000 per year in lost output and ultimately $3.5 trillion in annual output in current equivalent GDP.

The US Department of Labor estimates there will be 151 million jobs with 141 million applicants. The National Association of Manufacturers forecasts a skilled worker gap that will grow to 5.3 million workers by 2010 and 14 million in 2020.

Business 2.0 reported in September 2003 that a quarter of Cigna's IT workers will pass 55 in the next 10 years. More than a quarter of software maker SAS's staff will retire by 2010.

A McKinsey study, "The War For Talent," predicts that the demand for talented employees will rise by 33% over the next 15 years. It also predicts a 15% drop in supply.

And with once-impoverished countries and regions like China, India, Korea, The Philippines, and parts of Eastern Europe, Africa, and Central America booming or beginning to grow, there is less incentive for skilled immigrants to come to the US and Canada. Ironically these countries' and regions' existing and future prosperity derives from US and other Western nations' companies offshoring customer service, support, sales, and computer programming abroad to access cheaper skilled labor.

Traditional "Carrots" are Rotting

The traditional retention tools no longer cut it. Forget about the corner office. Promotions are rotten carrots. With the pancaking of middle and senior management layers

THE BEST RETENTION TOOL—HOME WORKING

Most employees want to work at home, even if it is only occasionally. Studies have shown that home working ranks top of the list of benefits employers can offer.

Here's a list of some data gathered by the Canadian Telework Association, one of most comprehensive sources of home working information. This is up to press time—visit www.ivc.ca for the latest:

InfoWorld, in its year 2000 survey, ranked telecommuting as #1 on the wish list of IT workers

Information Week (a CMP publication) surveyed 11,000 IT professionals. 29% rated telecommuting as something that matters most to them about their jobs

Monster.ca polled 900 Canadian workers. 38% said telework would influence their job choice

Ekos Research conducted a survey that found that 55% of 3,500 Canadian respondents want to telework; 43% would switch companies to telework; 33% would choose telework over a raise

Washington Post reports that 3,500 hi-tech workers rated telework as the most sought-after job perk

Rutgers University presented a study that found that 59% of those participating in the study said that they would telecommute if given the opportunity

Information Technology Association of America, in its July 2002 report, which cites its "Anytime, Anyplace, Anywhere" survey, found that 54% of American voters felt home working would improve their quality of lives (this jumps to 66% for those who have a one-plus hour commute), 36% would choose home working over a pay raise, 43% felt they would be a better spouse or parent if they were able to home work, and 46% think that the quality of work would improve if they were able to work at home.

People are placing greater priorities on their personal and family lives. They want to enjoy the time they have, by participating in outside activities, by taking their kids to sports practices, by evenings out with their loved ones, by not spending valuable time stuck in traffic or on trains or at airports.

Employees also know that vertical promotion is less of a possibility these days. To avoid the real chance that these qualified individuals will leave your organization entirely, offer exclusive home working—it just may help to retain them. After all home working provides them with a better quality of life, and saves them money on commuting and clothing costs.

You also can keep those valued employees whose spouses have to move by offering exclusive home working. That preserves the investment you put into those employees.

from cost cutting, how many positions, and real premises offices (with doorways!) are there left for employees to aspire to and work their tails off to get?

Forget about 'incentivizing' programs with gift certificates and trinkets, especially if that's all you do to retain them. Deploying these programs treats workers like circus and "marine park" animals: do a trick and get a banana. Jump through the hoop and get a fish.

Employees, like other animals, can see through this: they work the system. And employees, like other animals, will bolt the first chance they get.

The "who must quit their job?" Dilemma

What happens to employers and employees when both spouses work (as is often the case) and one gets a new job or must move to another city? Who has to quit? Which employee (and employer) loses? What happens if the couple can't afford to live in the new location or remain in their existing locale? Unless they're earning megabucks many couples and families can't afford big costly cities like New York, San Francisco, and Boston or Canada's Toronto and Vancouver on one income.

Moving is also costly. The Canadian Telework Association cites Statistics Canada says it cost companies an average of $42,000 to relocate the average homeowner to another city.

Think about it. If your organization is increasingly using conferencing tools to communicate with clients and customers, why move that employee? Why endure the costs, headaches, and hassles such a move would entail?

Larger Available Labor Pools

Long commutes and traffic delays and hassles shrink labor pools. That means fewer potential workers to select from and lesser chances to find the best employees around.

People who work part-time and split-shifts are especially sensitive to commute times. If they have to spend the equivalent of half their shift commuting they won't want to work for that employer. The benefits of working are rapidly eaten up by gas, vehicle wear-and-tear, or carfare.

Home working frees employers from the constraints of commuting distance and time. Employers who offer it for those jobs that are home-workable have better odds of obtaining the best, most productive workers, almost anywhere.

Note: Even in circumstances where you require your employees to come to your premises offices, say once a week or once a month, workers are willing to travel farther distances: up to twice or three times the distance and time than they would if they commuted. Traveling is then a break, not a grind.

You also get to tap into an older workforce. Many in this group would like to work; retirees especially would like to work part-time. They possess a lifetime of experience and skills: they're nobody's fools. They often have better educations and superior work disciplines compared to younger folk. But most in this group don't want the hassle of commuting and travel. Been there, done that. And the older one gets the more inconvenient and dangerous, from diminished driving skills, commuting gets.

Carla Meine is president and CEO of call center outsourcer O'Currance Teleservices. She reports that home working enables her firm to tap into a high quality and more loyal workforce. More than 80% of her agents work at home.

Meine's employees or agents are in their late 30s compared with the early 20s typically found in call centers. She says they are better able to deliver excellent customer service and sales compared with younger, inexperienced agents.

Freed from the straitjackets of dressing for and commuting to work, her agents are likely to stay longer, reducing staffing and turnover costs. O'Currance's turnover is 100% per year, compared with 200% to 300% for typical inbound and outbound sales operations.

"Our agents are those who would never dream of driving to and working in a call center," says Meine. "They are a different caliber of people."

A Stopgap to Poor Workforce Quality

Sadly, in parts of the US (and other places), the public education system has become part-babysitting service/part juvenile prison, whose "products" would be hard-pressed to read this sentence. A study released Feb.9, 2004 by the American Diploma Project (ADP) and reported widely in media outlets like *The Washington Times*, said that more than 60% of employers rate high school graduates' skills in grammar, spelling, writing, and basic math as only "fair" to "poor." The report also reveals 53% of college students take at least one remedial English or math class.

Employers have had to pick up the slack. One study estimated remedial training cost one state's employers nearly $40 million a year.

In a February11, 2004 *New York Times* op-ed piece, Nicholas Kristoff called for more math and sciences education to keep American jobs onshore. He cited a recent study of eighth-graders that was published in the "Trends in International Mathematics and Science Study." That study placed the US at 17th, just ahead of Latvia.

Jeff Furst, president of staffing firm FurstPerson, reports that in many communities, call center applicants are less skilled than in the past. Reading, writing, and comprehension abilities are diminishing; so is the work ethic.

In a typical community, only 30% to 45% of candidates tested will meet the abilities and behaviors to do the work, "which really impacts the ability to hire employees that meet the job requirements," he says.

Out of 100 interested job candidates, typically only 10 will be hired. This takes into account pre-screening, selection testing, and background checks. But, the new hires will stay longer and perform better.

Yes, this situation stinks. Employers should hold the school systems to account: they are financed by the taxes you pay. But that requires a revolution bigger than that from premises offices to home offices. Most companies are revolting, instead, with their feet—moving or outsourcing their operations offshore (outside of the US) to countries that have a better quality, as well as lower-priced, labor force.

Home working, however, enables you to compensate for declining education standards by widening the labor pool. The Internet is an excellent recruiting tool that lets you find people—and they find you—from anywhere. You can use the Web to prescreen and screen applicants, train and supervise them: all without seeing them.

Reduced Health Problems and Costs

Home working minimizes exposure and transmission of diseases that, at best, cause sick days and resulting productivity losses and, at worse, serious illness and death. The suffering also extends to the bottom line in the form of higher employee and employer healthcare costs.

Premises offices are "bug spreaders." One person comes in sniffling and three days later everyone's in the running for the "Booger Olympics." The "Gold Medallists" are those who have to call in sick; those who come back while the illness is still contagious get to train the "competitors."

Diseases also spread on mass transit. For that reason, experienced commuters quickly shuffle away from passengers that are sneezing, or if crammed nosehair to nosehair, they literally turn the other cheek.

That's no (hack!) laughing matter. The Seattle *Post-Intelligencer* reported March 9, 2003 that up to 40% of the strains of Streptococcus pneumoniae, which also causes meningitis, sinusitis, and ear infections could be resistant to penicillin and erythromycin in the near future.

The Associated Press, in a story posted on the (Toronto, Ontario, Canada) *Globe and Mail's* website, Sunday, Feb.29, 2004, reported that British deaths from an increasingly drug-resistant superbug, methicillin-resistant Staphylococcus aureus, are 15 times higher than they were a decade ago.

Some strains of staph have also acquired resistance to vancomycin, said the article. Vancomycin is considered by medical professionals as the "last line of defense" when all other antibiotics have failed.

"Although new antibiotics are constantly being developed, some experts fear it is only a matter of time until virtually every drug is useless," warns the article.

The severe acute respiratory syndrome (SARS) outbreak that hit China and Canada is an object lesson. During the outbreak authorities resorted to the disease control method developed in the Middle Ages: quarantine.

The Canadian government recommended that employers have their employees work at home. Employees exposed to SARS did work from home, remaining productive until the risk had passed. Some companies also made use of conferencing tools instead of having workers take business trips.

One of the many events cancelled in Toronto, Canada on account of SARS was, ironically enough, a global corporate site selection and real estate conference that I had been planning to attend (I also cover site selection for *Call Center Magazine*). Instead, I made my contacts virtually: by phone, e-mail, and the Internet from my home office.

A 2003 survey of CEOs by PriceWaterhouseCoopers reveals that healthcare benefits costs companies nearly $5,000 per full-time employee. Any step employers can take to lower those costs, like home working, goes right to the bottom line thus enabling organizations, like yours, to do more.

Unscheduled Absences Cost Money

A February 2003 report by Mercer Human Resource Consulting and Marsh found that employees took an average of 12 unscheduled days off. Assuming $40,000 in wages/benefits per employee working five days a week, each employee is nominally available 240 days a year [50 weeks, 10 legal holidays] each employee/day costs approximately a $167/day, or about $2,000 a year.

That figure is an *understatement*. Why? Because it does not include the missed employee's contribution: the reason why you hired that person in the first place. Nor does it include the additional burden on other employees and lost productivity and revenues caused by work not being done, like calls answered.

Significant too is the fact that more people are taking time off for personal reasons. A study by CCH, published in 2002, reported by the AP showed that personal as a cause for absences jumped to 24% from 20% in 2000 and due to stress to 12% from 5%.

Home working will minimize these absences. How? By reducing stress and by providing more family and personal time. The Kitsap County study revealed that the average commuter spends 790 hours a year in transit.

Employees could use that time relaxing, being with their family and friends, enabling family activities, and develop their careers, all of which makes them better, more productive employees. "Even a one-day a week reduction [in time/days spent commuting] is significant," says the Kitsap County study.

Premises Work Clothing is Unhealthy?

Everyone knows that the clothing many people must don for work, especially premises office work, is usually uncomfortable. Discomfort that means parts of the body is taking the brunt of the stress and that is not healthy.

High heel shoes that are de rigueur in many offices cause foot problems. Pantyhose and tights contribute to yeast infections. Research revealed by the British Institute of Opthalmology in 2003 showed that wearing tight neckties raises glaucoma risks—men don't call them "nooses" for nothing.

Home Working Promotes Fitness

Fit employees mean reduced sick days and health costs. Some employers provide gyms or pay for gym memberships—which cost money. But, even when an employer provides a gym or gym membership, they are often not used, because the employees want to get home.

There are fewer excuses for employees not to exercise and stay healthy—if they are working from home. After all, they can't use traffic and stress as justification for dropping out of the fitness war. That saves organizations money; you don't have to pay for costly fitness programs and gyms.

Home working encourages physical fitness because those employees have fewer excuses *not* to exercise; they have the time to actually use the dust-gathering rowing machine in the basement. The employees buy the equipment and work out at *their expense, not yours*. They exercise on *their time, not yours*. They select gyms convenient to where they live.

When I home worked on Staten Island, I worked out in the mornings during the time I normally wasted commuting to Manhattan. Where I live now, I take brisk walks at the end of my workday; there are plenty of hills and trails to keep me fit.

Because home workers are not driving, which most Americans and Canadians do to reach premises offices, they won't get injured or killed in accidents caused by, of course, "the other guy", or "the tree that jumped in front of me." That, too, lowers health costs; and keeps your workers alive, and productive.

Less Environmental Damage

Burning just one gallon of gasoline produces 19.64 pounds of carbon dioxide: a leading greenhouse gas, plus nine pounds from upstream refining, transporting, and refueling says the Surface Transportation Policy Project. By reducing the need to commute (over 75% of Americans drive to premises workplaces alone), home working helps to clean the air. It also removes traffic from the roads, leaving room for the commuters and truckers, who need to be on the pavement.

Just like commuting, the pollution resulting from requiring employees to come into premises offices is not someone else's problem—it's yours. You—and the rest of us—pay the consequences of commuting through environmental damage, higher health costs, and resulting lower productivity.

Business Continuity

Home working is the best strategy when it comes to business continuity. Having part of your workforce off premises, and having plans to make on-premises staff virtual workers in an emergency is your best hope of surviving natural and man-made disasters, unless the events are truly cataclysmic like an asteroid hit, the Earth's magnetic field collapsing, the Sun exploding, or nuclear war. In those cases business continuity is probably the *last thing* on anyone's mind.

Having employees work from home enables survival by dispersing your workforce across a broad area. So if a bomb, a power outage, or a tornado hits one locale the others will still survive and function. If a snowstorm hits, and the electric and telco lines are up, employees can work without sliding into a snow bank.

Employees also will not be trapped on the roads, on transit vehicles, in airports and train stations trying to get out in case of a disaster. Such sudden evacuations and panic often only make disaster response worse by clogging roads that impede emergency response.

If the power goes out in one neighborhood where one group of workers lives, chances are it will be on in another community or city where another group resides.

Proof that Home Working Works: September 11, 2001

Proof that dispersal of technology and workforces can keep companies up and running came on September 11, 2001. That day the Internet did what it was designed to do: protect the nation's information network in case of an attack. Its military-industrial creators understood the risk if a bomb wiped out vital computer infrastructure, so they created a network of distributed computers operating virtually.

When terrorists attacked the Pentagon and the World Trade Center on that day, those employees who worked at home, or who could make it home, or had laptops and could connect into a network did so; I was one of them.

My wife and I were on the New York City Transit Authority M6 bus going to midtown Manhattan; I on my way to *Call Center Magazine's* office then located at 12 West 21st Street, my wife to hers on West 31st Street. I had been working from home a couple of days a week from home on Staten Island, but that day both of us had taken the ferry in.

The bus rolled by the east side of the WTC as normal: crowds of commuters scurrying out from the subway and PATH train lines under the huge complex to Wall Street on a clear blue morning. A few minutes later, as the bus rolled through Greenwich Village the driver looked through his rear-view mirror and screamed; all of the passengers pressed their heads against the left-hand side windows and saw the smoke pour out of the north tower.

I got out at my stop, watched the black clouds billow out and listened, like the dozens of people on the street, to radio reports and speculation. Then I saw the flames erupt from the south tower.

My company shut down our Manhattan offices and the staff evacuated; I took my laptop, emergency pack (it had been for earthquakes—we had been under parent United Business Media's San Francisco-based Miller Freeman unit) and briefcase.

I ended up staying with a friend of my wife's in New Jersey for a couple of days; the authorities would not let people back onto Staten Island. But I logged into the CMP server through a virtual private network (VPN). I learned later that VPN usage jumped by 40% after 9/11.

And yes, I did get, and I did send "Are You OK?" e-mails. I knew someone who worked in the WTC, but he had been let go and was in Tampa that day...

But my wife's employer required her continue her daily commute into their premises office. The cleanup, bomb threats, and security checks made her commute nightmarish: sometimes two to three hours *each* way. Was the commute worth it? My wife didn't think so. Her job: programming and analysis was virtually workable, but she had supervisors who were leery of it because another employee who had been given home working privileges had allegedly goofed off while at home.

⊙ HOME WORKING DOWNSIDES

There are costs and issues with home working. Let's examine the leading factors that should be taken into account when considering the implementation of a home working program.

Need for In-person/premises Interaction

There are many tasks that cannot be done at home. Chapter 1 illustrates a number of such tasks including handling and treating people in-person, seeing clients and customers in person, physically manipulating equipment and objects, face-to-face meetings and networking with others, and negotiation.

But these tasks may not take up an employee's complete workday or workweek. There may be home workable duties, like paperwork, that eat up two or three hours a day. Jack Heacock illustrates the example of dockworkers who need to file paperwork at the end of their shift: that task could be done at home.

Field salespeople and support staff are primarily mobile workers. But they may not need to come back to premises offices to finish their tasks.

The Costs

Equipping employees to work from home is not a free ride. If you are replacing premises workspaces with exclusive home workspaces then you and your employees must agree on paying for equipping them with voice, data, equipment, and furniture. Your per-employees voice/data costs are higher that at premises—though with new technology such as VoIP and intelligent network routing, plus greater competitiveness amongst carriers, the differences are quickly shrinking.

You may need to provide hot desks, which cost about the same as regular workstations. But you can make them smaller than regular cubicles (no one is keeping files or family pics on them). More importantly, you spread out the cost over as many as five or more workers, depending on how often you require your home workers to come into the premises office.

To make exclusive home working work, you will need to research and put together a program, and/or hire a consultant to do it for you. Occasional home working is less extensive and does not need to be as formal.

You also will need to spend time in putting a home working program together, selling it to senior executives and adjacent departments, such as HR and IT. You may need a separate IT home support team and dedicate or hire a Home Working Program Manager to oversee the program. You also will have to select, qualify and possibly train people to work from home.

Existing Property

If your organization has just signed a long lease with a landlord for premises offices, then there may not be any real estate savings. Or if there are too few employees qualified to exclusively work from home, you may see little savings. Also, some landlords will penalize you if you let space "go dark," i.e. vacant, even when you're still paying rent. However, you may be able to sublease—if you are in an attractive location, with a good building, in a prosperous market.

There may be deals that could offset real estate savings with home working. Hungry landlords may offer zero-net-leases where in effect you're paying just for the overhead, with very little or no profit. In some cases local economic development agencies are willing to give buildings and other incentives that reduces taxes and training costs.

You may be able to negotiate leases as short as five years and exit in three years. But you may not be able to exercise such options if you received government incentives requiring you to be in a building for X number of years, unless your lawyer is bigger than the government's and you don't care about the bad publicity.

Supervision

Far and away, and since time immemorial, employers have instructed, corrected, rewarded, and punished employees to their faces. Employers have measured employee performance by their direct senses, against what their supervisors expect.

Home working, especially exclusive home working, requires that you implement a different way to supervise workers, and that is by performance (see Chapter 9). That means supervisors have to trust employees, and exercise people-skills like interacting with them rather than talking down to them. Some supervisors may get with the program; others may drag their heels to sabotage it.

Internal Communications

Communications is another major hurdle. There is still nothing like talking to someone face-to-face, or hearing others talk firsthand. Employees who work at home are "out of sight/out of mind" *unless* supervisors and employees make it a point to communicate with each other. When you need to talk to employees, either one-on-one, in a group, or to make an announcement it is often easier or quicker (and the message is delivered with more impact) in-person than it is over the phone or via the Web.

Also, when employees are on premises they usually feel more connected to what is going on; they hear the gossip and rumors, and ask questions. There are greater opportunities to chat with individuals outside of their own work groups, such as those in other departments. For employees wishing to advance, they often need that kind of information to plan their career moves.

Home Environment Distractions

Home-based distractions are another concern of employers. Few people live like hermits, though loners usually do make good programmers (and writers).

Chances are your employees will have two-legged and four-legged (or more) critters, and often both. Chances are they won't respond immediately or well to "shut the QWERTYUIOP up!" or "get the ASDFGJKL:" out of here!" Sometimes if you call your workers at home you may hear the 3:15 train to Poughkeepsie or the 3:55 flight for Port Angeles or a "music" performer or TV set hitting equivalent decibels.

The last thing you want is your customers, clients, and colleagues to know that your employees are working in anywhere but a quiet, professional environment. In reality, many premises are anything but, judging from high noise levels when I've talked to people in offices and call centers.

Then there's personal computer and the personal phone, which uses you can't trace. Or the afternoon ball games your employees have always wanted to watch; they can bring the TV into their workspace and you'll be none the wiser.

Data Theft

There is a risk that home workers that they have access to customer or other confidential data could steal it. Data theft is becoming a very serious and widespread crime and there are now laws to stop it, including provisions that penalize companies that don't protect it adequately.

The data theft and law violations risks increase when the data leaves the premises, such as to at-home employees. Employers have less de facto control then because there is no one on-site to make sure rules and regulations are complied with. On the flip side, however, at home offices, unlike at premises offices (where people other than the individual employees have direct access to their computers and files), the home office worker typically has a great deal of control over who gets into their equipment and information. They can lock the doors and secure the computers, including if necessary, disconnecting the laptop and locking into a file cabinet.

Career Development

Out of sight is, for many people, still out of mind. People are, for the most part, reluctant to promote or to hire someone sight unseen, especially when there are tasks that involve interacting face-to-face, such as on a business trip.

But this is changing. Mainly due to the growth of online recruiting and hiring, more careful screening, stricter background checks, and employers placing more focus on performance.

Corporate Culture

Many executives shut their minds when you mention home working because they feel that employees who work from home miss out in being part of that organization's "corporate culture." Countless outfits have built an identity around their corporate culture (e.g. company events including the dreaded "office parties" and "picnics", field trips, getaways, and community works) and use that to attract and keep employees. These organizations stress loyalty and service, with the "office" as the secondary or primary "home."

Is corporate culture dying? There is a lot to be said for caring about employees, and encouraging an atmosphere of togetherness within an organization. But many of these attitudes and trimmings of corporate culture were dispensed with during the recent downturn. Employer and employee loyalty are a part of history.

Employers reorganize, downsize, outsource, and move operations to lower cost-locations, including offshore; employees go job shopping on the Web. Socialization is disappearing: in striving to become more productive there are fewer and shorter lunches and after-hours "get togethers," and no one wants to risk dating in offices. Strict impaired driving laws are doing away with "happy hours," making lives truly happier for those who would have been injured or killed by drunk drivers. With an aging, more settled workforce, fewer people have the need to hang out with their co-workers.

A competitive workforce means few friendships; you may have to walk over your colleagues just to keep your job. Nothing personal: just business.

Couple those items with nightmarish commutes, and you find that most of your employees just want to get to "the office," do their jobs, and get home.

Workers also are understandably more cynical. They will put in 150%. But they don't believe that you actually "care" about them.

Nor should you care about them. Your organization exists to meet objectives, like making money, satisfying legislated goals, and winning causes, *not* to run a social club. You have an obligation to pay and treat workers fairly in exchange for good hard work. That's all *any* employee expects of you.

Top Management Buy-in

Just as home working is a different way of managing for supervisors it is equally a radical means of governing for *your* superiors. The organization's fiefs have the right to inspect their fiefdoms and many enjoy doing so. They want to see and want to preserve the right to see that everyone is working hard and properly.

Premises offices are becoming frills. If employees own computers, have voice/data connections and rely on them to work, then why pay money out of *your* profits and *your* budgets to duplicate this in buildings?

Your management should turn the paradigm around. Managers should instead be forced to make the business case for premises offices because they add costs and take away from the bottom line.

○ HOME WORKING RECOMMENDATIONS

Many of the objections and issues raised with exclusive home working can be worked around. Take a close look at the situation. You may find home working doable even after you've run into what may seen like insurmountable obstacles. While other chapters offer suggestions on how to cope with most of these matters, here is a summary of potential solutions:

Winning over top management

You will have to make a strong business case for top management to get them to approve home working. Chapter 13 provides planning points to ways how.

The main to emphasize is the bottom line. The aforementioned sections had explained that home working could save you and your organization *serious* money. Money that had been spent on unproductive premises offices could go more productively instead to boosting profits, getting more done, and enhancing the value of stock options.

Deciding and analyzing which tasks are/are not home workable

Take a close look at the functions of your employees. Determine such as from the guidelines in Chapter 1 which positions are home or not home workable. But in doing so, be mindful of rapid technological changes that are making many functions now doable from an employee's home. Keep it open. For example, I saw an ad for a virtual museum in Canada that shows a lonely house on a windswept cliff but inside the children were seeing and learning about nature. There are now "virtual tour guides."

Productivity, other benefits, if real estate benefits not feasible

Examine closely all the factors that impact on productivity: tardiness, early departures, sickness/injury, turnover, and quality of work. Take a hard look at the facilities costs, e.g. workstation installation, depreciation, moving, and maintenance costs per employee. You may find out that the benefits from home working, minus the costs of setting up and managing home workers, may exceed the costs of letting existing or just-leased space go dark.

Real estate professionals say there are many millions of square feet of leased commercial space that is not being used, but also not on the market, especially in large metropolitan areas like Dallas-Fort Worth. If you want to see for yourself drive through any office park complex at night and glance through the windows. Chances are you'll come across buildings without "For Lease" signs with floors that have lights on but there may not be any desks or workstations, or signs of use.

Your lease may permit you to sublease that space made vacant by home working. Or if leases come up on other property you have in the same locale, you may move into that building those employees who cannot work from home.

Jack Heacock urges that you examine how much it costs to fit out and maintain premises offices. In one of his examples, it costs $35,000 to put in a workstation. But real estate only accounts for $7,000-$8,000 of that. That leaves $26,000 to $27,000 the organization has to fork over. After subtracting the amount needed for hardware/software/voice and data for all employees the organization is still faced with spending $17,000 to $19,000 for each workstation setup at the premises office. That does not include maintenance—which home employees do—and moving workstations around, which is rare in home workplaces and is done by the employees themselves.

Should you ever have to move or consolidate your premise offices you should examine how many employees can work from home and set them up long before the move date. That reduces the amount of space you need, hence your costs and minimizes disruption. Setting up individual home offices is relatively easy: cut-over from premises offices can be as simple as unplugging the laptop and taking it home, for good. You can get the workstations and lines installed and checked days in advance, and check them out.

Voice/data/equipment technology advances

Voice/data/equipment technology is advancing so rapidly, while costs are plummeting, that it may be worth your while to make supplying and supporting home workers as cost effectively, IT-wise, as hooking them up and outfitting them at premises offices. If you run the numbers, you may find it worth your while to junk the old phone switch.

Network/intelligent routing has become so cheap to enable you to provide services like "being there" call monitoring at the same price as long distance rates. VoIP offers one-wire routing that avoids long distance charges entirely. Laptops have become so inexpensive that they are displacing desktops; tablet PCs may displace laptops. Storageless PCs have emerged as a low-cost, low-footprint and much more secure option to desktops, enabling organizations to comply with strict data privacy laws.

Besides, with properly selected home workers, you can lower your IT costs. Because they know the help-desk is many miles (or kilometers) away, they're more likely to solve their own problems—as long as you teach them how—than those workers in premises offices.

Supervisor buy-in

Concepts such as performance management and efficient and effective interactions with employees may be new to supervisors. You will then have to sell them on home working and management by performance, including what is in it for them; offer *them* the opportunity to work at home. Of course, then you will have to train them how to manage that way.

Communications techniques advancements

You may have noticed that increasingly internal communications involve less shoulder-tapping and more "mail-flashing." Methods such as the traditional phone, e-mail, and instant messaging (IM) have replaced the time-wasting and disruptive method of walking to the cube. Also, more organizations are mass commuting online than with in-person presentations.

All these advances and changes bode well for supporting home workers. All of these methods can be, and are being, used successfully by managers to communicate with their home workers and to keep them in the loop. Colleagues and team members also communicate electronically, especially with IM, and with more advanced methods like collaborating and whiteboarding through data conferencing.

Home environmental management

In many cases you can manage noise and other distractions at home offices by working them out with your employees. If they want to work from home badly enough they will make sure they will not get bothered by rugrats, on real life or on TV, or by the "real thing" on four legs.

If you manage by performance, so what if your employee knocks off early to watch TV—as long as they don't hurt their performance. If they goof off too often remind them that working from home is a *privilege* that can be easily withdrawn.

Data protection techniques

There are many methods for protecting data at home. They include requiring your home workers to have their offices behind locked doors, providing lockable filing cabinets, keeping your data on servers behind firewalls, gating access through methods such as virtual private networks, authentication devices like USB plugs, smartcard readers and fingerprint and iris scanners. You can also equip home workers with storageless PCs on which spyware cannot be installed.

Career development?

Deploying modern communications techniques and managing by performance keeps in the loop those employees who are interested in a career, but still want to work from home. As more organizations and departments offer working from home, "being there" is becoming less of a barrier. AT&T found that home working *improved* promotability; nearly one in three managers report that teleworking (including home working) has had a positive effect on their career.

There are good reasons for this. To work from home you must be self-disciplined, conscientious, a great communicator, and a problem-solver. You must also take the initiative. Aren't these the same characteristics that enable great employees to move up the ladder?

Also with today's pancaked management structure and impending labor shortages, you may find that more employees, especially older, settled workers seeing few advancement possibilities, prefer to improve their lot horizontally by working from home. That way you can keep these excellent workers that you have invested time and resources in, and they have paid you back by creating net value for your organization.

Expressing corporate culture—at home

If having a strong corporate culture is important to your organization, you can have that culture expressed at home. E-mail newsletters, conferences, collaboration, and websites are great tools for this. They also cost less than putting on picnics and parties.

If the culture stresses family and employee well-being, make it clear that home working meets those goals because employees can spend more time with their families instead of being stuck, snarling in traffic. If the culture encourages community involvement, set up programs and procedures by which home workers can become involved, on your organization's behalf, where they live. That has the added benefit of giving your outfit more, and wider-spread positive publicity.

Considering occasional home working as a first step or alternative

Occasional home working—having employees work from home one or two days a week— confers most of the benefits of home working: reducing tardiness and early departures, increasing productivity, hiking employee retention, lowered healthcare risks, and business continuity. The results are scalable: the more people who work from home and the more often, the greater the gains.

Occasional home working is a great technique for maintaining performance from those employees whose tasks require them to be in premises offices, or whose jobs are physical and/or face-to-face most of the time. They can undertake those duties that can be done at home, such as writing reports.

Occasional home working misses out on the big real estate and building operational gains of exclusive home working. There are nettlesome issues like who pays for what, reasonable requirements for temporary home workspaces, and whether to permit employees to use their own computers to work on and access your network to be worked out.

But occasional home working minimizes the investments, commitments and staffing, training and management changes required for exclusive home working. And you should make occasional home working part of your business continuity plan.

Moreover, occasional home working is a good first step on the road to exclusive home working: walking before running, for you, your employees, their colleagues, your supervisors, and for your top management. Chapter 13 on planning outlines this further.

Chapter 13: Putting Home Working Together

You may have decided that home working will help your organization by, for example, improving productivity, reducing costs, attracting and keeping valuable employees, and enabling your outfit to withstand disasters. Or that your organization can no longer afford to keep the "tried-and-true" premises offices because your workforce is continually stuck in traffic or forced to work in buildings threatening to become the next "Flu Freddies," "Typhoid Marys," or "SARS Sams."

Your initial analysis helped you to make up your mind that the benefits of home working outweigh the initial costs and effort entailed in setting up and managing a home working program, be it an exclusive home work program, or a more informal one involving only occasional home workers, or a bit of both. You are satisfied that you can handle issues such as ensuring your home workers have the right facilities and environment, that you can manage them remotely, and that your supervisors can do the same.

You also have decided on the goals of your program. These targets can range from keeping those employees who must occasionally be at home productive, to accommodating and retaining good employees, to meeting defined objectives, such as productivity improvements, real estate savings, cleaning up the environment, and reducing traffic congestion.

In addition, you have figured how you are going to manage the home working program. You have looked at and decided on qualifications, screening processes, and how you will ensure program compliance. You have thought out how you are going to have your home employees supervised and have trained both the employees and supervisors. You have decided on and put together a draft home working policy.

At the same time you may have decided that your home working program needs a separate set of metrics to measure its performance. You have figured out that you need a manager to implement, run, and be an advocate for home working. The next step is to put it together.

✪ PLANNING: THE HARD PART

Now comes the hard part: planning your home working program and marketing it successfully to your colleagues, to counterparts in other departments, and to senior executives.

Remember: home working by now may seem common sense to you and many of your employees, but it is a different, revolutionary, and out-of-the-blinders way of working for most managers, department heads, directors, and vice presidents. Even though, let's admit it, many of them like working from home too.

Planning can be as simple as getting permission from your superiors (if you need it) to let employees occasionally work from home on their machines, stipulating that they notify you, Reception, and their colleagues their location, and provide a contact number. Or the program can be a full-scale business plan with details of the program laid out: cost-benefit analysis with costing; data collection, and benchmarking to flesh out the figures; pilot projects; training and education projects; and roll-out timetables.

You also will need to discuss the program's plan with your supervisors and with your counterparts in affected departments, especially HR and IT, to get their input; then shape the plan accordingly. In putting together your plan you *must* consult with your employees—after all the plan is for *them*. They will let you know what is doable and what is not.

If the plan is fair-sized, you most likely will market it to senior management. Once you have top management approval, the next step is to initiate it (next chapter), with feed-back cross-checks from all parties so you can tweak the program as it is put in place.

✪ TYPES OF HOME WORKING PROGRAMS

The book has discussed what goes into home working: voice, equipment, data, locations, facilities, employer investment, administration, and management based on the types of home working. Those types, to repeat (this is a big book), are *exclusive* and *occasional*.

Exclusive home working is where the primary workplace of the employee is their home; there is no dedicated spot for that employee at a premises office. Occasional home working is where the secondary workplace of that employee is their home; the primary workplace could be premises offices or mobile, including operating vehicles.

But there are also two types of home working programs, *formal* and *informal*. Here is an explanation of them:

Formal home working

Formal home working is just what it says: a well-planned home working program that meets specific objectives, that has dedicated budgets, administrative and management procedures, and staff. This book covers the key elements that go into formal home working. Formal home working includes exclusive or occasional home workers.

In a formal program, employees are typically educated about home working, the requirements for becoming a home worker, and how to go about applying for the program. Also, when job openings come open and they are for a position where part or all of the tasks are done from home, employers post the qualifications/equipment and facilities needed.

With formal home working there is typically a home working policy that states the rules. For example, who is eligible, what the employer will supply and pay for, what the employees are expected to supply and pay for, home office requirements, and grounds for ending home work. Employees are required to sign a home working agreement.

Formal home working can be (and should be) part of an organization's business continuity plan. The plan could set out, for example, who gets the laptops in an evacuation; the laptops would be preloaded with the programs these employees need. The plan could have notification procedures of prequalified or essential employees who are at home or who are on their way in to tell them that they are working from home instead. There will have been network and e-mail access prearranged.

Informal home working

Informal home working is also what it means. There is no established program, budget, procedure staff, or policy, except perhaps with IT, if network access is involved. Employees work from home occasionally or exclusively.

With informal home working, employee approaches their supervisor saying that they want to work from home, usually occasionally, and the supervisor says "fine." Supervisors approve home working on a case-by-case basis; they can end the home working arrangement without notice.

The employee may have a laptop or they may use their own computer; there may be Webmail set up if e-mail access is important. Costs like long distance and Internet charges are expensed. If there are IT issues with the organization's gear, the employee deals directly with the IT department; if there are issues with home workers' own machines and voice/data networks they deal with it.

✪ THE HOME WORKABLE "INVENTORY"

The feasibility of home working, how many employees will benefit from it, and how much it will cost depend on what your organization does and the functions and tasks your employees carry out. By modeling your existing workforce, i.e. seeing how many would be eligible to work from home by determining if their functions are home-workable, you can get a reasonable indication of the size of your home working inventory. That analysis also will help to determine the cost savings that will result from a home working program. For example, with a home working program how many premises office workstations will be eliminated, what will be the reduction in the size of property needed and the amount of parking needed. This will, in turn, provide a savings in overall premises cost.

Assessing Home Workable Functions

The first step you should consider taking is to take inventory of your employees and their tasks, then determine which of those tasks are home workable and which ones are not. You can determine that from reviewing each employee's duties, what equipment each uses, and from observation of your employees at work, (perhaps they already do occasional home working). You also can get advice from your HR department and from home working consultants.

Now, I am going to repeat and paraphrase some suggested suitable tasks for home working as laid out in Chapter 1:

Editing and writing
Graphics design
Proofreading
Transcription
Translation (spoken and written)
Legal
Credit/background checks
Recruiting/interviewing/testing
Research (chances are that most material produced since the mid/late 1990s have
    been archived electronically)
Business/program analysis and planning
Computer programming
Data and report entry
Database management
Website hosting
Customer service, reservations, support and sales
Internal help desk
Accounting and bookkeeping
Programming
Engineering and design
Sound recording
Teaching (including lecturing, and correcting students' work)
Tutoring
Training
Management
The list goes on...

There also are many tasks that can't be readily worked at home. Examples include:
* Where manipulation of material, equipment and machinery is involved. For example, assembling items, repairing by hand and with tools, driving.
* Where you must physically interact with people. Examples include medical, retail sales and service.
* Seeing clients and customers at their premises, at yours, or at events such as trade shows.
* Training when direct touching and operation of equipment and the handling of tangible products is involved, including training others face-to-face.
* Teaching, where teachers need to read their students' in-person behavior.
* Negotiations and dealmaking, where it is vital to read others' full body language.

As a general rule, any task that involves creation, handling, using, and processing information received electronically (and by that I also mean by phone) is home workable. And, as a general rule, any tasks where employees have to physically interact with others (e.g., handing another person an item from a shelf), or with equipment (e.g., operating a lathe or driving a truck)—in other words, employees who must eyeball another, network, or touch objects cannot be home worked.

But these are general rules. There are special in-person tasks that can be done at home, such as accounting, bookkeeping, and law; clients and customers can go to home offices.

And chances are more non home-workable tasks will become home workable as technology improves (and becomes less expensive), enabling more information is stored as bits and bytes rather than ink-and-paper. In publishing, printout galley proofs or "blue lines" where you really needed to be there to review have been replaced by .pdf files, which can be reviewed and marked from home or on the road.

Assessing Home Workable Time Periods

You will also need to assess how much of a workday or work week a potential home worker performs each task, and if the tasks can be readily compartmentalized. Once the determination is made, then you can decide if, and how much of a day or week, those employees can work from home.

Jack Heacock's example of a dockworker, discussed in Chapter 1, illustrates this nicely. For most of that shift that dockworker is loading and unloading vessels, but for part of that shift that employee needs to file paperwork, which can be done from home on a laptop and/or home fax machine.

In my field, an editor or managing editor for a monthly publication may need to come to the premises office one or two days a month to go over layout with sales staff. But as all content becomes digital, including ad submissions from agencies, there may be no need for that commute; the documents can be shared in a dataconference.

Determining Home Workable Workforce Size

If you know the tasks that can be worked from home and the time periods, you can figure out from the size of your workforce how many employee positions are potentially home workable. That will determine the size and scale of benefits.

Notice that I said *potentially*. There are probably many existing employees in those jobs who lack the qualifications to work from home: the inability to work without supervisor or without needing others around them (Chapter 9) or they lack the appropriate facilities, environment or location (Chapters 5-6). Or they are not willing to work from home, period.

But that can change. Many organizations that began offering home working started to attract a different and often better caliber of employees.

The ARO Experience

ARO, a Kansas City, MO-based call center outsourcer provides a stark example of this. Its labor force completely changed from young 20-somethings to 40-plusers. When the company began advertising for at-home agents, Michael Amigoni, ARO's chief operating officer, found to his surprise that instead of tech-savvy young adults the respondents were largely baby boomers. These workers are educated, experienced, highly skilled, loyal, and responsible, he says. They also want to work part-time to supplement existing incomes.

The home-working employees enabled ARO to cut the supervisor-to-agent ratio to 1:20 from 1:10, thereby reducing labor costs. The baby-boomers also matched the demographic of ARO's clients' customers, which improved sales and satisfaction.

The combination of home working and a more mature workforce enabled ARO to slice turnover to 7% from 25%, saving $126,000, while productivity rose by 15%, netting $375,000.

"I could not believe that I could get this caliber of people working for us," says Amigoni. "These are people that would never step foot into a call center."

◎ HOME WORKING TIMETABLE

Home working can be implemented as quick as a "yes, you can work from home," i.e. an informal home working agreement. But chances are there will be some formality to these arrangements and that requires lead time—how much time depends upon the scale of investment and implementation, number of employees and staff involved, and budgets.

Implementation

A formal home working program will need about 6-10 months of planning and implementation. For example, plan on a few months to source, buy, install, and debug hardware and software like standalone off-premises extensions or new switches with IP gateways to reroute calls to home workers.

A typical informal home working program takes a few weeks, depending on how informal you want to make it. It can be as simple as waiting for the telco to install additional lines into homes and ordering or picking up a few laptops.

Seeing results

The results of home working can begin as soon as employees start working from home. Employees see immediate gain, like setting the alarm for 8am instead of 6am and donning sweats instead of suits. You can begin tracking their performance from the get go.

If the reason for home working is real estate avoidance, the savings begin when you are out of your old, unneeded building. Or when your workers' phones go live at their homes instead of inside new, costly, and useless additional offices.

Based on Jack Heacock's model, with 100 home workers, which as presented in Chapter 1 and Chapter 12, you will experience modest gains in the first year, but there will be greater benefits thereafter. For example, the first year average cost for a home worker is *$9,740* but that drops dramatically to *$2,640* in the second year. The net annual recurring savings in the first year is *$1,276,000* but that grows to *$1,986,000* in succeeding years.

◎ MEETING GOALS—WHAT'S ENTAILED

What you want your home working program to achieve defines the scope and size of benefits and investments—time and resources—to make it happen. The goals you pick determine the scale of planning.

You then need to determine whether, and how much of, the tasks and functions your employees carry out are or *will be* (with technology changes) home workable, i.e. the "home workable inventory." That will draw the field of the results you can expect from home working and in doing so enable you to plan how much effort and expense (time-wise and staffwise) it will take to make those goals.

You also will need to look at the timeframe to meet those goals. How much are you under the gun to cut costs, improve productivity, and/or meet government clean air regulations? Are you facing a major expansion of staff, but do not have the room in your existing buildings, or the lease expiries on your current premises?

Here are a few goals to consider, and a brief discussion of what is involved in meeting them:

Enabling work when employees have to be at home

Scale of planning and investment involved: To enable employees to work from home does require some planning and expenditures. Do they call in for messages and give you and their colleagues their home phone numbers? Or do they have calls transferred to their home or wireless phone, which you pay for, directly and indirectly, by picking up the employees' costs? If they are making work-related expenditures, i.e. long distance calls and faxes, you will need to pick up the costs; your expense procedure and policies will have to permit this.

You and your IT department will have to decide on equipment and network access. Do your employees use their own computers, and load your software and access your network? Or use their software and access their e-mail through Webmail? Do you issue them a pool laptop that they can use to access your network? Or do you decide to bite the bullet and replace their desktops with laptops?

You may need to consider whether to pay for voice/data lines and whether to pay the charges directly, especially if employees work at home fairly frequently, like once or twice a week. There may be substantial cost savings if you have agreements with your carrier, and your employees make a lot of calls.

Maternity leave or other long-term leave employees are exclusive home workers. The home offices will have to be set up accordingly.

Attracting, meeting needs of, and retaining, staff

Scale of planning and investment involved: The planning and implementation to meet your goal is more involved than with enabling work only when employees *must* be at home. In that situation, selecting staff is informal; they come to you or to your supervisors, then you and they decide on a case-by-case basis whether home working will be allowed.

With a formal home work program, you and your supervisors will need to develop guidelines to ensure employees are selected fairly. Those guidelines include performance without supervision, setting out the conditions of home working, and the employees assenting to the same—e.g. no background noise, lockable doors and windows, and home office inspection with notice.

You will need a formal policy that sets out who pays for what. If you have exclusive home workers you probably will need to invest in voice and data connections to the employees' home offices. You also will need to decide on equipment and network access and have trained IT staff to support them. You will need to set out administrative matters such as sign in/out procedures, data security, and how to handle terminations.

Business continuity planning

Scale of planning and investment involved: In incorporating home working into your business continuity plan you will need to decide ahead of time which employees can, or need to, work from home in emergencies. You will need to determine which employees are the most critical to your enterprise and examine which of them can carry out just a few of their tasks from home, have at minimum dial-up Internet access, and find out which ones will be willing to do so.

You also will have to make decisions about computing, phone, and network access. In disasters, PSTN lines are usually more reliable than voice-over-IP; dialup is less likely to go down than broadband, which requires a commercial power source. You should require all emergency home workers to have surge protection to safeguard their equipment (and yours, if they have your laptops in tow). You also should consider equipping with battery-powered uninterruptible power supplies home computers for any worker who is critical to your business's continuity. Better yet, if they have laptops, buy or ask them to have charged additional battery packs handy.

You will need to set up procedures to activate these workers when disaster looms or has happened. For example, in the event of a disaster warning from the National Weather Service that arrives prior to the opening of your office, your plan can stipulate that those critically designated employees stay at home. In the event of an evacuation, you will issue laptops only to those workers; otherwise the machines will remain under lock and key.

Meeting defined objectives (productivity, real estate, environmental benefits)

Scale of planning and investment involved: To reach these goals requires the most planning and resources. There are objectives to be met; the home working plan must be designed to meet those objectives, i.e. voice, data, and equipment at employees homes; hot-desks, if need be, at premises or satellite offices; and equipment and access set up at the employer's end, including IT support.

Organizations will need to set out in a detailed home working policy, facilities, locations, environmental and furniture/ergonomics and, if need be, privacy and insurance requirements. Employees may need to approve, as a condition of home working, permission for employers to check to see if they are complying with the policies.

If employers need to have employees come in periodically, the policies must stipulate that and set out the equipment to be provided to the worker while at premises: hot-desks for laptops, spare cubes with desktops, and phone systems modified to accommodate them. Normal management and HR procedures, like sign-in/out and termination must be adapted for home workers.

Such employers also must formalize home worker selection, training, supervision, and management, including training supervisors. They need to set up home working screening criteria, and rigorously assess existing and new hires for them. They should consider establishing means to regularly communicate with home workers, such as through instant messaging, along with the ability for home workers to easily communicate with colleagues.

Note: *Consider beginning your home working program with modest goals, e.g. enabling work when employees have to be at home FIRST. Many organizations want to see whether you can crawl before you can walk, walk before you can run. Otherwise you run the risk of falling flat on your face.*

✪ ASSEMBLING THE PLAN

You now should know the goals you want your program to achieve, the scale of the program, which functions and job titles are home workable, size of eligible workforce, and some idea of expected results. You know to measure it and which methodologies you will use. Now you need to start putting together your plan.

Let's go over the key components of a home working plan:

Technology (voice, equipment and data)

Your plan should set out what technologies your home workers and your organization need and will need to enable home working at a performance level that is equal, if not better than, premises working. That can include voice lines, Call Waiting, voice mail, computers, fax machines and printers, Internet connections, Wi-Fi and VPN lines, and authentication devices. That can also include off-premises extensions and VoIP gateways. Chapters 2-4 and Chapter 7 cover these tools in depth. Notice the words "will need." You need to keep options to grow and change technology.

Locations/facilities/environment

The plan must outline your physical home office requirements; this includes location—if you need your home workers periodically to come to your premises offices. It should spell out your workspace, environment (noise, human and otherwise visitors), furniture, and privacy policies. Chapters 5 and 6 cover these details in depth.

If your plan calls for having your employees come in to your premises office periodically, then you need to consider hot-desking: shared work areas. Chapter 7 looks at hot-desking.

IT Support

Equipment and networks go down on occasion. Your plan needs to set out how you will support home workers. It should set out how you plan to deliver field support, such as with third-party vendors. That can include having a dedicated IT team, investing in remote diagnostics and repair software, and enabling home employees to troubleshoot. Chapter 7 touches on these issues.

Administrative/HR/Legal

The home working plan needs to be in compliance with your HR and legal policies. That includes availability (e.g. sign in/out, absences) data protection, insurance, disability legislation, unemployment, and taxation. One example: letters stating that your employee is working from home for your convenience; you require them to work from home either exclusively or occasionally as a condition of employment. Chapter 8 examines these issues.

If your home employees have other employees working there, you need to set out requirements akin to small or satellite offices. If the home employees see clients and customers or handle goods on site, and otherwise fall under local bylaws, then your policies must set out these conditions and how your employees should be responsible for them.

Employee/applicant qualification and training

The home working plan must present qualifications for home workers. It must set out selection and, if necessary, termination procedures. It needs to outline how you plan to train home working employees.

But training also includes schooling on home working. Key details: how your employees should set up their home offices and how your supervisors should manage home workers. See Chapter 8.

Home worker management

This is key. The home working program needs to show how you will supervise, manage, stay in touch with, correct, and assess your home workers—you can't tap them on the shoulder. That can include deploying instant messaging, peer collaboration, having conference sessions and extending call/contact monitoring to home workers. You also need to look at how you will develop home workers' careers.

There are special circumstances you may need to look at, like international, small town home, and in-person (i.e. clients/customers/other employees at employees' home offices) home working. Chapter 10 explores these matters.

Management also includes measuring performance. It also entails measuring return on investment. Chapter 9 covers management.

Planning for alternatives and adjuncts

Your plan will need to cover any alternatives or adjuncts as explored in Chapter 11. For example, the extent of your home working program may be to outsource it to a self-employed home worker contractor or outsourcer.

You also may wish your employees, premises or home-based, to go on the road. There are distinct advantages, costs, and disadvantages to that. Conferencing: by audio, data, video, and Web can offer cost savings and productivity improvements (similar to home working), without having employees work out of the premises office. Conferencing can be a great tool in enabling home working, too.

Satellite working, including hoteling, as detailed in Chapter 11, may provide an option to home working because it minimizes commutes, thus saving time. But satellite working also can be an adjunct to home working for that same reason; it lessens the travel time for home workers who need to meet with your premises-working staff. Satellite working also extends the distance by which those home workers can live from the nearest premises.

Note: *If you use self-employed home workers you will need to lay out what you want them to do, what you will pay them and the terms. Then you have to seek, assess and decide on those vendors. Remember that the self-employed are not your employees. You don't control them. You can only manage the results, not how they do it.*

Employee/employer responsibility

Home working is a partnership between employees and employers. The plan must lay out who pays for what: voice, equipment, data, and possibly furniture. Employers must present, and employees who wish to work from home must assent to, a Home Working Policy that includes stipulations such as employees granting employers the right to assess and approve their facility for home working.

Plan implementation

You will need to set out how you will roll out the plan: from research to procurement, education, training, employee notification, and screening. The plan must include a timetable for each part to be in place by. The plan should also outline how you will modify it. Chapter 14 covers this topic.

The plan also needs to cover how to end home worker participation or end home working altogether. The last chapter, Chapter 15, illustrates ending home working.

Plan management

Every program needs someone to take responsibility for it. You may need someone, and possible support staff for that someone, to put together and manage your home working program. Their duties might include working with other departments, *especially* HR and IT; monitoring the program; suggesting changes to it, etc. That's where having a Home Working Program manager (Chapter 9) comes in.

The home working manager individual/position should have the same stature as facilities manager since the jobs are parallel. Just as facilities managers put together building policies, home working managers should write up home working policies.

○ DATA COLLECTION/PROGRAM VALIDATION

To flesh out your program, you need to gain the facts to show that the goals are valid and the program is feasible, and to make the financial case for your cost/benefit analysis. Here are some steps and tips:

Obtaining outside cost estimates

You will need to obtain cost estimates for hardware, software and connections; voice and data services like phone lines, local and long distance service and broadband Internet; laptops or desktops, printer, copiers, fax machines; and if you supply furniture: chairs and tables. Also include in your estimate technologies like VPNs and instant messaging to enable everyone to easily stay in touch. If required, you will have to get quotes for expensive technologies like OPXes and VoIP gateways. Also don't forget seemingly mundane, but quick-to-add up items, like office supplies.

Note: *There are still economies of scale resulting in lower prices in buying items like voice/data and office supplies for premises offices compared with buying them for home offices, or having employees expense them. But home workers can more quickly take advantage of deals, sales, and giveaways and do comparison shopping for services and items that meet their exact needs than corporate departments locked into purchase agreements with selected suppliers.*

Obtaining organization cost estimates

Equally important, you will have to find out the organization resources you will need to support home working. That includes additional (if any) HR and IT staff or staff-hours to support the home working program, and space, workstations and/or voice/data for hot-desks. It also includes home working training for your supervisors, a home worker manager, employee education, and other roll out costs, plus budgets for visiting employees.

Case studies/benchmarks

There are many organizations similar to yours that have done home working and quite successfully. You can learn from these outfits why they decided to go with home working, for what functions, the goals they set out, how they set it up, what metrics they used, how they made their business cases, and how they rolled it out. Many case studies offer lessons that can be applied to your organization.

Most senior executives want to see what others have done, especially name-brand outfits (like AT&T, Merrill Lynch, etc.). They need to know if what you are doing is proven, hence has minimal risk of failure. Most executives are skeptical as it is about home working. Providing that information about other organizations' home working programs gives them (and you) a benchmark to gauge your program against.

There are consultants and organizations that can assist you in finding the case studies that match your organization (see the Resources Guide at the back of this book). There are conferences and trade shows you can attend where you can meet others who have established successful home working programs.

When doing comparative research make sure that you equate the costs, especially real estate and labor, in the analysis as they change from organization to organization, location to location. You don't want to be left mouthgaping when told: "but that's California wages, they're much higher than our wages here in Kalamazoo."

Consider piloting

Piloting a home working program enables you to test it on a small scale to prove the concept. You set up the same methodologies as for a full-scale program. You learn from it what changes you will need to make to your main program.

There are arguments both for and against piloting. The pros are that piloting may give you that vital information to refine your program; it also may ease home-working-skeptical senior management into home working. But consultants say home working has been sufficiently proven to omit piloting: if the full program is designed right.

The cons: Piloting adds costs and time to home working and increases the time period before you reap the benefits of a home working. For that reason, some cynical executives that do not support home working will use piloting to defer home working. In doing so, they create a self-fulfilling prophecy: when the pilot's owns costs are higher and its benefits lower. They only went along with the pilot to shut you up.

If you go with piloting, Bob Fortier, president of the Canadian Telework Association and InnoVisions Canada, recommends that you include a varied cross-section of jobs, people, and business units. It's also a good idea to look at the impact of home working work on non-home-working working colleagues. There is no minimum or maximum length of a pilot program, but most typically run from six months to three years.

✪ THE BEST GUIDE: EXPERIENCE

When planning a home working program, no matter the goals it is to meet, you must thoroughly assess any experience with home working your organization has had with *any* type of *virtual working*: home working *and* mobile working. Find out what worked and what didn't and why.

Few reports will shoot home working programs or proposals down faster than those of employees goofing off when they work from the road or from home, even occasionally; critics will lump the two together. When the cat's away the mice will...

Chances are your department or another department has had people working from home on an informal basis. That experience is a gold mine of information that can make or break a formal or expanded home working program.

The best data is at your fingertips. If there is informal or occasional home working in your organization, learn all you can from it. Put in metrics to measure it; identify and isolate workers by where they are working.

In devising your program, begin by paying careful attention to and notating performance—is there a change (positive or negative), or no change at all—when your employees work from home occasionally, like waiting for the plumber. Benchmark this against when they work at your premises. Compare home workers' with premises workers' performances.

You also should interview those occasional home workers to find out what they thought about working from home, what they liked or disliked about it, and how can their experience be improved. You also need to find out what caused or didn't cause any performance changes.

Chat with the HR and IT departments and get their feedback. By listening to them *first*, at the early stages, you can build a solid program that has their support or at worst, non-opposition.

You also can correct any problems *immediately* and build those modifications into your home working program. This reaction also demonstrates your capability as a manager to be on top of and to fix issues quickly, which will help you sell the program to your colleagues and superiors.

✪ COST-BENEFIT ANALYSIS

Your home working business plan needs to contain and quantify all the features, costs, and benefits.

Let's look at these issues in depth:

Big (and little) picture impacts

There are two vectors that your cost-benefit analysis must focus on. These are (a) overall department net revenues as a result of introducing or adding home working, and (b) home working as a standalone component.

Senior decision-makers want to know how home working will affect departmental costs and output. They also want to see how home working as an investment could improve financial results and enable more profits or, in the case of governments and non-profits, permit more activities and programs for the money.

Cost overview

You will need to lay out and price out all the components, e.g. voice, data, and equipment (at your end and your employees' homes), facilities, environment, furniture, and if salient, location requirements, HR and IT requirements and staffing, training, and management procedures for your plan. The cost overview should lay out the **total** costs: who is paying for what: employers and employees.

Also, your program may entail employee hot-desk and/or satellite working and self-employed home workers. Set out those features and costs. If there are, or might be, special circumstances such as international home working, or customers and clients visiting employees at their home offices, set out any additional costs and, again, assess what the organization pays for.

If you are seeking either self-employed home workers, freelance or contract, or outsourcers that have home worked employees you may need to put together an RFP (requests for proposal). The RFP would outline your requirements. If you haven't decided for sure to go this route, consider preparing and sending out a request for expressions of interest (RFEI). A RFEI will give you an idea what services are available.

You also will need to budget for unique-to-home working management requirements. That includes training supervisors on how to manage home workers, communications investment (instant messaging, off-premises meetings and any travel allowances to the premises) and a Home Working Manager, with staff if necessary.

Cost analysis

The cost analysis should show how your home working program stacks against premises offices (see Chapter 9, and next section). I devised a calculation: infrastructure cost per employee (ICE).

When undertaking such cost analyses like ICE you have to account for whether you are in a position to dispose of the property. That will affect the analysis.

If you are in the middle of a 10 year lease, have no early termination, go dark, sublease or other exit clauses, your operation is not growing or there is no department inter-

PREMISES, MOBILE AND HOME WORKER MANAGEMENT

It's 10 AM (or PM). Do you know where your employees are? Are they on your premises, on the road or at home?

Do you have space and workstations that are unoccupied, costing you money? Do you have employees that are set up to work from home, only they are presently doing it just occasionally? Could they work from home exclusively, enabling you to reduce your facilities costs?

And in case of a disaster, do you know and do you have plans for filling excess capacity in existing premises offices if one should be shut down? Do you have an inventory of current home workers and workers who could work from home if they can't get to your premises?

John Vivadelli is CEO of Agilquest, which manages conference rooms, hoteling, move management, shared offices, and office space. He reveals that there is high underutilization of workstations in many buildings: from 15% to as high as 50%, caused by vacations, sick days, training and travel.

That's a lot of workstations to go unoccupied: wasting property and furniture. To shrink that wastage Vivadelli recommends managing the workplace as a "shared office environment," i.e. not permanently assigning workstations—a concept others, not him, refer to as hot-desking.

"You should allow employees to select an available workspace, either in advance or on the day of arrival, that meets their needs," he says. "Management can control access, giving each agent the proper workspace permission and restricting access as necessary."

Managing an office this way requires the proper workspace management software (like that provided by AgilQuest) that can track workstation availability for a given day or shift and then makes the assignment. That limits the number of empty workstations.

Such software can be tied into workforce management software that shows supervisors who is coming in and going out. If an employee books out sick, that workstation becomes available in the pool.

There is another benefit to AgilQuest's approach: business continuity. By knowing where your employees are: in premises offices and at home you can shift "resources" in case of a disaster.

For example, if you have a call center in the L.A. area and an earthquake hits it, you can see what space you have available at your Tennessee facility. Knowing you have availability of workspaces here allows you to call in off-shift employees. Or if you have a certain percentage of employees who have home offices you can ask them to come on stream.

"You need to have a system where you can look at your total workplace resources, traditional offices, satellite offices and home workers, "says Vivadelli. "If one of your buildings becomes unavailable, employees can immediately find available workspaces in other facilities, management can determine which employees were affected. Plans can be made to shift work to other locations and employees."

ested in your space then you can't gain real estate savings from home working: the premises ICE applies to *all* employees. But if you can alter who uses that property, or get out of the occupancy or the lease comes up, or if you own the building and you can sell it then you can break out the ICE between those employees who must work at premises offices and those who do not.

Resource identification

Your cost-benefit analysis should look at available resources you have in order to minimize additional expenditures. They include voicemail ports or VoIP gateways on your phone switch, stock of laptops, quality stored furniture you can ship to employees (which enables them to get on board with excellent and sound chairs and tables allowing you to cut down on costs), screens, docking stations, headsets, and phone sets. If you have spare workstations at your facilities you can identify and designate them as hot desks.

Resource identification also includes people. Find out from IT if the department has any staff members who are occasional home workers and would like, if they have time, to support others; they will know and feel home workers' pain. Find someone, again who is a home worker, even if occasionally, and who is willing to become a Home Working Manager.

HR/legal overview

If need be, you will have to present legal concerns, such as complying with data privacy laws, jurisdictional issues, and how you will cope with them. If there are taxation issues, like in Canada where employers have to fill out special forms outlining what expenses they pay for, then list and briefly explain them. Also, if your home employees need to see clients and customers, or have other employees, at their home offices, and that requires legal and regulatory compliance on their parts and yours, lay that out too. Senior decision-makers don't want any surprises.

Benefit analysis

You will need to quantify projected benefits from the program. You also must propose how you will measure them and with which methodologies.

The measurable benefits include real estate and furniture (infrastructure) savings, greater productivity (e.g. from internal metrics like customer satisfaction and sales), greater availability (like from longer working hours), lower tardiness and early departures, and fewer personal absences from sickness or other causes. But they should also include environmental impacts such as lower vehicle emissions, which we *all* gain from.

The methods of calculating those benefits are available either through means such as customer satisfaction surveys, sales per employee, tracking employee log-ins/outs, and/or through new measures such as ICE, which measures premises and home working employee infrastructure costs and per-commuting mile emissions. As noted in Chapter 9, all of these measures must identify employees by where they are working to obtain these benefits.

In showing the benefits, you need to assign the value of employees. That value is typically their compensation (wages/benefits) divided by hours worked.

But as consultant John Edwards pointed out in Chapter 9, the analysis also should include the employee contribution, expressed as net profit or budget value per worker. After all, your employees are providing value for your organization. Otherwise why have and pay for them?

✪ ALTERNATIVES (TO HOME WORKING) ANALYSIS

The US federal government has long required transportation agencies seeking funding for projects (like new highways and mass transit lines) to undergo an "alternatives analysis". The goal is to try and make sure that the preferred project is the best one. The alternatives always include doing nothing, minimal changes, and different modes (e.g. bus, rail, and routes).

The agencies examine the data, and analyze and give their judgments on the projects. Then it is up to Congress to fund them or in some cases reject the recommendations.

The local community, the federal government, and many in Congress are often skeptical about what transportation agencies propose, especially rail lines that some influential people are ideologically opposed to on principle. If the opponents don't like a project they act above-board and often underhandedly to derail a project—even one that has been approved by the federal government.

By the same, ahem, "token," many department heads are often similarly skeptical about home working. Those that don't like it, for whatever reason, may want to pull the plug on it.

Therefore, you must take every care to show (a) that your goals are valid and (b) home working, as you have planned, is the best way to meet these goals.

You should perform the same alternatives analysis with home working. You must apply *the same rigor,* **not more** to other options, as you hopefully have done to home working. You must illustrate their components, both on the employee and employer sides (like parking and space), employer investments, and the advantages and challenges that each option possesses. Then you need to compare the alternatives with home working.

Here are some cases in point:

Automation

Automation, such as IVR and Web self-service, may cut costs radically over having customer service handled by people. But only if self-service is done right—and in many organizations it isn't; and if customers want to use it—some don't. Otherwise you may lose customers. Also the labor saving software can take, in some cases, up to two years to get running. You could end up spending almost as much in high-priced labor: in consulting, systems integration and help desk costs than you may have saved.

Mass transit

Mass transit is an option to driving to work; it is more reliable, somewhat less stressful, and it minimizes pollution. But many premises offices are not situated on mass transit lines; they are in edge cities where the car is king. Many organizations have abandoned or avoided transit-served central business districts. To relocate a premises office to a transit line is usually more costly and hassle-ridden when compared to deploying a home working program.

Don't count on proposed new bus or rail transit lines to slice tardiness overnight. These projects can take many years: 5 to 6 years *at minimum* from conception to operations in design, engineering, environmental assessment, funding and in political fights

For example, Seattle, Washington is one of US's most badly congested cities. Yet it has taken over *35 years* to get a rail rapid transit system from its original 1968 proposal to the start of construction—with huge political fights, setbacks, voter approvals, legal action, funding holdups, and near-countless design changes on the way—and *hopefully* no more delays between now and 2009 when the trains begin running.

Carpooling/road improvements

Carpooling is also an option, especially for organizations with premises offices not located on mass transit routes. It saves parking spaces and cuts down on emissions. In several major cities there are high occupancy vehicle (HOV) lanes that permit carpoolers to speed past bumper-to-bumper traffic.

But carpooling has declined rather than taken off. The reasons are hassles in coordinating employees' work, pre-work, and post-work schedules, changing schedules and locations, whose turn is it to drive, and what to do in emergencies. There are sometimes disputes between employees, e.g. the driver or their vehicle might be wretched. Some carpool lanes have now been abandoned because of poor use and because drivers objected to being stuck in traffic while others in nearly "empty" HOV lanes whizzed by.

Don't expect announced new highways, with or without carpool lanes to alleviate congestion. They rarely do—new metro areas highways soon become as clogged as the old ones. Even when there is a justifiable case for road expansions, environmental assessment, funding and political resistance can take decades to overcome. When home working can be set up in less than 12 months why wait up to 12 years for traffic to "get better"?

Space analysis

It makes sense to better utilize what you have. Space analysis is an excellent tool in conjunction with home working and with business continuity planning. You will know who is where at any time in case there is a temporary shutdown; you can also see the impact of an expanded home working strategy on your space needs.

But sometimes strategies like hot-desking, without home working, could be disruptive especially for higher-paid more professional employees. Everyone likes to have "their" space. At least at home they have it. Literally.

Consolidation/relocation

You can save money by relocating your premises offices to locations with lower wages and real estate costs. Yes, the timeframe to set up a home working program: up to a year may seem long, but it is short compared to other strategies.

There is a big trend to move call centers, back office processing and, increasingly, higher end accounting, architecture, engineering, and programming work offshore to countries like India, The Philippines, China, Eastern Europe, Central and South America, and Africa where highly educated people work for significantly less money than Americans. Forrester Research predicts as many as 3.3 million white-collar American jobs will be lost offshore by 2015.

But Jack Heacock argues that home working is a viable option to that trend because of the downsides of offshoring, which includes cultural affinity problems between foreign workers and American clients and customers, and loss of quality control. There is also a greater risk that data protection and other laws could be violated. Americans who travel and work in developing countries face greater risks of crime and terrorism. Finally, the US loses innovation and security with key engineering work being exported.

"Home working can close much of the cost gap between offshore and the US and Canada," Heacock points out. "Plus it avoids the consumer and political resistance, personal and data risks, vendor assessment, training, and management issues with going offshore."

Also, relocation to a community with lower labor costs can take months, especially if it is offshore to another country. Relocations typically require the services of professional site selectors that track labor and property supply and prices.

Then there are the moving costs, and the costs of time and loss of productivity caused by planning, packing, takedown, moving, takeup and burn-in time in the new location. You may have to pay to move staff or hire new people and bring them up to speed.

Moving is *extremely* disruptive to operations. You have to disconnect and reconnect voice and fax lines, notify everyone of the move, update websites, employees and customers and change stationery. Then there are the staff gripes to contend with, with employees jockeying for cubes, complaints about the desks, chairs and phones...you begin to wonder if it was worth it.

Maybe you should have let everyone work at home instead. Compared to the hassles of moving, the effort involved in setting up a home working program is a breeze. At least half of the work will be done out of sight, in their home offices.

○ OBTAINING BUY-IN

You're convinced home working is right for your organization. You've put together your business plan. Now you need to convince your counterparts in the other departments and your superiors. And that won't be easy.

Remember, for many within your organization, home working is considered a revolutionary and, by inference, a risky new method of doing business. Especially since it changes the main gears on how the organization manages: from traditional line-of-sight to new performance-based.

Most managers and executives are rightfully conservative; most do not want to change, and risk wasting time and money on new approaches that may fail. Shareholders', donors', and the public's money must be carefully handled.

If a program like home working fails and word gets out to the media, the public, and to the owners, you will have to hire spin doctors to cauterize the resulting wounds to the organization. And there will be top managers cut out of the organizations, i.e. you, to save the patients.

Moreover, you may be able to win over one superior, but *their* superior may overrule them. Or you may get a new superior who says 'no way!' and your efforts will be for naught.

So how do you win over top management? How do you ensure that the home workplace program you have devised has staying power, beyond one such individual?

Here are some suggestions how:

Make a conservative case

This strategy goes back to alternative analysis. In a revolutionary and controversial change like home working you need to strike a careful balance between underplaying the benefits, to be on the safe side with a skeptical audience, and diminishing the benefits to the point of "why bother."

You should take the tack that you have gone to home working *as a last resort*, by focusing on the goals, and then structure your analysis and arguments accordingly. By using such an approach you can show that no other method has proven, either through experience or upon study, to work as well as home working.

You should also stress to all players that your home working program enables better performance than if the employees worked in premises offices. Where your organization is skeptical, work into your proposal measures such as locked offices, rigid login/out procedures, instant messaging, readerboarding data, monitoring, and authentication to reassure them that the home workers will be doing their jobs without risk to the organization.

Offer to begin the program via a pilot or test home working program. The more you show your willingness to be prudent, and to change when change is needed, and if necessary, to drop the program altogether, the greater your credibility in delivering the program.

Getting Facilities, IT, and HR buy-in

As has been recommended throughout this book, you will need to contact and touch base with Facilities, IT, and HR to get their input when devising your program. Your program will need to have separate budgets for those departments' contributions so that any additional resources the program requires will *not* come out of their existing budgets. These departments are very conservative, as they often are overworked, under compensated, and cash-strapped.

Facilities, IT, and HR can make or break a home working program because it needs their support for technology and for staffing, training, and management. The more your program takes their needs and wishes into account, including finding out and budgeting for additional staff-hours and other assets, the better the odds of getting their buy-in and obtaining authority for the program's roll out.

All of these departments are treated by organizations as costs. Consequently, they not only are usually overworked, but also underfunded and most departmental heads are in no mood to take on anything that adds to their burden. But at the same time, they might be receptive to you if you can prove that home working will make their lives easier.

For example, if you show how home working will cut turnover and reduce personal absences and medical costs, HR might buy in. You just made their jobs a lot easier and improved their numbers dramatically in exchange for their assistance.

If you designed your home working program in a way to minimize additional IT costs, such as network/intelligent routing, VoIP, storageless PCs, and propose internal FAQs to cut support calls, then you may get IT's buy-in. You should also propose that the IT staff might work from home—if they are not making field repair calls.

It is ***especially important*** that you get your Facilities people on board. That department will be the most threatened by home working and may fight you on it. You must show to Facilities that your program will not lead to job losses.

Instead, home working will give Facilities more options to handle future expansion, or if they see the possibility of some kind of relocation down the road, property selections. Discuss with them what they would like to see in existing or new facilities that having people work from home could afford—and then suggest to senior executives that part of the saving from home working should go to better-quality premises.

Obtaining supervisory staff buy-in

Your supervisors are your organization's NCOs: they are the ones who directly manage your troops. They too are on the front lines. You must consult with them when devising the program, such as finding out what concerns they have, what their experiences have been, and what suggestions they have, and then test them in existing informal or case-by-case programs, or offer to test them in pilots and rollouts.

Supervisors can make or break a home working program either by supporting it enthusiastically, or by dragging their heels, causing it to fail meeting the objectives. The more you listen to them in putting together the program the more likely they will help make it a success.

Showing compliance with corporate culture

In presenting your case for home working you will need to demonstrate how home working fits within, but more importantly *enhances,* your organization's corporate culture, with the values that your outfit expresses to its employees, the owners, and the public. Fortunately, home working's attributes and flexibility enables these goals, even when the culture encourages premises working.

For example, many outfits preach collective working, as families within families. They manage to maintain, even in tough economic times, outreach to employees such as in-person and conferenced meetings, and community involvement.

Is home working—individual home workers scattered throughout a region or the world—in conflict with this culture? No, because home working enables employees to be happier and more productive because they can spend more time with their families and in their communities, wherever they are.

If your organization has a collective, family culture you may want to build into your program electronic newsletters, chat sessions, and regular conference sessions with home workers. Or, if feasible, having regular in-person breakfasts or lunches (on your tab) with your home workers.

You also may want to plan for a method by which you propose that your home workers engage in community outreach programs on behalf of your organization wherever they might live (and work). That turns a potential liability—employees not being in the premises offices and contributing to programs there—into assets by spreading the image and the word of your organization over many more locales.

Demonstrate how home working improves image

For a finish, show how your home working plan makes your employer look good, by doing their share to clean up the environment and promoting family values. As a bonus, mention that if more employers took your lead we could reduce traffic jams and government spending that costs everybody in taxes.

Make the case that premises offices should be an exception

Let's turn the home working argument on its head. The information that you've gathered should enable you to make the case that premises offices are more expensive than home offices. So argue the financial point that premises offices should be treated as an exception and should be allowed only for those functions that require in-person, hands-on interaction.

After all, if most employees work on computers and communicate by voice and data (and possibly video) they have computers, voice and data lines, and space somewhere in their homes for these activities, then why should your organization spend investors' (or taxpayers') money on supplying them with voice/data and space. When the act of going from A to B wastes productivity, enables the spread of diseases, puts the organization at greater risk, and ruins the environment?

⊘ ARE PREMISES OFFICES BETTER?

This book examines many of the issues entailed in setting up and managing home offices, like location, access facilities, environment, and potential data theft. Thus there may be a perception that because you are leaving it up to the employees to set up their home offices (in buildings that were designed and selected as homes), albeit to your standards, that the resulting offices will be less "professional" than premises offices.

But *are* premises offices, though they have been selected, outfitted, and managed by people who are being paid to do so, any better than home offices? The answer, in all too many cases, is **no.**

Don't get me wrong. There are many fine premises offices: state-of-the-art and fun/funky/productive ones. There are excellent architects, designers, and facilities managers, interfacing with executives and managers who know what they are doing and have their employees' interests at heart. No matter how desirable home working is there will always be the need for fine, functional, efficient and conveniently located premises offices.

But from what I've seen—covering site selection, design and facilities, construction and materials and transportation, having worked in many premises offices in three countries, having been responsible for those properties and their occupants, and having been active in community and transportation policy—there are woefully too many poor premises offices. Dorothy is right...there is no place like home...

Let's look at the key issues plaguing premises offices:

Poor Location

Site selectors admits that—against their advice—organizations all too often poorly locate their premises offices where the property is cheap, where there is a deal on rents, and the building is sited five minutes from the senior execs' mansions, kids' private schools, and their favorite 19th Holes. Employee access, evacuation, and business recovery are mouthed first, acted on last.

The decamping of premises offices from transit-accessible downtowns to cheaply priced middle of nowhere "office parks" worsened commuting in two ways. Employees are left with no other way of getting to the premises other than being stuck in traffic. They are often forced to buy additional cars, which costs them money and which break down and need repairs that takes time out of their workday. Also, they have to either travel farther than if they commuted downtown, or move closer to the premises.

Moving is easier said than done. There is the hassle of buying and selling homes, packing and unpacking, dragging kids from one school and breaking them in (in some cases literally) in a new school. Not surprisingly, when faced with a premises location change many employees have chosen to work somewhere else: maybe with a competitor.

Increased disaster risks

Buildings have been located and laid out sometimes to invite disaster. Site selection consultants like Graeme Jannaway, president, Disaster Recovery Institute-Canada, points out that organizations have put in premises offices near airports, defense contractors, and embassies, which ups the odds of protests, strikes and terrorism. Others have them near chemical plants, oil refineries, and freight carrying railroad lines with grade crossing. That risks fire, poisonings, and blocked evacuation routes.

Yet there are dangers locating near seemingly benign and convenient locations such as restaurants. That's because restaurants are frequently targets of arson.

But no matter how located the building, organizations often make matters worse for themselves by not planning for disasters. For example, they may not have battery-powered uninterruptible power supplies (UPSes) or do not have UPSes between backup generators and electrical panelboards to smooth out power.

Sometimes the generators and the UPSes are located in the basement: the *worst place* to put them because that is the most likely part of a building to get flooded either in a storm or by water from fire hoses. Sometimes the computer rooms are located under hot water tanks that leak and eventually burst, wiping out the electronics.

The phone room and the computer room should be in separate locations. If a flood, fire, or other similar disaster hits one room, the other is not affected.

The buildings' biggest weaknesses may be its people. Does the building have an on-site super or one that floats between sites? Ask about the super's qualifications and experience. Who handles problems when the super is not there after hours? Are there security guards and if so what are their responsibilities and how well trained are they? How thoroughly are the backgrounds of the building's employees and contractors, like cleaning and security companies and their employees checked?

"You may be getting a good deal on a building or a lease and it may be in a great location," says Kurt Sohn, principal consultant, business continuity planning firm 180cc. "But what is the price on your business if the building is not able to withstand a disaster and if the building staff has not taken steps to prevent and respond to them?"

Other safety and security hazards

Premises offices can be unsafe and poorly secured. Some employers have them located in dicey areas. Few locales are spookier than industrial areas and deserted office parks at night.

Security in many offices is so lax some buildings ought to have "Welcome" mats for outside and inside burglars. Employees make it even easier by propping open fire doors.

In a past life (i.e. when I was in my 20s) I worked as a security guard. I was once assigned to a defense contractor. Terrorists had hit another contractor nearby.

Was I trained and drilled on what to do if we had a bomb or a threat of a bomb? Or what steps I would take— besides getting the !#%^&*()) out of the way—if I saw a truck barreling toward the gate? What do *you* think? Anyone who is familiar with private security, such as at airports, knows all too well the answers to *those* questions.

The same goes for fires. Many premises, especially newer buildings, are built to withstand blazes at least long enough to permit their occupants to safely get out of them; many older premises (and homes) are not.

But employers and employees make matters worse by being stupid. They leave combustible material lying around and they block exits. They smoke inside and toss their butts into paper-filled containers. They are not trained and drilled on proper evacuation procedures. They sometimes try and go back during an evacuation. Net result: preventable injuries and deaths.

Employee crime

Employee theft is a big problem; it cost companies $200 billion annually. Too many employees feel, wrongly of course, that such goodies are an unspoken perq of their jobs. And they will intimidate others to look the other way. If your organization has announced that is about to downsize or there are rumors that a move is afoot watch out for the "early bird liquidation specials."

Having employees work at home controls that theft because there is only one possible suspect. In many cases home employees do not get reimbursed for office supplies; if they do, then you can figure out from the expense reports if the use is excessive.

There is also the specter of assaults, robberies, murders, and rapes on premises. You can't predict or realistically prevent some nutjob especially in today's gun-worshiping society from coming in and slaughtering targeted individuals or people at random.

Lastly, premises workplaces can be great places to abuse legal and illicit drugs. It is one thing for employees to cook their brain cells in their own homes: though they risk making idiots of themselves with your customers. But it is another when they're getting fried and stewed when they are working at premises where they could harm others and expose how much of a fool they are to more people. And then they drive home...

Premises health risks

Building premises often are not healthy places. As noted before, they are "bug factories." When one individual comes in coughing up a lung, others soon will be doing the same. With dangerous diseases like SARS and drug-resistant staph and strep on the rise, those coughs could be their last.

Employees also may suffer from what is known as "sick building syndrome": symptoms such as headaches, dizziness, muscle cramps, edema, and chronic fatigue that arise from sources found on premises like formaldehyde, cleaning solvents, and mold. These symptoms often fade away when employees leave the premises such as for their home offices.

Too often premises offices are poorly ventilated, with minimal air changes, keeping airborne diseases and toxins inside. The cover story from the June 5, 2000 issue of *BusinessWeek*, titled "Is Your Office Killing You?" reports that some buildings draw in only five cubic feet of fresh air per person per minute.

"'That is almost enough to keep people alive'", the article quotes New York architect Robert F. Fox Jr. The American Society of Heating, Refrigeration and Air Conditioning Engineers recommends 20 cubic feet, below which, the article says, sick building syndrome increases.

And too often that air is anything but fresh. Sometimes brainless—and that is the only polite term I have for them—"professionals" install air intakes by parking garages and loading docks, at street level, and by building entrances, bringing deadly pollutants from vehicle exhaust and sometimes cancerous cigarette fumes from nicotine addicts smoking it up *inside*.

Smoking exposes employees and others to cancer and emphysema. That exposure occurs in the all-too-common "gas chamber gauntlet" outside of building entrances and when managers must go into formal smoking areas to speak to employees.

Smoking has other consequences that cost employers' money. It causes damage, creates litter, and could cause fires. It only takes one jerk to flip a butt into the dry grass on a hot day during a drought and in seconds you have an instant inferno.

Sometimes the sickness comes from other businesses. The *BusinessWeek* story reports that some Levi Strauss employees in San Francisco had rigged umbrellas to protect themselves from soot from a nearby restaurant.

There are now effective methods to "treat" sick building syndrome, such as ultraviolet germicidal irradiation (UVGI) lights. A study by McGill University, reported by the *Washington Post*, indicates that such lamps placed in heating/air conditioning systems could kill mold and microbes in cooling coils and drip pans. When the lights were tested, employees reported a 20% drop in respiratory and mucosal symptoms.

But the study also pointed out that UVGI is not a cure-all. For example, they can't do anything about allergic reactions to chemicals.

Noise

Building premises, especially newer ones, are built with a fair amount of noise protection, especially from outside. They often have better noise insulation than that in many houses and apartments. The floors are solid, limiting noise from above and below.

But older buildings can be rattletraps. They could have creaky elevators and old ventilation systems. I worked in one old office with the air conditioning system from hell: the ventilation system was so loud and the chiller so poor I often wondered if it was revving for takeoff.

However, most of the time the noise risks are more to come from humans. Co-workers can be loud, especially those that prefer to use speakerphones and who love blasting their music. Office parties can be heard sometimes all over the office. I've had to tell people to quiet down, and they've told me to do the same. The cube panels and premises walls are often thinner than the panels and walls in per-hour motels.

Ergonomics experts say a typical office noise level is high enough to cause problems, like increased stress and errors, diminishing performance. The din of speech, especially in environments like call centers where everyone is on the phones, is comparable to power tools, says Dr. John Triano of the Texas Back Institute.

But often employees are reluctant to say anything about noise, especially if it comes from their boss or from a popular co-worker. And sometimes there is nothing that can be done about the noise. So they suffer.

Vermin

Premises offices, especially the older ones also get vermin. The former *Call Center Magazine* office at 12 West 21st St. in New York had for years a big mice infestation. On more than one occasion my Size 14s performed euthanasia on some hapless creature caught in a glue trap.

One evening I was working late and my computer mouse suddenly turned tail. I exclaimed: "Oh Diety! I got a dead mouse!" My colleague Joe Fleischer, who was sitting in the next cube, made the Olympic record for jumping straight out of his chair...

Mice, the furry variety, are not "cute." They carry dangerous diseases such as hantavirus pulmonary syndrome which infects the lungs and has a 50% mortality rate and rabies which has a 100% mortality rate if not treated with injections.

Socialization and harassment

Let's face it, the term "workplace," especially when it is applied to a traditional premises office, is a misnomer. Because when you get a group of people together, especially those who are supposed to work with their brains rather than their muscles, they are going to socialize.

That can be good because that interaction imparts information that workers need. But more often than not it wastes time—you're an employer not the proprietor of a café—

and too often the yap is about topics that have zilch to do with the job. Like 'who won the space polo derby, look at the fat cow they hired, did you see the skinny b**** in Accounting, what did she have for lunch, one leaf of iceberg lettuce?, who vomited in the file cabinet after the office party?, and did you see the size of that zit on that actor's tennis kneecap?' Real bottom-line improving conversations.

Too often such mouthing can lead to low morale, hence productivity problems caused by backbiting, false accusations, and malicious gossip. How much work can take place everyone's griping about each other, you, your boss, and others in the firm?

Then there's the sexual end: the meeting and mating game. Too frequently "premises partnerships" create problems: too much "togetherness" at the cubes, jealousies and tensions, and perhaps turnover and lawsuits when these romances end.

And what happens if one's advances are not welcome? You guessed it: sexual harassment suits.

With home working those employees are not running up the clock gossiping or leering to see if she or he is wearing a G-string. Absence does make the heart grow fonder; it also removes suspects in the perennial who has been slandering whom in office politics.

Other premises distractions, and hazards

Premises offices have other problems and hassles. Employers who permit employees to bring their rugrats on premises risk, disrupting other employees who might not think their co-workers' progeny is cute and who find them distracting. I once worked in a building that allowed a co-worker to bring in a child that could politely be described as having "strong lungs." Then there's the doggie lovers who thinks if they love the snarling, allergen-spreading, flea-riddled Poochiekins that they literally dragged into the building, leaving its literal 'mark' so should everyone else.

There is also another issue: liability. Employers' premises are not usually designed to be childproof; children are not normal occupants of that space. What happens if a child decides that a paper clip is a good tool to explore an electrical outlet? And what happens if an employee was bitten: by a dog, ferret, cat, etc.?

I love dogs and cats. But the only other species, besides those human authorized to be there, that belongs in a premises offices are seeing-eye dogs, to enable the visually-impaired to work there. If employees need to have children and pets around them then they should work at home, providing you ensure that the kids and critters don't interfere with their tasks.

Employees *don't fully control* access to their computers and work areas when they are at premises. There are co-workers and visitors (like other employees' computer-savvy children) who can scan screens, load data-stealing software, and surreptitiously go into cubicles and desks when the employees are not there.

Given the above, instead of justifying the case for home working, ask your colleagues to justify the case for premises working

Chapter 14: Bringing It Home (Program Rollout)

You've decided the goals of your home working program; and you've determined that home working will meet those goals. But only after analyzing the results of informal home working that is currently taking place within your organization, and using that information plus benchmarked data to flesh out your formal home working program. All of which led you to create a home working program. You then wrote your Home Working Policy and presented it to the "legal eagles" for sign off. By now you may have even conducted pilot programs and adjusted the program based on the results.

Or if you have decided to go with a less broad-scaled program, e.g. enabling workers to be productive when they have to be at home. This means you obtained the budget approval for laptops, permission to pay employees' long distance charges and other incurred costs, and worked out with IT how to handle network access.

Or if you chose to go with contracted home employees (Chapter 11): for which you have an approved budget, and have analyzed and selected a vendor, worked out a contract or agreement, and figured out how to work with those vendors.

Whatever method you chose to use to implement your home working program, you have the approvals in place, the budgets appropriated, and staff resources committed. You're now ready to roll out home working.

Now here's where the rubber meets the road. Namely, how to make the program roll and keep it rolling smoothly—how to bring home working on home.

If for any reason you have to put on the brakes on home working employees and/or on the program you will need to know how to do it smoothly. That is covered in the last chapter, Chapter 15.

✪ PROGRAM MANAGEMENT

If your program requires it, one of the first steps you will need to take is to recruit and select a home working program manager. That individual would be responsible for program rollout, implementation and management, including troubleshooting with line managers and with departments including Accounting, HR, and IT.

As explained in Chapter 9, that individual could be a highly capable, well-respected home-working-committed existing manager who is happy to undertake additional responsibilities. Or, if the program merits it, the position could be a separate individual, with a salary to match his or her duties.

⊙ ENABLING THE EMPLOYEES

Now it is time to enable your workforce to be ready to work from home. That includes new hires as well.

The enablement can be as simple as putting together a home working policy that includes your requirements. Then you can have it ready for your supervisors or yourself to discuss it with employees who want to work from home on a case-by-case basis.

Or the enablement can be a full-fledged analysis of your work force based on the tasks they undertake and whether those tasks and when they fulfill them are home workable. You then pre-screen those employees for suitability based on task and function and then screen the individuals based on whether they are interested in home working and whether they or their home can meet the requirements.

The enablement process should have the following features:

A formal Home Working Policy

With a formal Home Work Policy, even for informal programs such as covering absences, there is no ambiguity, although there is room for flexibility. Employees (and applicants) plus supervisors will know what responsibilities and investments are expected from whom. The Home Working Policy can be summarized for internal and external applicants, such as in job postings and advertisements. Then, if employees are willing to accept the terms, they sign the policy.

Home employee screening/assessment in place

Your plan may call for screening and assessing prospective home workers, especially exclusive home workers. That can take several forms: formal tests that assess the abilities of employees (and applicants) to work independently, without supervision, and to be problem-solvers (consult with your HR department on tests). Or they could be in-person interviews between your supervisors and your employees.

Note, however, that the latter (in-person) method should not substitute for the former, but the types of questions in the former method should be asked by only your supervisors. In large enterprises or for large-scale home working programs you should have both testing and interviews.

Home office assessment

If your plan calls for assessing home offices for suitability, then your plan should detail how to conduct the assessment. That includes checklists, some suggestions on how to bring a home office up to your standards (your assessors should be creative problem-solvers), deciding who will do the assessing, and estimated staff-hours and travel budgets for the assessment process.

Home worker training

Training prospective home workers may be just the key to making your program a success, especially if it is a large-scale program designed to meet specific objectives such as cutting real estate costs. Training can help if home workers need to learn how to set up home offices and cope with environment issues. Training also can minimize support issues, by teaching home workers how to troubleshoot their voice/equipment/data. But equally, if not more importantly, training can help your employees to learn how to work with colleagues and supervisors without "being there."

All home workers should be educated on the home working policy. You should make it available to all employees, either when requested by an employee, or by posting it on your internal website.

Should your program call for home worker training, the education should take place before your employees begin working from home. The curricula should be made part of new employee training if those workers are going to be home workers.

When conducting the training be on the watch for any signs of employees "not getting it." Some individuals may have difficulty working independently. The signs can include workers saying that they don't see how that can work, for instance remote IT fixing of equipment, how to conference call, or deal with environmental issues like kids screaming—not just asking questions.

You may use that opportunity to discuss the problematic issues with the employee and see if they can or cannot be resolved in their home office. If a resolution can't be found, then the workers will work from premises; if you have limited premises space and there are other workers whose functions require it or have more seniority or are better performers, then you have to let such workers go. You are justified in doing so, just like when you move premises offices to a location that some existing employees cannot get to or are unwilling to move to.

Supervisor/manager training

Your supervisors should be taught how to coach, train, and manage, i.e. manage by performance, without "being there." This must be done before employees "go live" from their home offices. Training should include how to build and maintain a virtual team, how to stay in touch, and how to word and read messages and e-mails correctly to elicit responses to indicate employee performance.

Supervisors and managers should be required to learn the home working policy. They also need to be informed as to whom to go to in other departments, i.e. Accounting, HR, and IT, or above them if they have questions or if they need approvals such as for cable modems or other investments.

If you are deploying tools like audio/data/video/Web conferencing at the same time as you are implementing your home working program, you may need to teach supervisors how to use these tools. That way they can conduct effective and smooth-flowing virtual staff meetings.

As with home worker training, your supervisor/management training must occur before the program goes live. And as with home worker training, be on the watch for signs those supervisors or managers don't or do not want to "get it": signs such as by statements—not questions—like "how do I know they are working, not goofing off?"

You can take the supervisors/managers to one side to see if their questions can be resolved. If not, then you can give them a choice: change, transfer out, quit, or be let go, with recommendations. You *cannot afford* to have supervisory/management staff who are unwilling to follow your policies and programs, and thus be the cause may cause for your program to fail.

❂ PROGRAM MEASUREMENTS

There will be many eyes on you—people will want to see how the home working program succeeds, or stumbles. In rolling it out, you will need to have measurement metrics in place. For example, means for identifying employees by location, and the ability to record and track productivity, availability, turnover, absence, environmental benefits, and infrastructure cost differences.

Now that you are ready to launch the program you should know the number of employees who are working from home, from premises offices, or a combination of the two; you also know most of the costs involved in operating the program. Now you are ready to run some models to test the measurements, before the program begins. That way you can make any changes to your measurements, calculations and models, and to your results predictions.

❂ MAKING THE INVESTMENT

If your home working program requires purchase of hardware and software ranging from laptops to new phone switches and VoIP gateways, or if you are outsourcing for service bureau-employee-provided or self-employed home workers, you need to begin your procurement process, as per your organization's policies. You should know, after consulting with IT what your requirements are, the prices, and the lead times.

The more complex the investments—or the scale of services you need, e.g. going to network routing— the longer the lead times. For example, buying/specifying, taking delivery of and installing and debugging a new phone switch now takes only a few weeks.

❂ EMPLOYEE SETUP

Rollout of your home working program must include making sure employees are on track for going live from their homes. This is especially critical if you are downsizing your real estate or have committed yourself to environmental goals and have deadlines.

The key time factors in setup are voice/data installations, which can take a week or two including debugging, and computer/equipment setup and debugging. Your employees may have to wait for furniture to be delivered, especially if you supply it. But oftentimes they can buy the necessary furniture from a local office supply place and take it home.

Your home workers may need to make investments, from their own pockets. That includes buying their computers and acquiring furniture.

You may need to give employees time—if they ask for it—to spread out the costs such as over two paychecks. If money upfront for the gear is an issue, consider lending it to them. Or if Accounting approves, have a buy-payback program where you purchase the equipment and furniture and a portion of the employees' paycheck is deducted in regular payments; if the employee leaves or is let go you deduct the balance from the last check. I once worked for a newspaper in Canada that bought me a camera that way; I still have it.

Other Departments' Assistance

In your home working program rollout you will more than likely need the cooperation of other departments. They will need to adjust their policies and programs to adapt to home working. If you have done your homework, you've already those departments' cooperation and sign-off.

The key areas include:

Accounting

If your program entails additional or changed expenses you will need to make sure the policies enabling it are in place. Examples include having home office expenses (such as home Internet access and supplies) as chargeable costs on employees' expense forms, and direct billing for home workers' voice and data lines.

HR

If the program requires HR participation and assistance, such as hiring/screening/training, and in ensuring employee availability through sign-in/out and timesheets, then you need to coordinate such procedures with HR. They will need these procedures in place before the program can begin; they even may need to be there during the training phase (both of the home workers and their supervisors/managers).

IT

Your IT department has to be with you well before the program is launched. You and the IT department will need to work out, for example, when they could have phone systems ready to take voice mail for home workers or their timetables for VPNs, gateways, and other investments, and/or having their people available, trained and equipped to handle home-worker-related support issues. IT also may have to arrange with their hardware suppliers or other independents for field support for the home workers.

If the program includes having a website FAQ to enable employees to solve some of the minor technical problems than the site should be ready—and tested by existing premises employees—before going live. Some of the managers also should test the adequacy of the FAQ section from their homes.

Even if your program entails home workers supplying their own machines and voice and data lines, chances are the IT department will be involved somewhere.

Legal

If there are legal issues, like over tax laws and jurisdiction, then you need to go over them beforehand and work out a policy, guidelines, and procedures. The more such potential obstacles can be identified and worked out the fewer the problems faced when the program goes live.

Facilities

If the program has hot-desks or satellite offices then you and Facilities must work out size and location; IT must be involved for the locations to have the proper connections. Also, if other employees are required to have ID tags, exclusive home workers who hot-desk or satellite work also will need ID tags.

If you use hoteling, then you must have the space set up for your needs and prepare your employees for them. If you contract for hoteling, you need to assess and select contractors. For example hoteling firms typically have Web-based reservations systems that your employees will need to use to make sure there is room for them when they need it.

✪ PRIORITIZATION

You may want to consider prioritizing your qualified home worker applicants. This especially applies when circumstances include a limit on the number of positions that will be home workable: say you are told only 20 jobs out of 50 of the home workable jobs within your organization will be available for the home work program. But even with a large program, you may need to prioritize the order in which home workers will be accepted into the program.

If you prioritize your employees for home working there are several ways of doing so. They include:

* Best-performing (home working then acts as an incentive)
* Employees with care responsibilities (child, elder, disabled long-term illness)
* Commuting time/distance (eliminating the longest and time-wasting commutes)
* Available transportation alternatives (the most environmental, traffic and stress-relieving gains are had by cutting back car commutes)
* Spouse/family relocating to community outside of commuting distance

When going through the home working applicants, you may want to move onto the "qualified list" those employees who have care responsibilities or who have spouse/family relocation issues. This gives those employees a break, enabling them to continue to work productively for you.

✪ EDUCATION/MARKETING

You have done all the work and the people, process, and technology are in place. Now you have to let your employees, the rest of your organization, and ideally the communities, know about it. Here are some suggestions:

Meetings (in-person and via conference technology) with entire departments

Announcing a home working program is a big deal, especially for employees. If your organization has a culture and practice of large in-person or Web- or videoconferenced meetings, then you should arrange one for home working.

To do that you will need to put together multimedia presentations (in-person or conferenced) on what the program is about, why you are doing it, what are the benefits to the organization, to its employees, and to its clients/customers. How it will work. Who is eligible, and how they can apply. You will need the home working program manager to prepare and give the presentation, and to take questions, supported by personnel from Facilities, HR and IT.

The home working presentation must be introduced by a senior executive. This is vital because it shows the program has top management blessing. That will enable program acceptance and compliance.

Meetings (individual workgroups)

You also will need to get up close and personal with your employees by arranging for in-person presentations to individuals, or a few workgroups, in a premises office. Your manager, who has been trained on home working, will give the presentation; if the program is large enough, the person could be assisted by the home working manager or their designate, backed up by HR and IT.

Individual employees will have many questions. This is the time to get them out and answered.

Internal communications

All departments whether directly affected by the home working program or not, should be informed of the program's attributes. You should have a formal presentation, or highlights plus the formal Home Working Policy on your website. The program's specifics, plus key points of the policy, should be in any employee newsletter. Your employee handbooks will need to be updated for home working.

You also may want to consider showcasing some of your home workers. After obtaining the workers' permission, see about having senior executives visit the home office setups themselves so they can see the workers doing their thing; invite the executives to ask questions. This is an especially good strategy for a pilot. The more informed people within the organization the more the staff overall will accept and believe in your home working program.

Outside media

Home working is a big deal. The benefits are win-win for the environment, for traffic congestion, and for family values.

You should plan a media information campaign announcing that you are beginning the program. The campaign material should provide the number of home workers involved and list the benefits of home working. You will need the assistance of your organization's marketing/public relations department to ensure that the information presented complies with your organization's image and message. Your answers to inquiries must be preplanned and also should comply with your organization's policies, in order to ensure you hit the right buttons.

The key hotpoints

With the introduction of the home working program you show that your organization is doing its share to lower pollution and reduce traffic; and in doing so, your outfit is helping to avoid wasting more tax dollars on needless road and transit construction. The program also enables employees to spend more time with their families.

Don't forget to mention business recovery and reducing the spread of flu and other illnesses. Make sure you also mention the investments made to enable home working, including spinoffs in the communities where workers live, like furniture and office supply purchases.

You will also need to prepare for negative questions such as impacts on local business. For example, there are studies showing no net effect.

Point out that by cutting organizational and employee personal commuting expenses you free up more money for more productive business investments, or programs (if your organization is government or non-profit) in the area. Highlight the benefits in the community where the premises office is located. Home working will provide employees the time, money and free them of commuting stress to drive in or ride in on the train, ferry or bus to enjoy the culture, shopping and other amenities.

When planning media coverage make sure you have one or two home working employees ready to be interviewed, in their offices. Prep them on questions and answers. Reporters, especially TV crews, will want to check out "the office of the future."

⊙ MONITORING AND TWEAKING

The first few months of the home working program are the most critical. The program will make or break during this time.

You and your managers need to watch for, and immediately act to solve, any problems—voice/data access, transferring calls, equipment malfunctions and complaints, or finding out that home working employees aren't working when they should. These problems, if left unresolved, could cripple your new program. You also should be strongly proactive. Talk frequently to the home and premises workers during the startup period to uncover any issues that may pop up, e.g. poor communications with colleagues, and hinder the progress of the program.

This period also will inform you whether you need to modify any of your home working criteria: e.g. qualifications and facilities. For example, to Kansas City, MO-base outsourcer ARO's surprise they found that their younger very-tech-savvy workforce did not take to home working as well as older workers; gradually the 40-pluses *became* the labor force.

Here are four potential flashpoints:

Compliance

Employers need to be sure that employees have set up their home office program in compliance with your directives, which, when given the proper attention and emphasis, could serve to avoid regulatory and safety hassle, e.g. OSHA. Employers should furnish a comprehensive suite of ergonomically designed home office furniture (or at least full information on how to purchase such a suite), along with formal guidelines and training for the operation of a home office. By taking such steps, you can greatly reduce the risk of injuries or potential actions downstream.

Consultant Jack Heacock recommends having the employee take 4-6 photographs of their home office. The employers' home work program management team, along with risk management and the furniture manufacturers' representative can then review the photos and make any recommendations for any identified changes. The photos become a part of the home workers profile and record—further insulating the employer from future claims.

This strategy may avoid the need to physically inspect home offices, which may be nearly impossible, if they are located many miles from the nearest premises offices. That, in turn, saves you time and money.

"Once an employee experiences the joys and fulfillment of working from home, they rarely will want to return to the traditional office or fill a claim that could require them to go backwards to their daily commuters' grind," says Heacock.

Harassment

Sexual harassment is only a remote possibility with home working, but it can occur when supervisors visit the employees' home offices. It would only take one incident to ruin a home working program: the bad publicity and lawsuits would force senior management to unplug the program—for everyone. That's why you should not have employees of the gender opposite to the home working employees visit them unless accompanied by employees having the home workers' gender.

Employees goofing off

Human nature being what it is, people will goof off—to their own destruction and to everyone else's. That includes home workers. There are incidents, fortunately very isolated, of home working employees not being in their home offices, treating those days like paid vacations.

Employees that are not working when they should be are one of an organization's biggest fears about home working. Such actions play to the negative images of home working and will kill it, *fast*. I know of organizations that quickly clamped down on home working when one employee abused it.

The best responses to this are prevention, monitoring, checkup, and quick action. You can prevent abuse from occurring by thoroughly screening home working candidates. Employees who goof off on the job in premises offices are not good candidates for home working. You can monitor and check up on their performance, such as by asking for a synopsis of their work.

Also you can have employees work from home occasionally, on a trial basis, to measure their performance, backed up by login/logout plus occasional calls or e-mails during the day. That way if there are any problems with the arrangement you can resolve them quickly. Employees must be told that if their performance drops while they are home working that they must return to the premises office.

There is a good reason for having a formal home working policy that employees must formally sign. It allows managers to discipline and if necessary fire employees who do not work during the hours they are expected to work.

If you catch an employee goofing off, and they have no good reason for doing so, you may need to make an example of this individual. Canceling their working at home privileges and/or putting them on probation will send a message real quick—to employees and to senior management.

Organizational conflicts

Despite the best of planning, and involvement of other departments, like Facilities, HR, and IT, from the get-go, there still may be ongoing disputes. There are budgets and egos involved. For that reason Jack Heacock recommends hiring a neutral third party (e.g. an outside consultant) to help set up and facilitate the home working program. That is also a great reason for having a separate Home Working Program manager—who has excellent internal political skills—with status equal to those other ancillary departments.

In any organization, status and influence, more than budget, matters. A home working program will succeed or fail not just on the results but on how managers whose departments the program has been implemented and on how home working program managers present, run, and troubleshoot it and in doing win and keep top management blessing. Once your home working has proven itself repeatedly and has built a strong following in the organization, it will become part of the norm in your organization.

Chapter 15: Ending (if need be) Home Working

Yes, everything comes to an end. That includes this book, this writer, and home working.

Home working can end for individuals or for organizations. Individuals may not work out either as employees or as home workers; they may lack the competencies, discipline, and/or the skills to succeed or to function at home. There may be a crisis, like a divorce or a fire that destroys the home environment. Or employees are laid off, organizations are downsized, liquidated, subsumed, or sold; the subsuming or acquiring organization may have a different way of doing home working or may not want to do it at all.

If the end is preventable, then you must be on the watch for trouble signs in the employees and in the program, and take steps to correct them. If doom cannot be forestalled, then you must take the necessary steps to conclude the home working program for the individuals and for your organization—steps that are fair to all in order to prevent further loss and to allow you to protect your assets.

There are two main categories by which individual employees can have their home working terminated. These are *functional*, where employees are failing to work satisfactorily at home or no longer wishing to work at home, and *institutional*, where employees are having their home working ended for internal reasons. There is also ending home working by *organization*.

✪ FUNCTIONAL ENDING

The *functional* reasons for ending individual home working include:
* Inability to carry out their responsibilities at home
* Desire to return to work at premises offices
* Disciplinary, including failing to abide by home working policies
* Personal issues (children, divorce, changes in homes)

You can prevent many, but not all, of the functional reasons through proper planning, screening, selection, and management of home workers, and setting out and monitoring compliance with home worker policies. That is what this book has been all about.

Many of the circumstances behind functional reasons can be and *must be* identified and rectified quickly. You must monitor performance and procedures like sign in/out carefully, especially in the early days of the program and in the early days of employees' participation in the program.

For example, if you have a home working employee who is signing in late, find out why and remind them to sign in at the time required. If your policy forbids excessive background noise and your supervisor hears it, or more seriously, there are customer or staff complaints, then you must take that up with the employee and come up with ways to reduce the noise.

The experience of organizations is that their employees want the home working program to work. If they are warned that the arrangement is not working, or that they need to buck up their performance or it's back to the nooses and the !$%&*-hose, then most will shape up—especially if they put their own cash into the furniture and the equipment.

Remember, everyone is watching your program. There are some old-line managers and executives that would like to see it fail and go back to the old, safe (but inefficient) way of working.

At the same time some employees may find that home working isn't for them. They may miss the social interaction, or more importantly (from the organization's viewpoint), they need more supervision and better training that they feel can best come from being at the premises office in-person. Or they brought a new person into the world and the infant has an *extremely good* pair of lungs.

Then there are personal issues. A divorce, a roommate or family member moving out or getting laid off, or a terrible incident like a fire or tornado, or unsavory (and possibly noisy) characters moving next door or underneath may mean a change in home office locations. But those new "offices" may be less suitable. At the same time a spouse/loved one may be transferred, or has agreed to take a job offer in another community.

Recommendations

The tricky part with these functional reasons is what to do if the matters *can't* be solved. For example, if the home working arrangement is no longer working out but these individuals are good employees what happens if you no longer have space for them at your premises office?

Here are some choices:

* Make them "permanent hot-deskers." But if you use that solution, will that interfere with other employees who hot desk it?
* Try to find a corner for them. (Most buildings have some room for one or two people).
* Work with the home employee to accommodate changed circumstances, such as by being flexible with your rules. For example, if their spouse is moving to another part of the country or to another country see about having that employee work from there—even at possibly higher costs. Remember, great employees are priceless.
* Drill down to the next set of employees, existing or new, and see which ones will be willing to work from home.
* Let those employees go. As in "sorry...

None of these choices are easy. But you will find a solution amongst them that meets your needs. Remember, you are required to compensate employees for their work. But you are not required to provide them with a formal premises office.

✪ INSTITUTIONAL ENDING

There are a number of institutional reasons for ending individual home working. This section will examine some of those reasons.

Changes in Work Function

Examples of this type of institutional ending include adding in-person customer service, a step up to a more hands-on managerial position.

But think hard before consigning or allowing others to consign such functions to premises offices. Many managers and executives who are skeptical about home working will use that rationale as an excuse to slow down or derail the program.

At the same time you might lose employees who prefer to work from home and will quit if they have to commute again. Can you afford that?

Recommendations

If there are functions and tasks that must be done outside of the home office, consider or make sure you arrange to permit the former home work to occasionally home work, so that you and the employee still benefit from it. One example would be a sales rep or technician who has been told they are being moved to field sales or support, but arrangements have been made so that they would not have to perform all of their duties at the premises offices.

Technology is advancing so rapidly functions that what only a few years ago had to be done on premises can be now be done at home. One example is checking galley proofs—they were once sheets of paper that had been proofed off the press. Now they are portable document files, checkable and correctable anywhere.

Downsizing

If employees are being let go there is little that seemingly can be done. Budgets are budgets. Money is money.

Home office employees are too often the first to be let go. Out of sight/out of mind. But this is prehistoric thinking that could hurt your organization financially. Why? Home workers are often the *most* productive and the *most* efficient employees in the organization. Keeping premises offices staff that cost more to accommodate and who are less productive does not make financial sense.

If you are letting home workers go they may be at a disadvantage compared with premises offices staff in that they can't exactly find another position within your organization. But because these about-to-be former employees are not on premises they will have less compunction going outside, such as to a competitor, taking their skills and knowledge with them.

While you may offer to bring the home worker back to the premises office, in some cases, that is impossible for the home worker. They may live too far to commute or even to hot-desk into the premises, or they may have personal or family obligations keeping them at home.

Recommendations

When staff is being cut, managers, supervisors and executives have a fiduciary obligation to the organization to ensure that the best people, with the lowest costs to the organization, are kept, no matter where they are working. Therefore, you need to track performance and cost closely, and put forward those candidates who are lowest-performing, do not have much potential to grow, and who cost the organization the most in infrastructure and support.

There are also steps that you can take to ensure laid off home workers will stay within the organization. Chief among them is that in the event of a layoff talk to HR to see if those departments that have openings could accommodate home workers.

Home working could be a hidden asset that way: because the employees are working from home they can be assigned to another department in another location elsewhere in your organization —and not have to make a costly, time-consuming, and productivity-wasting move. The hiring department does not have to find space for them, or juggle the voice/data or supply equipment. The home employee is ready to go.

New Hostile-to-home-working Management

This section covers what happens when you get managers and executives who are skeptical or hostile to home working and want it downsized or scrapped. Not everyone is sold on home working.

Again, like downsizing, there may not be much that can be done. Especially if the dictates are coming from above and/or if the managerial personnel are otherwise well-qualified and/or they have advocates and mentors at the senior level.

Recommendations

You can take steps to try and protect your home working program. They include educating the naysayers on the program's benefits, what goals you are attempting to achieve through the program, why you selected home working to achieve these goals, and how well you are achieving your objectives.

In other words, you have to sell these reluctant managers and executives like you sold the managers and executives on the home working program in the first place. Make a business case. Only this time you have a track record, and you must defend it. That goes back to what was pointed out earlier—you *have* to make the program work by quickly identifying and resolving problems.

Also, if you have many people working from home, you may want to outline to these managers and executives the hard costs of rolling back home working. Those costs include losing valuable employees who are making contributions to your organization—many may quit and work elsewhere rather than commute again.

Many of those employees can't commute if they live outside of travel range of your premises offices. Relocation is horrendously expensive and you lose productivity in the moving process. With skilled labor becoming in short supply, driving away good people, and having to recruit, screen, train and bring up to speed, replacements have big price tags that can't be ignored.

Moreover, if your home working program has enabled you to downsize your real estate you may not be able to bring those employees "back" because you don't have the room. If that is the case, you will need to demonstrate the added property facilities costs and complexity. If the only space that is available is not contiguous with your existing offices then you run into the nightmares of managing additional facilities. The workers are better off working from home.

✪ ORGANIZATION ENDING

Organizations may want to end home working for several reasons. Let's examine a couple of the more common organization endings.

Desire to Reduce Total Costs

Home working is easy to pick on because it is "new" and may not be seen by many executives as "essential" to the outfit. Or they may eliminate home working as part of the department they are chopping.

Recommendations

You may not forestall a total downsizing: business is business and budgets are budgets. But you should point out and demonstrate that home working can and does save money by avoiding site selection real estate and furniture costs, and it boosts productivity.

You may be able to present enough convincing data to save some jobs, while enabling your organization to cut costs. That would be quite a feather in your career cap, while making you a hero amongst your staff.

If staff recommendations are inevitable, you can point out that having and keeping as many employees working from home as possible makes downsizing easier. That's the flip side of out-of-sight/out-of-mind. Because employees don't "see" each other there will be less rumor-mongering, loss of productivity, and less equipment and services theft that too often occurs in a downsizing.

Finally, with home working you have cost-saving options, such as hiring self-employed home workers. Former employees can freelance. But again be careful how you do that; taxation authorities are keen to nail firms for substitution scams to avoid paying Social Security and unemployment taxes.

Hostility to Home Working by New Owners

If your organization have been acquired or merged into other organizations, home working may not be part of the "new" corporate culture. There also may be technical complexities in meshing voice and data systems that the new management does not want to deal with, especially if they don't see it as being "essential."

Recommendations

If the acquiring organization is skeptical or hostile to home working then you will have to remake your business case for it. But in doing you should research the new parent and management carefully and present home working in terms that are relevant *to them*: how home working can and will mesh very well with *their* operations and *their* culture.

Show the new team the benefits of home working to *their* bottom line. But also show to these individuals how much dropping home working will cost *them*: financially, in employee contribution losses and turnover, and in image. Eliminating home working is turning back the clock. It adds to pollution and ruins family life.

Besides one of the nightmares in downsizing and mergers is what to do with the real estate, furniture, and people. Where do you put the smaller teams and the "new people"? How do you mesh formal and informal dress codes and cultures? Or do you open a new building and throw everyone together? Not to mention the horrors of cleaning up, packing, moving, unpacking, and setting up that throws your productivity into the toilet.

You can show that home working avoids such costly productivity-killing messes altogether. After all, employees already have "offices": their homes. There will be less "A team versus B team friction" if they are communicating via voice and online—which can be monitored—than in-person including sniping at the water cooler.

In contrast to all the hassles introduced by a move, merger, or acquisition, any technical changes to the voice and data systems to accommodate home working will be a piece of cake. This also may be an opportunity to *introduce* home working: by giving a rationale to buy new laptops, install VPNs and new VoIP-enabled switches, off-premises extensions, or on network call routing.

But you must also argue how expanding home working to cover the acquiring/subsuming firm will help them in saving costs and boosting productivity. That means you have to do your homework on them.

Merged organizations always try to stress "unity." If members of the A and B teams both home working that will show a "new beginning" that is beneficial to all.

○ WRAP-UP

Inevitably home workers will have to be let go, sometime, somewhere, for the reasons illustrated above. While organizations have long had policies in place to cover employee terminations and severance—such as canceling network access—there are special considerations involved with home working, especially exclusive home working employees:

Unemployment

Because exclusive home workers work from their homes they usually, but not always, come under the jurisdiction of where they live, *not* where they report to. Check with HR on that. If that is the case they come under their home jurisdiction's labor laws.

Voice/data connections

When you terminate or lay off a home working employee and you supply the voice/data connections you will need to make arrangements with the providers to sever them. If you have disciplined an employee, or an employee suspects they are to be let go inform your IT department to make sure they are not making calls to Cousin Bruce in Australia, i.e. "unauthorized" use of the phone. Just as you do or should do if the employees in question work from premises offices.

If you are disciplining or about to fire a home working employee and they have network access from their laptops consider suspending it and have them work from Webmail instead. That way you avoid any chances that the employee may maliciously harm your network, such as installing spyware, spreading viruses, and stealing data.

That's another great aspect of working from home. If there is an attack you know where it is coming from; it isn't some employee at another's desk.

Equipment return

You will need to get your equipment back. You also will need to make sure employees don't abscond with your data. Your Home Working Policy should require equipment and data or file returns.

Getting equipment "back" is easy for premises offices workers; the gear is there. But it might not be as easy for occasional or exclusive home workers to whom you supplied laptops and other gear such as copiers/printers/faxes. If the employees live outside of traveling distance, you must determine if it will cost you more in shipping than what the equipment is worth: standalone or as trade-ins.

If that employee is being fired or is upset that they are being laid off, they may want to steal the gear to "get back" or as "compensation" for your actions. It may or may not be worth your while to pursue legal action for a depreciated-to-$500 computer.

If you are terminating an employee, some jurisdictions require you to pay them on the last day of working; you are not allowed to hold that wage pending return of equipment. But if you are laying off an employee(s) and you have a severance agreement that employees must sign then they must return the property. Check with HR and Legal.

ADA compliance

If you have physically impaired employees, under the Americans with Disabilities Act accommodation telecommuting agreement if you end that agreement you are terminating that employee. You can end that agreement if your organization ends the home working program, believes the accommodation is no longer fair, or if your needs are not being met.

Recommendations

Most employees, even those who are poor performers, realize that it isn't their best interests to abscond with gear; burning bridges also burns careers. Someone who didn't work out at one place may be a star performer somewhere else.

If you know you are going to be terminating an employee or ending their home working agreement, and if that employee lives within commuting or traveling distance, ask that employee come in for a meeting (which your HR policy may require in any event). Make a reason that they need to bring in their laptop—like you don't have any available computers. That way when you tell the employee of that decision you can get that asset back.

If you are laying off an employee, consider making the equipment as part of the severance package—but on the condition that they hook it up to the network for your IT staff to wipe out any proprietary data and programs; employees must turn in any data and programs that belong to you. If the employee has been working for you for a couple of years chances are the hardware has been sufficiently depreciated to make it worthless. This avoids the returns and shipping issue.

Should you need that former employee as a freelancer or if business turns around, or there is a new business within that organization and that laid-off employee is suitable for one of the new positions, they will have the tools to work for you again. These workers may be willing to do so because you did treat them well.

During the course of writing this book I was downsized by my employer, CMP Media; there are sweeping changes in the publishing as well as in other industries. I cleaned off my laptop and sent it in; the company paid for the shipping. Because I had a great working relationship with the people at all levels in CMP I would not have hesitated accepting an offer for a similar position from the company; CMP also has a great IT Home Support department.

⊙ HOME WORKING REBIRTH?

If your organization rebounds, or grows another department, consider home working from the start; begin by rehiring those home workers who had been let go. You can add employees and benefit from their contributions, one by one, without over-committing your scarce assets.

Because home working avoids investments in real estate and facilities and it is more productive than premises offices working, it provides a much more fertile ground for a rebirth of your enterprise or department. The best way to prepare for possible future regeneration of your organization is to lay the seeds: by considering, planning and deploying a successful home working program.

Epilogue

"The Sky is Falling!"
—Will home working hurt cities? On the contrary

There are fears by many, especially older cities, that encouraging people to work from home will harm their communities. They say they depend on property tax revenues from the premises office, and the rush of commuters into and out of their business districts each weekday to support restaurants, shops and other businesses, like gas stations. Some of their revenues come from local sales taxes.

Those are legitimate concerns. And as a result, many cities may not want to support home working, such as by ending double-taxation of home workers' income, incurred if their employer has a premises office in one locale and the home worker lives in another community.

But this is short-sighted. Because if the cities, especially those with traditional walkable downtowns, take a hard look at home working they will find the benefits outweigh the losses. Here's how:

* The key to the central city is having people living there. The model of a central business district that empties out at 5pm and on Fridays, leaving an eerie, dangerous ghost town, is obsolete and does not work. Home working will encourage people to live in the downtown and to enjoy the enormous cultural amenities that cities large and small provide: the arts, museums, theater, plus restaurants and shopping are at their doorsteps. Yet those workers can "commute" to employers in the suburbs or to any part of the world at the speed of light—utilizing the fat fiber infrastructure that services cities. They do get the best of both worlds.
* Home working opens opportunities for inner city residents. By working from home they can tap into jobs anywhere in the world. By the same token, employers can tap into this little-tapped but hard working and very willing labor force with home working programs.
* Reducing commuting into the cities via home working saves them money and their taxpayers' money, making their communities more affordable for residents and businesses. Local governments avoid having to spend enormous sums to build and maintain roads and mass transit, plus sewer and water to handle rush hour and employer-driven weekday loads.

* Fewer commuters will lead to a cleaner environment and better quality of life for urban residents, visitors and those who need to commute in. Less traffic means reduced air, noise and water pollution, and lower numbers of fatal or debilitating so-called 'accidents'.

* With home working cities can also scale back on emergency services: police/fire/ambulances/medical facilities and staff to handle accidents, fires and crime. Few events screw up a city like a freeway pileup at 6:30. Everybody gets fouled up: from commuters, to truckers, to people needing to get to the hospital or catch a flight.

* Having fewer premises offices, especially prominent ones, reduces security and terrorism risks. Prestige buildings are targets, as has been tragically demonstrated (as noted elsewhere in this book I witnessed the 9/11/01 attack on the World Trade Center and my wife, sister-in-law and I evacuated Manhattan in the wake of it).

  Also, fewer commuters means infrastructure like bridges, tunnels, transportation terminals, and bus/rail lines are less inviting targets. I traveled on business into, and later commuted into, central London during a series of terror attacks and threats. The fear is there, even under the thick British mantle of "muddle through/stiff upper lip/not going to let them get to me."

  You can't as easily shut down a city if many people are working from home. By distributing portions of workforces to their homes cities gain survivability and reduce their vulnerability, at enormous potential savings in lives as well as money.

* Home working works to the *advantage* of downtowns and older suburban hubs. By shrinking the number of employees needed to work in premises offices, to those that need to interact face-to-face with others organizations can fit into high quality, fatwired small floorplate Class A space or attractive, enjoyable renovated older buildings that are only steps from the best restaurants in town.

With home working then, all sizes of cities get the best of all worlds. Home working, after all, is just a return to the model of the late 18th century where there was a balanced mix of small premises offices, urban residents who could "commute" to anywhere from their homes, albeit visitors or local businesses that serve them, and not just 9am-5pm, Monday-Friday.

Urban individuals and families will be safer and live healthier. They will have more employment opportunities. They also will have more money to spend—which helps local businesses and city coffers—because they are freed from shelling out fat chunks of their wages in income, property and sales taxes to pay for transportation and emergency services.

The same goes for residents in smaller communities too. They will have to worry less about pollution, traffic, and accidents, and have more time to be with their families, and to get involved with their communities. Having workers closer to home means more of those wages spent locally while reducing tax burdens caused by transportation, and the consequences.

Ultimately, for large and small communities alike, for the residents and employers alike, home working means a better quality of life. And isn't that worth achieving? The points, information and suggestions in this book are intended to help you do just that.

Resources Guide

Here is a list of companies, consultants and organizations that have graciously assisted me with this book and/or which I cited in this tome, either directly or as background. The suppliers covered have been selected because they provide products and services that especially apply to home working; it is therefore by no means a complete list.

Please note that the phone numbers and sometimes their website URLs change. If you reached a non-working number or website, you can find current listings on *Call Center Magazine's* website, www.callcentermagazine.com, or by doing a Web search.

✪ CONSULTANTS

AT&T
866-409-7051
www.att.com/teleworking

The Boyd Company (site selection)
609-890-0726
www.theboydcompany.com

Contact Knowledge
972-781-0994
www.contactknowledge.com

Gil Gordon Associates
732-329-2266
www.gilgordon.com

Jack Heacock & Associates
303-841-8799

InnoVisions Canada (Bob Fortier)
225-5588
www.ivc.ca

Joanne Pratt and Associates
214-528-6540
www.joannepratt.com

Laura Sikorski (space planning and design)
631-261-3066
www.laurasikorski.com

The Tanner Group
800-429-8550/801-994-5000
www.tannergroup.com

Teletrips
888-882-09426/202-434-8913
www.teletrips.com

TELEWORKanalytics
888-353-9496/571-434-7444
www.teleworker.com

Trammell Crow (real estate/site selection)
215-863-3000
www.trammelcrow.com

Wave, Inc. (design services)
972-387-7555
www.wave-fcm.com

✪ ATTORNEYS

Hale, Hackstaff and Friesen (John Paddock)
720-904-6000
www.halehackstaff.com

✪ CONTRACTORS AND OUTSOURCERS

Alpine Access
866-279-0585/303-279-0585
www.alpineaccess.com

ARO
800-722-8827
www.callcenteroptions.com

IntelliCare
800-524-1484
www.intellicare.com

O'Currance Teleservices
888-OCURRANCE/801-736-0500
www.ocurrance.com

Virtual-Agent Services
847-925-2343
www.vagent.com

West
800-762-3800
www.west.com

WillowCSN
888-899-5995
www.willowcsn.com

Working Solutions
972-964-4800
www.workingsol.com

✪ EQUIPMENT BACKUP/BUSINESS RECOVERY

American Power Conversion (APC)
877-800-4272
www.apc.com

Briggs and Stratton (also owns Generac portable line)
800-743-4115
www.briggspowerproducts.com

Generac
800-333-1322
www.guardiangenerators.com

Kohler
800-544-2444
www.kohlerpowersystems.com

Liebert
800-LIEBERT
www.liebert.com

Onan
800-888-ONAN
180cc (business continuity planning)
800.550.4180/972.814.6772
www.180cc.com

✪ FURNITURE AND ERGONOMICS APPLIANCES

Allied Plastics
800-999-0386
www.alliedplasticsco.com

Dowumi
866 811 7771
www.dowumi.com

Haworth
800-344-2600
www.haworth.com

Herman Miller
888-443-4357
www.hermanmiller.com

Keybowl
877-363-7774/407-622-7774
www.keybowl.com

LapWorks
877-527-9675/909-948-1828
www.laptopdesk.net

Plantronics
800-544-3660/831-426-5858
www.plantronics.com

⊙ HOTELING AGENTS/
OPERATORS/OFFICE SPACE
MANAGERS

Agilquest
888-745-7455/804-745-0467
www.agilquest.com

HQ Global Workplaces
888-OFFICES
www.hq.com

Regus
877-REGUS-US
www.regus.com

⊙ EQUIPMENT/OFFICE SUPPLIES/
SERVICES/FURNITURE DEALERS

Basics (Canada)
519-893-3039
www.basics.com

Best Buy
1-888-BEST BUY
www.bestbuy.com

Circuit City
800-843-2489
www.circuitcity.com

ComputerRepair.com
561-988-3008
www.computerrepair.com

Fedex Kinkos
800-2-KINKOS
www.fedexkinkos.com

Future Shop
www.futureshop.ca

The Home Depot
800-553-3199
www.homedepot.com

IKEA
www.ikea.com

Island Inkjet
877-4INKJET/250-897-0067
www.islandinkjet.com

Lowe's
800-44LOWES
www.lowes.com

Office Depot
800-GO-DEPOT
www.officedepot.com

OfficeMax (Boise Cascade)
800-283-7674
www.officemax.com

Radio Shack
800-843-7422
www.radioshack.com

Staples
800-3STAPLE
www.staples.com

WalMart
www.walmart.com

⊙ ORGANIZATIONS

Canadian Telework Association
613-225-5588
www.ivc.ca

**International Telework Association
and Council**
301-650-2322
www.telecommute.org

Telework Coalition
202-266-0046
www.telcoa.org

○ SUPPLIERS

AT&T
866-409-7051
www.att.com

Aspect
408-325-2200
www.aspect.com

Avaya
866-GOAVAYA/908-953-6000
www.avaya.com

Cisco Systems
(800)553-NETS/408)526-4000
www.cisco.com

Citrix
800-424-8749/954-267-3000
www.citrix.com

Concerto
800-480-2299
www.concerto

eOn Communications
800-955-5321
www.eoncommunications.com

Envision Telephony
206-621-9384
www.click2coach.com

FurstPerson
888-626-3412/773-353-8600
www.furstperson.com

Gatelinx (conferencing)
877-669-4283/910-695-0000
www.gatelinx.com

GemaTech
619-283-3765
www.gematech.com

Gryphon Networks
781-255-0444
www.gryphonnetworks.com

Interactive Intelligence
317-872-3000
www.inin.com

Literati Group (eGroupware-conferencing)
312-482-9229
www.literatigroup.com

MCI
800-465-7187
global.mci.com

MCK
888-902-6223/781-343-6338
www.mck.com

Microsoft (Live Meeting)
800-426-9400 [US] 877-568-2495 [Canada]
www.microsoft.com/office/livemeeting

Net2Phone
973-438-3111
web.net2phone.com

Neoware
800-neoware/610-277-8300
www.neoware.com

Nortel
800-4-NORTEL
www.nortelnetworks.com

OnRelay
011-44-1732 371-590
www.onrelay.com

Primus Telecommunications
888-899-9900 (US)/888-501-8430 (Canada)
www.primustel.com

Shaw Communications
403-750-6990
www.shaw.ca

Siebel
650-295-5000
www.siebel.com

Siemens
800-765-6123
www.siemensenterprise.com

Spectrum
800-392-5050/713-944-6200
www.specorp.com

Sprint
800-370-6105
www.sprintbiz.com

Sun Microsystems
800-555-9SUN/1-650-960-1300
www.sun.com

Symantec
408-517-8000
www.symantec.com

Telephony@Work
858-410-1600
www.telephonyatwork.com

Teltone
800-426-3926/425-951-3398
www.teltone.com

Telus
888-811-2323
www.telus.com

UCN
888-UCN-5515
corp.ucn.net

Vonage
866-243-4357
www.vonage.com

WebEx (conferencing)
877-509-3239/1-408-435-7048
www.webex.com

West
800-762-3800
www.west.com

White Pajama
877-725-2621
www.whitepajama.com

Appendix I: Telework Employment Agreement[1]

THIS TELEWORK EMPLOYMENT AGREEMENT is entered into between The Company International, Inc. (the "Company") and _____ (the "Employee") and is effective on the date it is signed by the last signatory. The Company and the Employee are referred to in this Agreement together as the "Parties," "we," "our" or "us," or individually as a "Party."

1. **Employment.** In consideration of our respective rights and obligations set out below, the Company employs the Employee for the position and to provide the services described below, subject to the terms, covenants and conditions of this Agreement.

2. **Contract Consideration.**[2]

for a new employee – The Company is hiring the Employee based on the Employee's agreements stated below.[3]

for a current employee – In return for the Employee's agreements stated in this Agreement, the Company is permitting the Employee to work via telecommuting and is granting the Employee the other additional benefits stated below.

3. **At-will Employment.** Neither this Agreement or any other agreement between us states how long the Company may employ the Employee. Therefore, the Employee's employment with the Company will be indefinite and **AT-WILL**, which means either of us may terminate that employment with or without cause at any time, and without advance notice, procedure or formality.[4]

4. **Employee's Duties and Worksite.** The Employee will provide the following services under this Agreement: *state the duties the Employee is agreeing to perform; alternatively, the employer could attach a separate addendum or exhibit where there is a detailed description or where the duties may change.* The Company may change the Employee's duties at anytime. The Employee will provide such services at [*check all that apply*]:

☐ *the Company's offices*   ☐ *the Employee's home*   ☐ *Other: specify any other applicable territory or location* and at such other locations as the Company determines from time to time. As used in this Agreement, the "Remote Worksite" refers to: _____ the Employee's home or: _____.

The Employee's services performed for the Company and its supervisors, employees, customers and others, and the Employee's representation of the Company regarding customers and the public, will in all events be consistent with the Company's best interests and with the Company's policies and standards.

5. **Telework**. Beginning on _____ [*date*], the Employee will work via telework as specified below.

Use any of the following that suits the particular employer and employee.

   a. **Telework Personnel Policies**. The Employee has reviewed and will comply with the Company's policies concerning telework, as stated in the Company's Personnel Policies dated *effective date*, and as may be revised later. The Company's Personnel Policies provide guidelines and are not a contract; so, if there is any conflict between a policy and a provision of this Agreement, the provision stated here will control.

   b. **Telework is Voluntary**. The Employee understands and agrees that telework is voluntary. The Employee also understands and agrees that the Company may at any time change any or all of its telework policies and the conditions under which the Employee is permitted to telework, and may withdraw permission for the Employee to telework. Nothing about our agreement the Employee may work via telecommuting will alter our at-will employment relationship; so neither of us has any obligation to continue that relationship if the Employee or the Company chooses not to have the Employee telework or to change the amount of time or the conditions under which the Employee teleworks.

   c. **Evaluation Period**. The first three months after the date Employee starts telecommuting will be an evaluation period during which we will each evaluate the effectiveness of this working arrangement. For example, we will make special efforts to monitor how the telecommuting is working and to communicate with each other about productivity, procedures, equipment, etc. During this evaluation period, either of us may decide for any lawful reason that the Employee should not work via telework and may cancel the telework provisions of this Agreement with notice to the other party.

Subparagraph (d) is recommended for nonexempt employees; it can also be used for exempt employees where the employer tracks work time for customer billing

   d. **Reporting Work Time**.[5]

      i. **Nonexempt Employee** – The Employee is a nonexempt employee under applicable wage and hour laws and, as stated in the Company's Personnel Policies and other provisions of this Agreement, should keep work time records and report work time for purposes of determining the Employee's compensation, and for tracking leave and other employee benefits [*optional*—and for customer billing].

ii. **Exempt Employee** – The Employee is an exempt employee under applicable wage and hour laws and, as stated in the Company's Personnel Policies and other provisions of this Agreement, should keep work time records and report work time for purposes of tracking leave and other employee benefits [*optional* – and for customer billing].

Subparagraphs (e) and (f) should be used only for nonexempt employees

e. **Work Hours**. *choose the method that applies*[6]

  i. The Company will be responsible for establishing the Employee's specific work hours, including those completed via telecommuting.

  ii. Until further notice, the Employee's work hours will be as follows:_____

  iii. The Employee will telework using the Remote Worksite _____ days per week, or as otherwise scheduled/needed, and will be in the Company's customary offices/work areas (or making calls on customers) during all other work hours.

Use the following for nonexempt employees only

f. **Overtime**. The Company may need the Employees to work overtime on occasion. The Employee will receive advance notice of necessary overtime whenever possible. Time reports will be used to compute the Employee's pay. Time reports should be completed each day and submitted to payroll at the end of every work week [*or state alternate schedule*]. The Employee's time report should indicate for each day worked: the total amount of time in hours and quarter hours the employee worked based on the time the employee actually began or stopped working at the start or end of a scheduled work period. No overtime will be allowed unless the Employee's supervisor approves it in advance.

g. **Billable Time**. The Company will track the Employees' time spent on projects for customer billing and/or internal accounting purposes. When the Employee's time is tracked for these reasons, the Employee's time report should contain appropriate records of the time spent on each billable project or task.

h. **Remote Worksite Equipment, Services and Software**. The Employee [*or—The Company*] will be primarily responsible for equipping and maintaining the Remote Worksite. As indicated in the following table, the Employee or the Company will provide the identified equipment, services and software for the Remote Worksite to enable the Employee to perform the Employee's duties under this Agreement: *list any needed equipment and software, such as:*

Equipment	Make/Model/Other	Serial No.
Computer, keyboard, mouse		
Monitor		
Modem		
Printer		
Fax machine		
Phone		
Phone/DSL Line No. 1		n/a
Phone/DSL Line No. 2		n/a
Cellphone		
Pager		
Long Distance Carrier		n/a
Software:		
Desk/work station		
Other:		

i. **Insurance, Maintenance, Supplies**. The Party providing the equipment, service or software identified above [*or*—The Company] will provide insurance, maintenance, repairs, upgrades and supplies as needed. The Company's provision of any equipment, software, services, supplies or facilities to the Employee will not give the Employee any ownership interest in any such equipment, etc. and will not obligate the Company to continue providing any of them.

j. **Security, Safety, Access**.[7] The Employee will comply with all applicable telework security, safety and access policies the Company implements to protect the Company's and its customers' and suppliers' interests, assets, information, trade secrets and systems.

Note: If the Company uses (j), its personnel policies should cover these areas. Alternatively, the telework contract can state all or any of the following:

k. **Safety, Ergonomics**. The Employee will be responsible for operating all equipment used in performing the Employee's duties in a safe manner and consistent with all applicable manufacturers' instructions and warranties and the requirements of any applicable insurance policy. The Employee will set up and maintain the equipment in a manner that maximizes productivity and comfort and that minimizes the possibility of injury, such as from strains or repetitive motion.

l. **Use of Company Equipment**. The Employee agrees to use the Company's equipment for its intended business purpose only, not to damage or abuse it, and to allow only other employees to use it. The Employee may not use any Company equipment for personal reasons without the express prior permission of the Employee's immediate supervisor.

m. **Software**. The Company supports legally obtained software and shareware. The Employee may use authorized software only according to the applicable software manufacturer's licensing agreement and any patents or copyrights applicable to software the Company provides or the Employee uses in performing the Employee's duties under this Agreement. The Employee will not make any unauthorized copies or alterations of such software.

n. **Notice of Malfunction**. The Employee will promptly notify the Employee's supervisor of any failure or malfunction of any Company equipment, service or software.

o. **Access to Data**.[8] The Company may review any data, software, program or other information stored on any of its computers, in its network or on disks, tapes and other means of storing information and data, including any program or information associated with any unauthorized software or shareware. The Company may remove any unauthorized software or shareware and has no responsibility to save or to preserve as confidential any data, program or other information stored on any of its computers or in its network. Only authorized passwords, access codes and other security measures are permitted for use with the Company's computers, network, software and programs, and no personal password, access code or encryption may be used to prevent management from using or reviewing any part of a hard drive, disk, tape or other electronic storage medium.

p. **Access to Equipment**. Management at anytime may inspect any area within its offices and workplace and any Company equipment at the Remote Worksite or in the Employee's control.

q. **Equipment Return**.[9] The Employee will return all the Company's equipment in good, unaltered condition on request and on or before the Employee's last workday with the Company. Subject to applicable laws, without giving prior notice to the Employee or posting a bond, the Company may obtain an order from a Court permitting entry into the Remote Worksite to recover any of its property that is not returned; and all amounts owing for damaged or unreturned Company property will be deducted from the Employee's pay.

r. **Confidential Information.**[10] The Company's passwords, access codes and Internet server account information are all confidential information which the Employee must treat with care and that should not be disclosed to any unauthorized person. Access cards, passwords and account records should not be posted, entered into automatic sign-on protocols or accessible computer files, or left in open drawers or other places where they could be seen by visitors or other non-employees.

s. **Internet and E-Mail.**[11] The Employee will access the Internet during work hours or using the Company's equipment only for business purposes directly related to particular research, projects or other purposes consistent with the Employee's duties. The Employee will not send confidential information over the Internet without using precautions, such as pass words or encryption, to ensure the information remains confidential. Consistent with the Company's Harassment and Sexual Harassment policies, the Employee will not access, download or send to anyone via the Company's Internet access or E-mail any offensive or inappropriate material or message.

t. **Communications Monitoring.**[12] Subject to applicable law, management at anytime may monitor and/or record any communication made or received using the Company's computers, network, modems or telephone equipment, including any e-mail sent or received within the Company or to or from third parties and including any information or materials reviewed, received or sent over the Internet.

u. **Utilities – Safety.** The Employee verifies that the Remote Worksite's utilities meet all applicable manufacturers' requirements to operate the equipment safely. Specifically, and without limitation: all electrical breakers or fuses meet applicable building codes and are labeled for the circuits they control; have adequate power and are properly grounded; and there is appropriate surge protection.

v. **Utilities – Cost.** The Employee [*or*—The Company] will pay the cost of all utilities needed for the Remote Worksite.[13]

w. **Long Distance and Other Phone Charges.** The Employee [*or*—The Company] will pay the cost of long distance services and all other phone charges incurred by the Employee making job-related calls from the Remote Worksite.[14]

x. **No Restrictions.** The Employee verifies that, to the best of the Employee's information and belief, nothing restricts using the Remote Worksite as a place to complete the Employee's duties, including any zoning, lease or community covenant.

6. **Compensation.** For all services the Employee provides under this Agreement, the Company will pay the Employee the following salary: *state the Employee's com-*

pensation, including any information needed to calculate commissions or other salary components

   a. **Overtime**. The Employee is an exempt worker under applicable wage and hour laws and regulations and will not receive overtime time.

or: _____ The Employee is a nonexempt employee under applicable wage and hour laws and regulations, and, for each hour over 40 hours the Employee works in any work-week, the Company will pay the Employee overtime at 1? times the Employee's regular hourly rate.[15]

7. **Paydays**. The Company's paydays are described in its Personnel Policies [*or, state how the employer normally pays*]. All compensation the Company pays the Employee is subject to employer withholdings, e.g., for FICA, Medicare/Medicaid, any applicable occupational privilege tax, and any court ordered deductions such as garnishments. Compensation may also be reduced by deductions the Employee authorizes for insurance, 401(k) contributions and other similar purposes. *Optional*: There will be no compensation advances, unless otherwise agreed in writing by the Company.

8. **Professional and Educational Expenses**. *Consider including a provision about the Company's reimbursement of professional and educational expenses.*

9. **Vacations, Holidays and Leaves**. *Consider adding a provision to confirm the Employer's vacation, holiday and leave benefits.*

10. **Voluntary Benefits**. *Consider adding a provision that describes any other voluntary benefits the Employer provides (i.e., health insurance or moving reimbursement, rather than benefits that are required by law such as workers' compensation and unemployment insurance.*

11. **Other Agreements Between the Parties**.

   a. **Preservation of the Company's Confidential Information**.[16] *Insert here a confidentiality provision, or use a separate agreement and state:* The Employee will preserve the Company's confidential information and customer records, as further specified in our Confidentiality Agreement dated:

   b. **No Competition with the Company**.[17] *Insert here a noncompetition provision, or use a separate agreement and state:* The Employee will not compete with the Company, as further specified in our Noncompetition Agreement dated:

   c. **Agreement to Avoid Conflicts of Interest**.[18] *Insert here a conflicts of interest provision, or use separate agreement and state:* The Employee will avoid conflicts of interest with the Company, as further specified in our Conflicts of Interest dated:

   d. **Work Made for Hire**.[19] *Insert here a work made for hire provision, or use a separate agreement and state*: All work made for hire the Employee creates or with which the Employee is involved will remain the Company's sole property, as further specified in our Work Made for Hire Agreement dated:

12. **Death [Or Disability] During Employment.**[20] *Consider adding a provision covering this contingency.*

Optional and requires caution if the employer wants to preserve an at-will relationship.

13. **Termination**. This Agreement may be terminated under **Paragraphs 1, 3, 12** or under this Paragraph. *Consider a provision specifying how the contract may be terminated.*

14. **Resolution of Disputes**. *Consider adding a provision to arbitrate disputes.*

15. **Entire Agreement; Amendment; Enforceability; Interpretation.** *Consider adding what is known as an integration provision and other "boiler plate" language to enhance the Agreement's enforceability: the wording of the necessary language depends on the laws and practices of the state or province where the employee lives and works and where the employer is headquartered and does business.*

Each Party has read and considered this Agreement carefully, believes that Party understands each provision, and has conferred, or has had the opportunity to confer, with the Party's own attorney before executing this Agreement.

IN WITNESS OF OUR AGREEMENTS, the Company and the Employee have executed this Agreement on the date(s) indicated below.

THE COMPANY INTERNATIONAL, INC.   THE EMPLOYEE:

By: _____   _____

Title: _____   Signature

Date: _____   Date: _____

DISTRIBUTION: THE EMPLOYEE; HUMAN RESOURCES; ALL APPLICABLE DEPARTMENTS

---

[1] This model agreement was written by John R. Paddock, Jr. and appears in his book, the *2004 Colorado Employment Law and Practice Handbook*, copyright © 2004 (West). The *Handbook* and Mr. Paddock's other text are available from West (www.westgroup.com.). The agreement is used here with the author's permission. John Paddock is a partner with the Denver, Colorado law firm, Hale Hackstaff Friesen, LLP, and can be reached at hrlawcolo@aol.com. Mr. Paddock was a member of the Blue Ribbon panel of national experts on telecommuting issues of the International Telework Association & Council, and wrote the chapter "Legal and Employment Issues" for ITAC's book *e-Work Guide, How to Make Telework Work for Your Organization*, (www.telecommute.org).

[2] Consideration: agreements must have consideration, or an exchange of performance, obligations or benefits between the contracting parties, to be enforceable as contracts. For new employees, being hired and receiving employment compensation are adequate consideration for a contract related to their employment. In many U.S. states, continued employment of current employees may be inadequate consideration for an employment agreement that imposes a new condition of employment because the employees get nothing more than they have received before and lose benefits or rights they have previously enjoyed. Therefore, to create an enforceable contract that changes the terms of a current employee's employment by imposing new conditions (e.g., drug testing, conflict of interest limitations or an arbitration agreement) that actually or arguably reduce an employee's rights, an employer usually must give the employee a cash bonus, raise or something else of value beyond the employee's present compensation

and benefits. Allowing an employee to telecommute should be sufficient consideration for this agreement with existing employees even if there are other new terms of employment since both the employer and employee will benefit from the new arrangement. An employer should decide whether it is wise to couple the Telework Employment Agreement with a raise, one-time bonus or other clear benefit to ensure the agreement may not be set aside for lack of consideration when the employer also wants to use this agreement to replace an earlier agreement that was not at-will, to impose an agreement to arbitrate disputes, to protect its confidential information or otherwise to diminish an employee's rights or benefits.

[3] While this Agreement can be used to permit a new employee to work via telecommuting, there are many reasons an employer may need an employee who will be telecommuting to be in the employer's offices for significant periods of time during the start of the relationship; e.g., for training, orientation, and to meet and get to know supervisors and other employees.

[4] The common law of U.S. states generally follow the principle that employees are presumed to be employed at-will unless: (a) they have agreements stating they may be terminated only for cause or that they will be employed for a specified term; or (b) their employer creates implied or express employment contracts through promises, policies or conduct inconsistent with at-will employment; e.g., policies that employees will be terminated only for cause or only after the employer follows certain disciplinary procedures. At-will employees are not employed for any set period of time, and they may resign or their employer may terminate them at any time, with or without cause, and without advance notice or procedures, provided, however, that employers may not terminate employees as a result of illegal discrimination or in violation of statute or public policy.

[5] The following provisions concern employers' obligations regarding time records, compensation and overtime calculations under the U.S. Fair Labor Standards Act (FLSA).

[6] Consider a rotating schedule for telecommuting employees so they can share equipment and a work area in the employer's offices and to even-out coverage of in-office work.

[7] This provision should be consistent with policies the employer has in its personnel policies regarding remote access, security and use of the Company's computers and network. Among the issues a security policy for remote access should address are: guidelines for controlling access between authorized users and the Company's network and other computers and networks; physical protection of the communications medium, devices, computers and data storage at the remote site; how the Company will have access to equipment and to proprietary information at remote sites; who is responsible for theft or damage of hardware, software or data at remote sites; appropriate use of remote equipment and the Company's right to monitor all use of remote equipment; and controls to prevent the uploading of unauthorized programs (e.g., virus programs) into remote site equipment and from remote sites to the Company's network and other computers. As part of formulating the security policy, employers should consult with knowledgeable legal counsel and with their liability, property and workers' compensation insurance providers about telecommuting employees and insurance coverage for the employees and for the employer and its equipment.

[8] Without clear notice from employers that nothing brought to work or created or received as part of work is private, employees may have an expectation of privacy in things they bring to work or make or store using Company equipment, and in communications they have or information they create or receive as part of work. The U.S. federal Electronic Communications Privacy Act (ECPA), 18 U.S.C. A. §§ 2510-2522, prohibits anyone from monitoring the phone communications of other people who reasonably believe their communications are private. The ECPA also prohibits electronic monitoring of employees' phone communications unless the monitoring is done in the ordinary course of business using a device the phone company installs. U.S. states' laws also offer restricted access to electronic data. For instance, Colorado's wiretapping law, West's C.R.S.A. § 18-9-301 – 312, prohibits monitoring any phone or electronic communication without the permission of a sender or recipient. Employers should also have personnel policies that are consistent with these provisions.

[9] Before taking measures that could violate a law or give rise to liability, employers need to verify their rights and their employees' rights with knowledgeable legal counsel. State or provincial real property and privacy laws will govern an employer's right to seek an order to enter an Employee's or third party's residence or property to recover the employer's property. State and provincial wage laws often limit deductions employers may make from employees' pay. For example, the Colorado Wage Claim Act limits deductions employers may make from employees' final pay to "lawful charges" and permits deductions for

stolen property only if the employer reports the theft to the police or sheriff. U.S. federal pension laws (e.g., ERISA) can come into play if an employer makes deductions that effect the 401(k) or other pension contributions. State and federal minimum wage laws should be considered before an employer makes deductions to compensation that reduce employees' pay that would take the effective compensation rate below the minimum wage. The interplay of a variety of laws and court decisions mean employers are well-advised to have clear agreements and policies that limit the use of their equipment to business purposes, that require all their property to be returned in good, unaltered condition on request or when employment terminates.

[10] *See also* the confidential information provision in ¶ 10.a.

[11] This provision should be consistent with the employer's personnel policies covering e-mail, Internet access, software, and computer passwords.

[12] Employees or private employers generally cannot assert they have an expectation of privacy in their work-related phone calls or electronic communications when the employer has given them notice it may monitor calls and communications. U.S. federal law, however, places some limits monitoring telephonic or electronic communications. The Omnibus Crime Control and Safe Streets Act (OCCSSA), 18 U.S.C.A. §§ 2510-2520, and the Electronic Communications Privacy Act of 1986 (ECPA), 18 U.S.C. A.)) 2510-2522, restrict the ability of both private and public employers to use surreptitious surveillance and monitoring activities. State or provincial law, such as Colorado's wiretapping statute, West's C.R.S.A. § 18-9-301-312, may also come into play.

[13] If the employer will be paying for utilities, it should be billed directly, or the parties need procedures for the Employee to pay utility bills and to obtain reimbursement.

[14] If the employer will be paying phone bills, it should be billed directly.

[15] It may be possible to use the overtime calculations for fluctuating work week or what is known as a *Belo* contract to pay certain nonexempt employees overtime at a lower effective rate than 1? times the regular rate.

[16] Trade secrets may be protected in various ways: (1) through the common law tort of misappropriation; (2) as a legitimate business interest protected by noncompetition clauses; and (3) using a state's version of the Uniform Trade Secrets Act; see, e.g., West's C.R.S.A. §§ 7-74-101-110.

[17] State or provincial law, either in the form of statute or court decisions, will govern the enforceability of limitations on an employee's freedom to compete with an employer after the employment relationship ends. While they are employed, employees generally have a duty of loyalty which prevents them from competing against their employers. If permitted by applicable law, a noncompetition agreement should state reasonable limitations on the kinds of activities employees must avoid, on the geographic locations in which they may not compete, and on the time during with they will not compete.

[18] Employees' duty of loyalty to their current employers generally means they may not pursue their own businesses or activities when they would create a conflict of interest in performing their obligations to their employers. This type of agreement is useful to clarify this obligation.

[19] U.S. Federal statute, 17 U.S.C.A. § 101, generally defines "work made for hire" as copyrightable work and products employees make within the scope of their employment. The statute allows employee and employer to agree in writing about the kind of activities that will be within the scope of employment and about who will own the work made for hire. Particularly with employees whose work involves designing or creating products, software, manufacturing processes and other information or objects that are important to their employer's business, the employer and its creative employees should have a clear understanding about who will own anything the employees' create or develop for the company or using the company's facilities and resources.

[20] The employer should ensure any provision regarding a disability complies with the terms of any applicable disability insurance policy and does not violate laws prohibiting disability discrimination, such as the U.S. federal Americans with Disabilities Act.

Appendix II: Checklist for High Level Home Working

Readiness Assessment:

* Corporate "Champion' Support
* Current Telecommunications Expansion Capability
* Home Office 'Policy', Training, Reimbursement, Help Desk Support
* Recruiting and Retention of Home based and Virtual Workers
* Ability to re-allocate or sell-off soon to be excess office space
* Risk Management
* Home Office furnishings
* Home Office Equipment (voice and data)
* Corporate Communications – Internal and External

Human Resource Consideration:

* Corporate Standardized Remote Worker Policies and Rules
* Screen for the ability to work remotely and alone
* Selections & Criteria for Virtual Employees / Remote Workers
* Can the prospective teleworkers recognize and solve problems?
* Management roles and responsibilities – Management by Objectives – Setting Goals and Objectives
* Hours of Work
* Promotions and Transfers
* Budget coordination and changes
* Interacting with others working from home
* Interacting with those in the traditional centralized work place – duties and responsibilities, expectations and relationships
* Does the prospect have necessary hardware, software, telecommunications, and internet skills?
* Can the prospect balance schedules, meet deadlines and communicate with co-workers, clients and customers?

* Should and existing employee wishing to work from home not meet screening criteria, can the deficiency(s) be over come and who decides?
* Productivity, supervision, metrics and accountability
* Local, city and state taxation
* Returning to the traditional central work place – under what circumstances and who pays for the move, transition, office space?
* Develop an enforceable Agreement between the employer and employee
* [One of the most comprehensive agreements is found in the 2004 Colorado Employment Law and Practice Handbook© by John Paddock, West 2004, www.westgroup.com/product/store]

Home Office Environment and Facilities:

* Organizational Responsibilities
* Space requirements and workstation design & layout
* Safety and home office inspections / verification of compliance
* Home office ergonomics – Company provided furnishing policy & guidelines
* Home Office HVAC (Heating, Ventilation, Air Conditioning) and lighting policy
* Moves, adds and changes - Terminations
* Maintenance, repairs / replacement, insurance for fire and theft
* Office supplies, postage, heating, cooling and ventilation – who pays?
* Physical security – safeguarding company provide equipment and documents
* Any jurisdictional issues involving leases, covenants or home office / business restrictions
* Employer ADA responsibilities for 'reasonable accommodation'
* Space management
    — Use of or disposition for space relinquished by Virtual Workers
    — Electrical and Cooling management
    — Parking Lot Space reductions
    — Mailroom and Package handling requirements
    — Cafeteria and other support services like daycare, health maintenance and common areas reductions

Information Technology:

* Standardized company provided computer equipment
* On-line training for equipment and software utilization
* Computer, data, and network security
* Levels of service, including help desk support (24X7 recommended) voice and data
* Moves, Adds and / or Changes
* Installation Support - Provisioning
* "Intelligent Network" services voice and data
* Replacement, up-grades and refreshing technology
* Hoteling Systems and Software
* Virtual / Dynamic Connectivity

Finance & Accounting / Program Planning:

* Budget one-time costs for each teleworkers
* Prepare tentative 'Journal Entries' for telework program financial benefits:
  — Reduced Facilities Cost
  — Smaller training, recruiting and retention budgets
  — Increased telecommunications / IT budgets
  — Capture reductions in traditional office space 'dedicated connectivity'
  — Reward productivity gains
* Metrics / measurements before telework
* Metrics / measurements post telework – On going objective program assessments and line management testimony
* Review standing preferred vendor contracts for suitability and capability to support the new remote worker environment... furnishings, hardware, provisioning, etc...
* Corporate Recognition, Awards and Contributions
* Governmental Program Participation, Grants, Tax Credits – Cleaner Air, Less Congestion, Enterprise Zone, Hub Zones, Rural Economic Development, etc...

Appendix III: Getting A Life

Almost by definition, when you are working from home you are working alone. That's why some occupations (like mine) that require concentration and the work to be completed by a single individual are tailored for home work.

I don't need other people to write my articles and books. When I finish my end then others take over: the Managing Editor, Editorial Director and Art Director.

But no one, not even writers, used to and who prefer to be solitary, can stay alone all the time. That's why it is vital for all of us home workers to stay in touch with others, to get a life, to have life, and to have some company. It helps us refresh our brain cells, to relax and gives us balance to our lives.

Here are my suggestions:

Chat with your co-workers

There's all good and nothing wrong in sending an instant message or calling up your co-workers, like close to the end of the day, to chit-chat, especially if you've worked with them face-to-face. That way you're keeping yourself in the loop and staying in touch with what's going on. You hear the gossip that is vital to your career, like staff changes, moves, and impending decisions. Only by conversing with others will you not be forgotten.

Ask anyone at our New York office and they know I'm around, even though I'm not there, because I stay in touch. I've talked to new staff people who said they had heard about me but they have never seen me.

Go out on your breaks, and after work

You go out for breaks and for lunch at premises offices. So why not at home? Become a regular at your local coffee shop, diner, café or pub. Take a jog or walk in a local park. Get to know the people in your community. Make some friends and develop interests outside of work.

Consider going to a gym in town. That way you get out of the home, get active and interact with others.

Get involved!

One of the great benefits of home working is that you are in your community and your neighborhood, not wasting gas, polluting the environment, and getting stressed out traveling two hours or so round trip to another community. So why not get involved where you live?

There are community, interest groups, nonprofit, charitable, political and religious organizations that cover practically every purpose, cause, viewpoint and belief. Many of these bodies are looking for able, eager and helpful new people.

Being involved can be fun. I've been a director and vice president of transportation and social advocacy organizations. I speak at public hearings and been on advisory committees. I've taken a run at city council and I am active in a Canadian political party. I've been on the other side of the microphone, notepad, and TV camera: being interviewed by the media instead of doing the interviewing.

If you work on the West Coast and report to people on the East Coast you get the added benefit of having almost a second shift for your outside involvement. My workday begins at 7am and ends at 3pm, which means I have half the afternoon and all of the evening. Eliminating the commute also gives me extra time.

Make sure that if you get involved that you try as much as possible to undertake activities that are not work-related. Like if you're an accountant, don't do accounting. Or if you're a writer, don't do articles or newsletters. The reason: it feels too much like work and you'll resent it. The key purpose of getting involved is to undertake an activity that is different from work so you can become refreshed and alive.

That's why I don't mind doing publicity for organizations. I like writing press releases because it is not the same as writing from them. The same goes for speaking instead of reporting. But I draw the line at photography and story writing.

Also, if you get involved with public policy make sure you separate your professional from your public work. And see if you have enough time for that involvement; never let your personal life run over your work or you'll find yourself commuting again...if not worse.

If you get elected to a directorship or to public office avoid any conflict of interest with your career if such a matter comes up by recusing yourself. Architects, lawyers and real estate people do this all the time. Like if I was on City Council and they were voting to approve subsidies for a call center or vendor who was advertising, or to buy an ad in *Call Center Magazine* I would have to step out of the chambers.

Adopt a pet from the SPCA

Pets help bring life into a home when you are alone, even temporarily. They are good companions; they remind you that there are other living creatures with equally valid needs as yours. Like being fed, looked after, loved, and having their toilet handled (usually literally!).

A dog will insist that you walk them, rain or shine or sleet. That will most definitely get you out of the house. A cat will demand your affection at the time of their choosing.

We have four cats, two of which have adopted me, the other two have adopted my wife. My "catparents," Casey, who is older and Dot, who is younger, will come into my home office periodically to see if I am working and ask that I pay attention to them. When I sometimes, and to them, inexplicably leave the room Dot meows for me: like "where do you think you're going, get back here!" Who needs an on-site supervisor when you have one with claws?

I also have to let Casey see me leave the office and the house. If I don't, she will yowl for me, and search everywhere for me. I left for Orlando once, without saying goodbye, and she cried for a day or so before settling down.

Just keep in mind with pets, like with people, that there are boundaries. When I am busy the cats either stay quiet or I put them out of the room. I will shut the door when I am on speakerphone.

The American Society for the Prevention of Cruelty to Animals (ASPCA) and the Society for the Prevention of Cruelty to Animals (SPCA) in Canada are the best places to find a dog or a cat to give a loving home to. We adopted all four of our cats from the SPCA.

These societies pay for the neutering, if they are not neutered. Some animals have their shots; you may have to get others. But that's a small price to pay for another creature that wants, and deserves, caring, love and attention.

Spend time with others

Perhaps the greatest benefit of working from home is the time you spend with others: your friends, family and loved ones. You're not stressed out from the commute. You can enjoy others' company.

Even though my work gets hectic from home (like writing this book), I see and talk to my wife more often than when we both worked in New York. This is despite the fact that we commuted together (we tried to leave at the same time). When we got home we were so tired that we ate, watched some television, and then went to bed. No life in between. We tried to make it up on weekends by day-trips and seeing friends and family: in between shopping and other chores.

When I was single and worked from home I went to dances and to other singles' events. Going into Manhattan from Queens, or central London from the outskirts, was a treat; my trains weren't as crowded as the rush hour expresses going out. Meeting someone at these events avoided the potentially career-damaging hassles of "office romances."

While living in New York (and commuting to *Call Center Magazine's* Manhattan office) I met my wife at a dance. Nine months later we were married, over three years later we moved to Canada. I'm working for the magazine from my home; she has a holistic healing practice (she is a Reiki master/ teacher), also from home.

In the words of famed radio newscaster and commentator Paul Harvey: "Now you know the rest of the story..."

Appendix IV: How To Go Home

— Advice to employees who want to work from home —

Strive and become the best you can be at your job. Show that you are a top performer. Arrive a little early and leave a little late, if necessary to complete your tasks.

Take on extra responsibilities, if offered. But before accepting, determine whether the effort will have any impact on how you handle your primary responsibilities.

Show that you are a team player. Praise and help your supervisors and colleagues.

Behave as a star performer. Be a professional. Look out for your organization and for your colleagues. Put forward constructive suggestions. If you have any criticisms, make them constructive, too — they should be well-thought-out, backed by facts, and follow the established channels to present them, beginning *first* with your supervisor.

Demonstrate your ability to work independently. Undertake your tasks successfully without supervision, and without constantly asking your boss or your colleagues.

Now, take a hard look at your tasks to see if home working is doable, even occasionally. The less need for in-person interaction the more likely your job is home workable. Also, see if there are substitutes for in-person communications, i.e. audio, data, video or Web conferences for face-to-face meetings. Call or e-mail instead of visit. The more work you can make virtual the more likely you will be able to work from home.

Plan ahead for home working. That means setting up a functional, ergonomically-sound home office with everything, including communications, paid for by you. That way you will be able to work effectively from home occasionally, like for a doctor's appointment, a plumber coming to the house.

If you have a home office set up and you have to be at home, let your employer know that you wish to work from home that day and here's where/how they can reach you. Ask to borrow a laptop or get set up on a Webmail account. That way you're showing that you are being productive outside of the premises office.

Also ask to borrow a laptop and/or obtain Webmail access to work from your home office in the evenings or on weekends for a special project or to handle pressing deadlines. This too can demonstrate the flexibility and versatility of home working.

Working from home after-hours lets you avoid the necessity of staying later or coming into the premise office on weekends. It is often not safe in many areas to leave the office late; you don't know who is lurking in the lot or parking garage, or at the bus stop or train station. Premises offices on weekends, even during the day, can get lonely and creepy. There is also the hassle of having keys, like for doors and elevators.

Also, volunteer yourself to be on your employer's business recovery/disaster response team. By having an already-equipped home office you'll be set up to handle contacts and continue to work in case of weather and other emergencies. You'll be very much appreciated if the inevitable happens.

In case of a snowstorm or other such anticipated event that makes travel impossible, offer to work from home ahead of time or notify your employer soon after the event happens if it cannot be predicted or the scale was larger than expected. You'll be safe, out of harm's way, *and* working.

Once you have proven yourself working from home occasionally, with enough of a track record for your employers, begin to make the case for you to work from home exclusively. But when doing so, make your case from the employer's standpoint, **not** yours. Here are some solid rationales:

Higher productivity

If your work extends beyond 9 to 5 you may have a strong case for home working. You can catch people that you would have missed while commuting, and you avoid playing message tag that eats up office time the next day. This is especially relevant if you work in the East and have a lot of business and contacts in the West, where 5pm is actually at 2pm, and vice-versa where 8:30am PT is 11:30am ET.

Be careful here. Employers expect that you complete your tasks within the time allowed. So make sure you are a top performer *first*.

Family/personal needs

Should you have child/elder/spousecare responsibilities; your spouse is being transferred to another community, is burned out and wants to set up a new career in a new location, or has an offer to work from a new home in a place that both of you love — in that order — then your employer may be amenable to letting you work from home. Provided you have proven to be a star performer. Good employers don't want to lose top workers; it costs them a lot of money directly, and in lost productivity, to recruit, train, and bring up to speed to good employees.

Corporate relocations

If your organization is moving, especially to another metro area or out of a downtown to a sprawling suburban "edge city" or back into the downtown from the suburbs, then your employer might be approachable to letting you work from home. Employers do not want to lose good people if they don't have to. Also, the fewer the employees at a premises office, the less money they have to spend in real estate and facilities.

End result for your employer: big cost savings in facilities, productivity and retention. Using the assumptions, calculations, and methodologies in the Chapter 1 example posed by Jack Heacock, after accounting for employer investments in home working you could be saving your organization *over $20,000 a year* by working from home.

You will need to demonstrate that you cannot move, i.e. your spouse has a career locally and can't work from home and your kids are in high school and will find changing schools disruptive to their education. Or in the case of moves within the same metro area, you must show that commuting to the new location will add significantly to your trip time, making you less productive, i.e. more message tag.

If you live in a big city with excellent transit, you don't own or can't easily afford a car, and your organization is moving to the suburbs, you could make a hardship case with a sympathetic employer. But don't count on it. Your commute is technically your problem. The added commuting time is a stronger argument. And it depends on how much they value you as an employee.

The best opportunity to work from home is if the employers offer it. If you follow the above advice — making sure that you are a top employee, a team player, and perform tasks that are home workable — the more likely you will be selected for an "in-house commute."

About the Author

Brendan Read is a business editor/writer and consultant who writes about site selection, real estate, facilities, staffing/training, outsourcing and home working. He has covered the call center, transportation, metals and electrical industries. He is also the author of *Designing the Best Call Center for Your Business* and co-author of *The Complete Guide to Customer Support*, both published by CMP Books. He works from his home in Courtenay, British Columbia, Canada. He is active in his community with land use, political, and transportation organizations.

Glossary

Here is a glossary of the new, commonly misunderstood and less common terms used in this book. I have also included, where appropriate, similar and related terms in the same glossary item.

If you work in and need to find out more about voice/data systems I recommend that you purchase *Newton's Telecom Dictionary*, from CMP Books. *Newton's Dictionary* goes into considerable depth and is updated annually. Many of the definitions in this book are drawn from this source, devised and written by technology visionary extraordinaire Harry Newton.

ABC

Authentication: Verifying users' identities via physical means i.e. USB plugs contact or contactless smartcard readers and biometric devices, like fingerprint and iris scanners.

Automatic Number Identification a.k.a. ANI: ANI identifies incoming phone numbers but unlike its more popular cousin, *Caller ID*, ANI cannot be blocked. ANI also enables *computer telephone integration (CTI)* screen pops that carry customer, other employees' or system data that arrive on employees' computers.

Broadband: A generic term referring for data transmission speeds typically above 100 kilobits per second; dial-up is limited to about 56 kbps. Broadband is supplied by several services including cable, digital subscriber line (DSL), ISDN, satellite and Wi-Fi.

Business continuity planning: Planning before a disaster hits on how to keep the organization functioning in that event. Business continuity planning includes risk minimization,

having and testing response means (e.g. home workers) and mitigating the effects of disasters.

Centrex routing: A generic term for contracting with your local exchange carrier to handle the inbound call routing, rather than using the long-distance carrier's or switches on your premises. Centrex routing supports ANI, CTI, and skills-based routing.

Conferencing: Interacting with more than one person without being their physical presence, usually as an alternative to traveling and meeting. Conferencing means include audio, data, video and Web conferencing and data collaboration.

DEF

Employee Contribution: The *value* per employee to the organization as opposed to measuring the cost of employees based on wages/benefits paid. Consultant John Edwards recommends this measure because it gives a truer mark of what an employee is worth. Remember that's why you have employees in the first place: you pay them compensation to give you value.

Ergonomics: The Concise Oxford Dictionary, Sixth Edition 1976 defines it as "the study of efficiency of persons in their working environment". That has come to mean ensuring productivity of your employees by ensuring their work environment is safe and effective, enabling them to perform efficiently, without injury that costs them and you. Ergonomics, by definition, includes every factor that impacts on that work environment: furniture, lighting, temperature and noise.

Exclusive home working: Where the employees' principal workplace is the home; employers do not provide dedicated space or voice/data connections for them in premises offices. Exclusive home working arrangements pass taxation authorities' home office tax deductibility tests.

GHI

Home-based business: Businesses that operate out of owners' homes; the actual work may or may not be carried out in the home e.g. carpentry, gardening, nursing, trucking, window installation. Home based businesses include self-employed home workers.

Home offices, also Home Workplaces: Where the home working employee carries out their tasks in the same occupancy that they legally reside at. The home office can be a corner of a room, a separate room or basement or converted garage. Home office employees can be home, condo, co-op owners, tenants, roommates or family members.

Home working: Where employees carry out their work tasks from their homes, either exclusively or occasionally, including in emergencies compared with working in premises offices. Home working includes conferencing i.e. interacting with others via audio, data, video or the Web as opposed to conducting work face-to-face.

Home working program: The plan that sets up and runs home working. It encompasses goals, policies, procedures and management, including employee and program performance management.

Hot-desking: Supplying temporary space, voice/data connections or in some instances equipment in premises offices, including satellite offices that is shared with other users.

Hoteling: Temporary premises office space provided by employers or third parties provide workspace and shared services such as printers, fax machines, and in some cases, meeting and conference rooms for home and mobile working employees. The "hotel" accommodations range from basic offices to executive suites. These spaces often provide conferencing services (audio, data, and video)

that reduce time-wasting and costly employee travel.

Infrastructure Cost per Employee (ICE): One means to determine, for comparison purposes, how much does it cost to support each employee, whether in premises offices (including satellite offices), or home offices. In the case of occasional home working employees, you need to know how much more will this cost to support them in home offices on top of what it costs you to support them in premises offices.

Instant messaging (IM): IM is software that when installed on all computers permit messages to be sent and received over the Internet. IM enables you, your colleagues, your employees, and others in the loop to send and receive messages in real-time, unlike e-mail, which can take awhile. IM users must all have the same software (there are no universal standards — yet) and must log in and out.

JKLM

Mobile working: Where employees undertake their work responsibilities while traveling or from non-fixed or dedicated locations e.g. cafes, hotels, others' premises or home offices, restaurants, waiting rooms.

NOPQ

National Electrical Code (NEC): A set of electrical installation safety standards, adopted into law by most American governments; some cities have their own electrical codes that may or may not be based on the NEC. The federal Occupational Safety and Health Administration's design safety standards for electrical systems regulation has been based on the NE Code. Canada has the Canadian Electrical Code; there are minor differences between the NEC and the CEC.

Network routing a.k.a. "intelligent networks": Calls that are destined for the organization and reaches their carrier's phone switch, which then reroutes the calls to anywhere the organization sees fit: to premises offices, mobile workers, or to home workers. The organization tells the carrier where to route the calls through a network automatic call distributor (ACD). The calls can pick up caller data

(CTI) or be routed to employees with specific skillsets (skills-based routing).

Occasional home working: Where employees only work from home occasionally, including part-time and in emergencies; employers usually provide them with space and voice/data connections for them in premises offices. I say unusually because mobile workers may work in the field a majority of the time and work from home occasionally: they do not have premises offices.

Off-premises extensions (OPXes): Separate hardware or software located on your premises and depending on the technology on your home workers' to reroute premise-based or Centrex-switched calls to home workers, whether wireline or wireless. If these employees live outside of the local calling areas you may need to pay long distance charges: those inbound calls became outbound calls.

Performance-based management: Determining how well workers accomplished or performed their tasks compared with expectations and standards. Related to this is *Virtual Management* where supervisors manage employees without being there in-person, such as through e-mail, IM, phone.

Premises-based routing: Where the switches, either PBXes or VoIP routers are located on the organization's premises and route communications to employees, at premises office desks, or to their homes or mobile phones or to other locations.

Premises offices, also Premises Workplaces: Where employees carry out their work tasks in space provided and subsidized by and under the direct control of their employers. This includes *satellite offices* and *satellite workplaces*. With rare exceptions e.g. resort hotels employees typically commute from their homes to premises offices.

PBX a.k.a. private branch exchange: A PBX is a phone switch typically employer owned/leased at the premises office that routes calls to individual phones. Those phones can be at premises desks, at satellite desks, at employees' home office desks or be on the employees a.k.a. wireless. Calls can be transferred automatically, or rerouted out of the premises offices as outbound calls. Traditionally PBXes have PSTN only but many newer ones are now VoIP-enabled.

PSTN a.k.a. POTS: Public switched telephone work or plain old telephone service. The traditional phone service that is mainly voice but also data via dial-up modems that comes into homes typically over copper wires. PSTN/POTS is circuit-switched i.e. a continuous dedicated connection between both parties over a circuit.

RST

Satellite working: The providing of mini-premises offices near where employees live. These offices have shared peripherals like copy machines, faxes, and printers.

Secured networks: Secured networks are those in which access is restricted either by private leased line or virtual, over the Internet, known as *virtual private networks* (VPN) that use encryption and passwords.

Self-employed home workers: Independent, e.g. freelancers or individuals who have formed corporations who work from home for clients. These workers are *not* employees.

Softphones: Software that emulates a phone; instead of hitting keys on a phone, users click on screen icons to take or make calls.

Storageless PCs a.k.a network appliances: Computers that lack hard or removable drives and chips; they also do not have ports like USBs that could be used to upload and download data and software.

Telecommuting: Similar to home working. Strictly speaking it does not include conferencing.

Teleworking: Home and/or mobile working either exclusively or occasionally, with or without conferencing.

Text messaging a.k.a. SMS: Text messaging or short messaging service (SMS) is "mini-e-mail." Instead of sending that message to an IP address you shoot it to a special address or to a phone number.

UVW

Unified (or integrated) messaging:

Voicemail, e-mail, faxes and video digitized into one form that could be read, heard and downloaded.

Uninterruptible power supplies (UPS) systems: UPSes are battery supplies that keep computers, peripherals or other appliances e.g. modems going during voltage sags or brief (15-30 minutes) power outages.

Virtual working: The carrying out of employment-related tasks that is not totally dependent on employees being at premises or being face to face with others.

Virtual workplaces: The space, stationary or mobile, where the work is being carried out. This includes non-dedicated workspaces such as hot-desks provided employees also work in other environments e.g. at their homes or on the road.

VoIP (Voice over Internet Protocol): The transmission of voice as data packets as opposed to dedicated circuits, typically over the Internet either naked or encrypted.

Webmail: Webmail is a password-protected access means that allow users to receive and transmit e-mail over the Internet without requiring them to gain access to the network. Webmail sessions are limited: usually 30 minutes and logout is often without notice. Webmail users can't access your servers.

Wireless: The transmission of electronic communications without wires. Wireless media includes cellular, cellular broadband, infrared, satellite, underwater longwave, UHF/VHF, Wi-Fi and WLAN.

Wi-Fi and WLAN: Wireless local area networks (WLANs) consist essentially into two groups: home networking and hotspots, for mobile working. Wi-Fi is the term used to describe the latter.

Wireline: The transmission of electronic communications with wires. Wireline media includes telegraph, PSTN, Telex, fax, dial-up Internet, cable, DSL and VoIP.

Workplace: Where the work is carried out.

Index